Environmental Sustainability - Preparing for Tomorrow

Edited by Syed Abdul Rehman Khan

Published in London, United Kingdom

IntechOpen

Supporting open minds since 2005

Environmental Sustainability - Preparing for Tomorrow
http://dx.doi.org/10.5772/intechopen.91568
Edited by Syed Abdul Rehman Khan

Contributors
Manikanda Bharath Karuppasamy, Srinivasalu Seshachalam, Usha Natesan, Karthik Ramasamy, Alexander Evdokimov, Łukasz Kruszewski, Banwari Dandotiya, Olga Lucia Quintero, Santiago López-Restrepo, Jhon Edinson Hinestroza, Elias David Nino-Ruiz, Jimmy Anderson Flórez, Nicolas Pinel-Peláez, Angela Rendón, Monica Lucia Alvarez-Laínez, Andres Felipe Zapata-Gonzalez, Jose Fernando Duque, Elena Montilla-Rosero, Andres Pareja Lopez, Jean Paul Delgado, José Ignacio Marulanda Bernal, Bibiana Esperanza Boada Sanabria, Juan Ernesto Soto, Sara Lorduy, Jaime Andrés Betancur, Alejandro Velez Ospina, Arnold Heemink, Andres Yarce Botero, Arjo Segers, David Galan Madruga, Karzan Mohammed Khalid, Mario Enrique García, Héctor Ulloa, Omar García, Hermes Ramírez, Aida Fajardo, Claudia Saldaña, Sarah Messina, Yamilet Rodriguez, Sunil Gulia, Saurabh Mendiratta, Prachi Goyal, S.K. Goyal, Muhammad Ikram Bin A Wahab, Tiziano Tirabassi, Daniela Buske, Mahima Habil, David D. Massey, Ajay Taneja, Tomaz Langenbach, Luiz Querino Caldas, Tacio M.P. de Campos, Syed Abdul Rehman Khan, Zhang Yu, Adeel Shah

First published in London, United Kingdom, 2021 by IntechOpen
IntechOpen is the global imprint of INTECHOPEN LIMITED, registered in England and Wales, registration number: 11086078, 5 Princes Gate Court, London, SW7 2QJ, United Kingdom
Printed in Croatia

British Library Cataloguing-in-Publication Data
A catalogue record for this book is available from the British Library

Additional hard and PDF copies can be obtained from orders@intechopen.com

Environmental Sustainability - Preparing for Tomorrow
Edited by Syed Abdul Rehman Khan
p. cm.
Print ISBN 978-1-83968-785-3
Online ISBN 978-1-83968-786-0
eBook (PDF) ISBN 978-1-83968-787-7

For EU product safety concerns:
IN TECH d.o.o., Prolaz Marije Krucifikse Kozulić 3, 51000 Rijeka, Croatia, info@intechopen.com or visit our website at intechopen.com.

Meet the editor

Syed Abdul Rehman Khan is a professor of Supply Chain and Logistics Management. Dr. Khan achieved his Certified Supply Chain Professional (CSCP) certificate from the United States and completed his postdoctoral fellowship at Tsinghua University, Beijing, China. Since 2020, Dr. Khan has been affiliated with Xuzhou University of Technology, China, as an associate professor. He has more than nine years of core experience in supply chain and logistics at industry and academic levels. He has attended several international conferences and has been invited as a keynote speaker in different countries. He has published more than 100 scientific research papers in well-renowned international journals and conferences. He is also a regular contributor to conferences and workshops worldwide.

Contents

Preface **XIII**

Section 1
Introduction **1**

Chapter 1 **3**
Poor Air Quality: A Dark Cloud of Filth Poisons
by Syed Abdul Rehman Khan, Adeel Shah and Zhang Yu

Chapter 2 **17**
Implication of Secondary Atmospheric Pollutants in the Air Quality:
A Case-Study for Ozone
by David Galán Madruga

Chapter 3 **37**
Particle and Carbon Monoxide Atmospheric Pollution in the City of Tepic, Nayarit, Mexico
by Mario García, Héctor Ulloa, Omar García, Hermes Ramírez, Aida Fajardo, Claudia Saldaña, Sarah Messina and Yamilet Rodríguez

Section 2
Air Pollution **59**

Chapter 4 **61**
Characterization of Atmospheric Mercury in the High-Altitude Background Station and Coastal Urban City in South Asia
by Manikanda Bharath Karuppasamy, Srinivasalu Seshachalam, Usha Natesan and Karthik Ramasamy

Chapter 5 **77**
Health Impacts of Air Pollution
by Muhammad Ikram Bin A Wahab

Chapter 6 99
Medellin Air Quality Initiative (MAUI)
by Andres Yarce Botero, Olga Lucia Quintero Montoya, Santiago Lopez-Restrepo, Nicolás Pinel, Jhon Edinson Hinestroza, Elias David Niño-Ruiz, Jimmy Anderson Flórez, Angela María Rendón, Monica Lucia Alvarez-Laínez, Andres Felipe Zapata-Gonzalez, Jose Fernando Duque Trujillo, Elena Montilla, Andres Pareja, Jean Paul Delgado, Jose Ignacio Marulanda Bernal, Jaime Andres Betancur, Alejandro Vélez, Arjo Segers, Arnold Heemink, Juan Ernesto Soto, Bibiana Esperanza Boada Sanabria and Sara Lorduy

Chapter 7 137
Data Assimilation as a Tool to Improve Chemical Transport Models Performance in Developing Countries
by Santiago Lopez-Restrepo, Andrés Yarce Botero, Olga Lucia Quintero, Nicolás Pinel, Jhon Edinson Hinestroza, Elias David Niño-Ruiz, Jimmy Anderson Flórez, Angela Maíra Rendón, Monica Lucia Alvarez-Laínez, Andres Felipe Zapata-Gonzalez, Jose Fernando Duque Trujillo, Elena Montilla, Andres Pareja, Jean Paul Delgado, Jose Ignacio Marulanda Bernal, Bibiana Boada, Juan Ernesto Soto, Sara Lorduy, Jaime Andres Betancur, Arjo Segers and Arnold Heemink

Chapter 8 157
A Synthesis of the Information Given by Temporal Data Series: The Representative Day
by Tiziano Tirabassi and Daniela Buske

Chapter 9 167
Why Airborne Pesticides Are So Dangerous
by Tomaz Langenbach, Tácio M.P. de Campos and Luiz Querino Caldas

Chapter 10 183
Air Pollution, Health and Perception
by Banwari Dandotiya

Section 3
Environmental Sustainability Practices 197

Chapter 11 199
Aerotechnogenic Pollution of Boreal Forests in Northern Europe
by Alexander Evdokimov

Chapter 12 213
Influence of Microenvironments and Personal Activities on Personal $PM_{2.5}$ Exposures among Children and Adults
by Mahima Habil, David D. Massey and Ajay Taneja

Chapter 13 229
Fossil Fuel Fires: A Forgotten Factor of Air Quality
by Łukasz Kruszewski

Chapter 14 253
Evaluation of Particulate Matter Pollution in Micro-Environments of Office Buildings—A Case Study of Delhi, India
by Saurabh Mendiratta, Sunil Gulia, Prachi Goyal and Sanjeev Kumar Goyal

Chapter 15 271
Correlation between Air Quality and Wastewater Pollution
by Karzan Mohammed Khalid

Preface

Due to the lack of environmental policies and unplanned industrialization, the world faces air pollution and climate change challenges. No doubt increasing air pollution is creating an alarming situation for the fauna and flora. In addition, manufacturing and logistical operations are heavily dependent on fossil fuels and energy consumption, fueling air pollution.

Air quality is measured through the Air Quality Index (AQI), based on the concentration of pollutants present in the air in a particular area. Poor air quality has adverse impacts on human, animal, and plant life. Further, air pollution is also a cause of acidic rain, which directly and/or indirectly affects the agricultural sector and converts into water pollution.

With the help of the Clean Air Act of the 1970s, the United States significantly controlled air pollution in their territories. However, 80% of the world's energy budget comes from burning fossil fuels, thus the world needs collective efforts to reduce dependence on burning fossil fuels and recover environmental sustainability by adopting renewable energy sources and green practices.

We would like to thank the contributors of the book for their extraordinary hard work. We would also like to thank IntechOpen and Author Service Manager Ms. Iva Lipovćfor her help in coordinating and facilitating this project. We also extend our thanks to IntechOpen publisher for all of their support.

Thank you,

Dr. Syed Abdul Rehman Khan, PhD
Xuzhou University of Technology,
Xuzhou, China

Beijing Key Laboratory of Urban Spatial Information Engineering,
Beijing, China

Section 1

Introduction

Chapter 1

Poor Air Quality: A Dark Cloud of Filth Poisons

Syed Abdul Rehman Khan, Adeel Shah and Zhang Yu

Abstract

Our home, the Earth, is the rarest planet in-universe the to sustain life. The thing which makes it unique amongst heavenly bodies is balance in the environment. This balance is the key to sustain life for millions of years. Air is one of the most critical components of mother nature; it provides oxygen for all species, both plant and animal, to live. Air not only provides oxygen but is also essential for keeping the human body cool. The advantages of air are countless, from the cloud, weather, humidity, dust, and pollen migration to burning fire; without it, life will not continue. Air is made up of chemical components, and if pollutants added, it would become harmful for all living beings. The chapter put forward is to highlight the importance of the quality of ambient air, standards to measure, and sources of pollution. Further in the chapter, the impacts of polluted air on human health and the countries' financial obstacles are discussed. The chapter concludes with a summary and recommendations for policymakers, NGOs, and affected people to better their lives and repair the damage caused to nature's precious gift, the air.

Keywords: Air quality, pollutants, and AQI

1. Introduction

Human beings need combustion energy to survive and live their daily lives. The energy needs are constant in the everyday affair, from cooking to use electricity. Out of the most abundant of all fuels in an urban and rural setting in developing countries are coal, wood, methane gas, and cow dengue. The heat produced from the fuel is used for cooking, keeping warm, and boiling water for sanitation. The combustion fuels mainly categorize as gas, liquid, and solids [1]. Solid fuels need an air mixture to fully combust and provide energy [2]. Such fuels produce heat for our well-being and produce gases that affect the air around us. A good quality air constitutes seventy-eight percent nitrogen, twenty-one percent oxygen, one percent other gases (National Geographic) for human beings to survive. In the one percent of other gases, 0.1% of gases are CO2, methane, water, and neon gas [3]. The combustion of solid fuel results in carbon monoxide emissions, particulate matter (PM), and organic compounds [4]. The emitted gases badly affect air quality and thus are termed as pollution [5].

There are many other activities in the daily walk of human life which are adversely affecting air quality. The leading causes are the construction of megacities, manufacturing of goods, usage of pesticides, automobiles, and airplanes [6]. The polluted emission from solid fuels and biomass in domestic use has declined since 1990 [7] due to the availability of liquefied petroleum gas (LPG) and methane gas. However, in developing countries, households in cities are still using solid fuel and biomass for heating and cooking [8]. The rural areas in developing countries are mainly using biomass and solid fuel for daily chores [9]. The pollution caused by domestic emission results in death, and it has increased since 1990 significantly in emerging economies [7, 10].

The natural sources are also affecting air quality in the atmosphere, and most significantly, they are volcanoes and wildfires. However, anthropogenic behaviors remain the most prominent [11]. The rapid deterioration of air quality started since oil discovery and byproducts due to rapid industrialization globally [12]. The consumption of fossil fuel has increased mammoth in the 21st century due to the human population explosion. It is a significant pollution source to negatively impact the ecosystem [13]. The air quality and pollution are not localized. However, a transboundary problem [14], the pollutant emitted in a country can transport through the wind in the atmosphere and thus need global attention and regional collaboration [15].

The air quality is measured through a numerical index is used know as the "Air Quality Index (AQI)," which gauges pollution concentration [5]. The AQI serves to identify air pollution levels impacting the health of the public. The assessment of air pollution by the measure helps policymakers strategize reduction by curbing activities serving as the source [16]. The data for the indexes gathers from worldwide city stations responsible for monitoring air quality. The AQI data is available online for the general public to use in studies and reports [17]. The "US Environmental Protection Agency (USEPA) " is first to use AQI in the year 1999, and then the index modified to use six chemicals measurement in the air: "carbon monoxide (CO), nitrogen dioxide (NO2), ozone (O3), coarse particulate matter (PM10), fine particulate matter (PM2.5), and sulfur dioxide (SO2)" [18]. The AQI unit ranges from "0 to 500," and the values are directly proportional to the air's chemical concentration; thus higher the value greater the effect on health [19, 20]. As the AQI value exceeds 100, the air quality becomes unhealthy for certain groups of people. The AQI's values are color-coded to reflect the severity of pollutant concentration (see **Figure 1**) [22].

Air Quality Index		
AQI Category and Color	**Index Value**	**Description of Air Quality**
Good Green	0 to 50	Air quality is satisfactory, and air pollution poses little or no risk.
Moderate Yellow	51 to 100	Air quality is acceptable. However, there may be a risk for some people, particularly those who are unusually sensitive to air pollution.
Unhealthy for Sensitive Groups Orange	101 to 150	Members of sensitive groups may experience health effects. The general public is less likely to be affected.
Unhealthy Red	151 to 200	Some members of the general public may experience health effects; members of sensitive groups may experience more serious health effects.
Very Unhealthy Purple	201 to 300	Health alert: The risk of health effects is increased for everyone.
Hazardous Maroon	301 and higher	Health warning of emergency conditions: everyone is more likely to be affected.

Figure 1.
The EPA AQI [21].

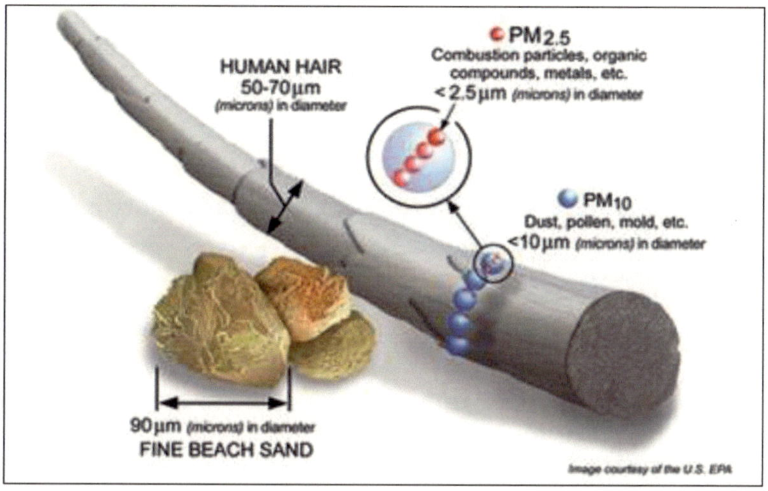

Figure 2.
Size comparison [21, 23].

The pollutant matter's size defines the PM; there are two types of sizes used in AQI: 2.5 and 10 microns (see **Figure 2** for size comparison) [23]. Many government agencies and NGOs like "IQAIR" collaborate with like-minded partners to report AQI to the public. The website also publishes reports yearly basis, and the result reflected are in PM2.5. There are many other websites which reflects air quality to view by tourists, policymakers, and certain groups with health conditions (see appendix 1 for websites).

2. Low air quality and society

The low air quality impacts plants, crops, aquatic water life, humans, and animals [24]. Globally, AQI standards protect the habitat; an example is the European Union's standards to protect vegetation. The transportation sector is also a significant source causing degradation, adding "acidification, eutrophication, and ozone." Acidification occurs due to the pH balance change of perception by the chemical reaction between "nitrogen dioxide, sulfur dioxide, nitric acid (HNO3), and sulfuric acid" to affect lake water, plants, and crops [24]. Carbon dioxide in the air reacts with water in the oceans to produce carbonic acid. The acid lowers the ocean's pH to impact sea life and is the primary cause of coral reefs [25]. The ozone is also a significant cause that harms human health and damage plants and jungles. The smog caused by ozone causes respiratory problems in humans and animals. It also interrupts the photosynthesis process in plants weakening to fight diseases, insects, and weather impacts [26].

Air pollution is responsible for killing millions; according to [27], seven million people die worldwide each year. In 2015, 2.8 million and 4.2 million people died from household and environmental air pollution [7, 10]. According to Hanson et al. [28], the same year's deaths categorize as "19% of all cardiovascular deaths worldwide, 24% of ischemic heart disease deaths, 21% of stroke deaths, and 23% of lung cancer deaths." Kioumourtzoglou et al. [29] studied the impact of air pollution and found evidence of the risk factor for neurodevelopment conditions in kids and neurodegenerative conditions in grownups. The effects of air pollution on humans depend on exposure, age, immunity, and predisposing conditions [11], resulting in nausea, respiratory problems, skin inflammation, organ failures, and cancer [30]. The epidemiological research on air pollution suggests that adverse effects primarily on cardiovascular and respiratory organs; however, the other organs may also damage [30–33]. WHO [27] published in their report below death percentages:

- "29% of all deaths and disease from lung cancer
- 17% of all deaths and disease from an acute lower respiratory infection
- 24% of all deaths from stroke
- 25% of all deaths and disease from ischemic heart disease
- 43% of all deaths and disease from chronic obstructive pulmonary disease."

Many studies suggest that air pollution long-term exposure to high and low pollutant concentrations affects the respiratory system [34]. The usual symptoms are irritation of nose and throat followed by bronchoconstriction and acute dyspnoea in asthmatic patients by exposure of sulfur dioxide [35] and nitrogen oxides [36]. The PM in air pollution enters the lung (see **Figure 3**) to affect alveolar epithelium [38], ozone starts lung inflammation [39], and nitrogen oxides increase the chances of lung infections [40]. The more prolonged exposure to ozone reduces lung function [41, 42]. Exposure to heavy metals increases the risk of asthma, emphysema, and lung cancer [43, 44].

Air pollution also affects the cardiovascular system as carbon monoxide joins with hemoglobin to reduce its ability to transmit oxygen [45]. It reduces oxygen availability in body organs, impacting their functions (see **Figure 4** for AQI chart and health recommendation). Blood coagulation is another problem that occurs due to PM presence in the body [47]. Similarly, air pollution may cause blood clotting, resulting in obstruction in cardiac blood vessels [48]. The nervous system in the human body exaggerates by inhalation of "heavy metals (lead, mercury, and arsenic) and dioxins" [11]. Researchers have found linkages of neurotoxicity and memory loss, sleep disorders, mood swings, emotional imbalance, impaired vision,

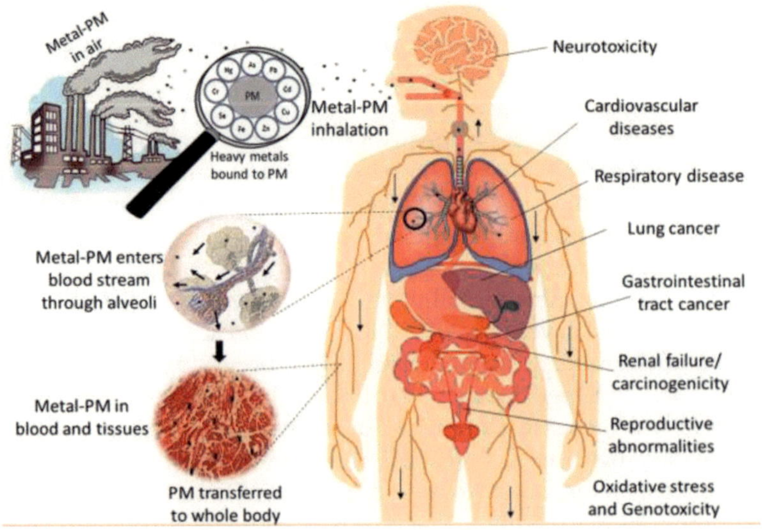

Figure 3.
Human diseases due to heavy metals in the air [37].

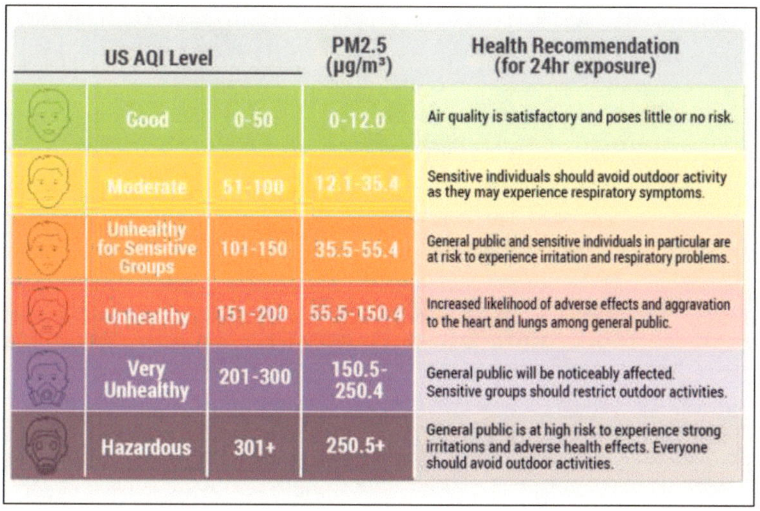

US AQI Level			PM2.5 (µg/m³)	Health Recommendation (for 24hr exposure)
	Good	0-50	0-12.0	Air quality is satisfactory and poses little or no risk.
	Moderate	51-100	12.1-35.4	Sensitive individuals should avoid outdoor activity as they may experience respiratory symptoms.
	Unhealthy for Sensitive Groups	101-150	35.5-55.4	General public and sensitive individuals in particular are at risk to experience irritation and respiratory problems.
	Unhealthy	151-200	55.5-150.4	Increased likelihood of adverse effects and aggravation to the heart and lungs among general public.
	Very Unhealthy	201-300	150.5-250.4	General public will be noticeably affected. Sensitive groups should restrict outdoor activities.
	Hazardous	301+	250.5+	General public is at high risk to experience strong irritations and adverse health effects. Everyone should avoid outdoor activities.

Figure 4.
AQI chart and risk to health [46].

and slur speech [45, 49]. Air pollution can increase the risk of pregnancy complications by impacting fetus development [50]. In the case of exposure to lead and heavy metal, it risks abortion [11].

3. Low air quality and financial obstacles

Air pollution also influences economic conditions in a country. According to McCarthy [51], air pollution in China is USD 900 billion, and in the USA, it estimates to cost USD 600 billion. The deaths by the burning of fossil fuel and coal are three times higher than road accidents. The estimated economic cost of air pollution is USD 2.9 trillion that equates to 3.3 percent GDP of the world [52]. According to the report, the disease and treatment due to air treatment cost disability from chronic diseases costing USD 200 billion in 2018. Sick leave of absence and premature birth cost around USD 100 and USD 90 billion, respectively [51].

Similarly, India's air pollution cost a staggering yearly USD 150 billion (see **Figure 5** for GDP comparisons). The labor hour lost due to illness and absentees due to health-related issues caused by air pollution directly impacts the labor market. It results in less productivity and reduced labor contribution to GDP [54].

The above are direct economic cost implications of air pollution; however, indirect costs also impact the country's economy (see **Figure 6** for direct and indirect cost) [55]. According to OECD [56], air pollution is " The indirect economic effects come from reallocation of the factors of production across the economy, changes in international trade and changes in savings, as changes in relative prices induce them." The report further explains that unlike direct cost, which affects GDP, indirect cost increases gradually. Consumers and environment watchdogs globally aware of air pollution on the environment and remain concerned for ethical and sustainable manufacturing practices. According to Qiu et al. [57], air pollution and consumers' negative sentiment correlate and impact buying behavior. The textile and apparel exporting sector in developing countries exports goods, which significantly impacts GDP. The industry is also responsible for air pollution in manufacturing countries.

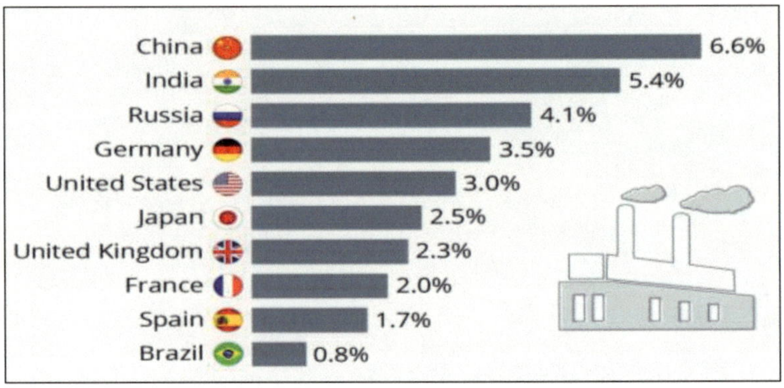

Figure 5.
Air pollution cost as GDP share (MacCarthy [53]).

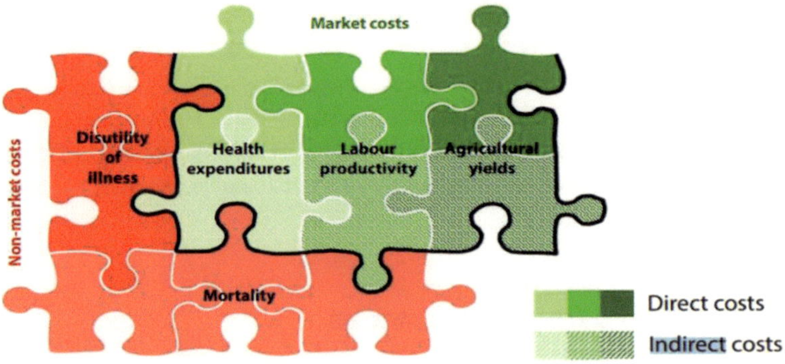

Figure 6.
Direct and indirect cost of air pollution.

In the textile and apparel sector, in which many processes use chemicals for treatment. Other than chemicals, the loss of short fiber called lint is air pollution hazardous to labor and people living near the factories [58]. According to Karthik and Gopalakrishnan [58], the following are the source of air pollution in the textile and garment industry (see **Table 1**): Similar studies by Haseeb et al. [59] and Heydebreck et al. [60] confirm the positive contribution of carbon dioxide CO2 and poly-fluoroalkyl substances (PFAS) in the textile and clothing industry.

An example of cotton fiber is relevant to the indirect cost of air pollution socioeconomically. An estimate of 25% of the pesticides used globally is the direct cause of cotton production. Further organic cotton is another demanded product in the west, used in fabric and garments label as "ECO" [61]. The water uses for irrigation is natural and rainwater. In air pollution, the rains will become a source of acid rain and damage the crop, resulting in economic loss to the country's farmers and exports [62].

The other primary air pollutant of the textile and garment industry is "toxic heavy metals, biochemical oxygen demand (BOD), chemical oxygen demand (COD), residual chlorine, dissolved solids, color, sludge, used oil, dyes, and non-biodegradable organics" [63]. Air pollution is a CSR activity in which organizations and governments are very conscious. Siri Lankas government in 2015 voluntarily designed and produce products that come under "ethical manufacturing" and to certify its industry to a complaint of "Garments without Guilt Certification" and

Due to manufacturing	Due to diffusion
"Boilers"	"Solvent-based"
"Ovens"	"Wastewater treatment."
"Storage tanks."	"Warehouses"
	"Spills"

Table 1.
Sources of air pollution in the textile and garment industry [58].

"Sustainable Environmental Friendly Manufacturing" to target EU markets and consumers [64].

4. Conclusion and recommendations

Low air quality is now a global phenomenon in which all countries are giving due consideration to improvement. Developing countries are the worst affected; according to [65] 21 out of 30 worst air polluted cities belong to India. New Dehli ranks at number 1 most air polluted city in the world (see **Figure 7**). Incontinent, Asia ranks no. 1 as most countries are developing economies and rely heavily on fossil fuel and coal to fulfill energy demands. Air pollution impacting on labor productivity and health care systems. China has turned around and is investing heavily to reduce air pollution [66]. The air transportation emission standards in China are at par with European standards [67]. Mainly china is investing in solar energy, and by 2024 it will be generating twice in comparison to the USA [68]. The world has to unite in the effort as air pollution is a transborder phenomenon and needs to combine effort from regional countries. By acting together, the world can safeguard the habitats that are getting toxic and affect billions of human lives, plants and trees, and aquatic life, essential for the ecosystem's substance.

The air pollution can reduce to a level if collective efforts are in the right direction. The awareness programs should start in developing countries, educating the masses about the effects of air pollution on human health and the environment. The primary cause of air pollution is due to transportation. The developing countries should invest in upgrade the transportation system in the cities. Local bodies and city policymakers should provide facilities to reduce traffics on the road. Mobile Applications should introduce in for pooling resources and transportation for passengers going to the same destination. Air pollution is because of the high usage of pesticides.

and stubble burning to vacate lands for the next harvest in rural areas. The government should make laws to restrict and educate the masses in this regard. The farmers should use the right amount of pesticides to avoid excess chemicals in the air.

People should avoid trash burning, and the government should implement a trash disposal system. Countries should invest in alternate energy, and countries with prolonged exposure to sunlight and coastal areas with wind should identify to generate energy from solar and wind. These are clean energy and can generate energy with long term wait, unlike hydraulic power. Hydral power generation is not possible everywhere. Tree plantation is also essential to reduce the greenhouse effect and decrease CO2 in the atmosphere. The government should invest and enact laws for the purpose. All fuels containing lead and heavy metal must put to a legal stop. Euro 5 standard fuels are the best examples; the compliant fuel has

Figure 7.
Ten air polluted city in the world [46].

shallow content polycyclic aromatic hydrocarbons and sulfur to reduce unsafe environmental discharges. Education is the best weapon to use in the fight against air pollution, and it needs government and joint public effort, but once achieved, the whole community will enjoy the fruits and generations to come. Earth is unique and the only home which could sustain living beings. It is human beings' moral obligation to protect them from degradation, only possible by stopping all types of activities that cause air pollution.

A. Appendix 1

Website	Web address
Air Now	https://www.airnow.gov/
World's Air Pollution	https://waqi.info/
World Air Quality	https://www.iqair.com/us/world-air-quality
Purple Air	www.purpleair.com
EPA United States Environmental Protection Agency	www.epa.gov/outdoor-air-quality-data
Air Pollution in the world	www.aqicn.org/map/world/
Green Facts	www.greenfacts.org/en/ozone-o3/links/index.htm

Author details

Syed Abdul Rehman Khan[1,2*], Adeel Shah[3] and Zhang Yu[2]

1 School of Management, Xuzhou University of Technology, Xuzhou, China

2 School of Economics and Management, Chang'an University, Xi'an, China

3 Institute of Business Management (IOBM), Karachi, Pakistan

*Address all correspondence to: sarehman_cscp@yahoo.com

References

[1] Morawska, L., & He, C. (2014). Indoor particles, combustion products and fibres. In *Indoor Air Pollution* (pp. 37-68). Springer.

[2] Paraschiv, L. S., Serban, A., & Paraschiv, S. (2020). Calculation of combustion air required for burning solid fuels (coal/biomass/solid waste) and analysis of flue gas composition. Energy Reports, 6, 36-45. https://doi.org/https://doi.org/10.1016/j.egyr.2019.10.016

[3] Nathanson, J. (2020, October 19). *air pollution | Effects, Causes, Definition, & Facts | Britannica*. Britannica. https://www.britannica.com/science/air-pollution

[4] Smith, K. R. (1993). Fuel combustion, air pollution exposure, and health: the situation in developing countries. Annual Review of Energy and the Environment, *18*(1), 529-566.

[5] Jassim, M. S., & Coskuner, G. (2017). Assessment of spatial variations of particulate matter (PM 10 and PM 2.5) in Bahrain identified by air quality index (AQI). *Arabian Journal of Geosciences*, *10*(1), 19.

[6] Landrigan, P. J. (2017). Air pollution and health. In *The Lancet Public Health* (Vol. 2, Issue 1, pp. e4–e5). Elsevier Ltd. https://doi.org/10.1016/S2468-2667(16)30023-8

[7] Forouzanfar, M. H., Afshin, A., Alexander, L. T., Anderson, H. R., Bhutta, Z. A., Biryukov, S., Brauer, M., Burnett, R., Cercy, K., & Charlson, F. J. (2016). Global, regional, and national comparative risk assessment of 79 behavioural, environmental and occupational, and metabolic risks or clusters of risks, 1990-2015: a systematic analysis for the Global Burden of Disease Study 2015. The Lancet, *388*(10053), 1659-1724.

[8] Molla, M. F. (2020). Assessment of the Carbon Footprint Across Urban Households in Kolkata. In *Habitat, Ecology and Ekistics* (pp. 115-131). Springer.

[9] Gupta, E. A., & Jain, A. (2020). *Role of Bio Gas Plant for Empowering the Women Communities in Rural Area of India*.

[10] Prüss-Üstün, A., Wolf, J., Corvalán, C., Bos, R., & Neira, M. (2016). *Preventing disease through healthy environments: a global assessment of the burden of disease from environmental risks*. World Health Organization.

[11] Kampa, M., & Castanas, E. (2008). Human health effects of air pollution. Environmental Pollution, *151*(2), 362-367.

[12] Al-Salem, S. M., & Khan, A. R. (2010). Monitoring and modelling the trends of primary and secondary air pollution precursors: The case of the state of Kuwait. *International Journal of Chemical Engineering*, *2010*.

[13] Tang, H., & Dhari, A. (2008). Introduction to Air Pollution and Air Quality Management. *Environment and Urban Development Division Kuwait Institute for Scientific Research*.

[14] Haryanto, B. (2012). *Air Pollution: A Comprehensive Perspective*. BoD–Books on Demand.

[15] Abdullah, M., Hamzah, N., Ali, M. H., Tseng, M.-L., & Brander, M. (2020). The Southeast Asian haze: The quality of environmental disclosures and firm performance—*Journal of Cleaner Production*, *246*, 118958.

[16] Ruggieri, M., & Plaia, A. (2012). An aggregate AQI: comparing different standardizations and introducing a variability index. Science of the Total Environment, *420*, 263-272.

[17] Van den Elshout, S., Léger, K., & Heich, H. (2014). CAQI common air quality index—update with PM2. 5 and sensitivity analysis. Science of the Total Environment, *488*, 461-468.

[18] EPA. (2009). *Atmosphere | National Geographic Society*. National Geographic. https://www.nationalgeographic.org/encyclopedia/atmosphere-RL/

[19] Gore, R. W., & Deshpande, D. S. (2017). An approach for classification of health risks based on air quality levels. *2017 1st International Conference on Intelligent Systems and Information Management (ICISIM)*, 58-61. https://doi.org/10.1109/ICISIM.2017.8122148

[20] Kumari, S., & Jain, M. K. (2018). A critical review on air quality index. In *Environmental Pollution* (pp. 87-102). Springer.

[21] EPA. (2020). *Wildfire Smoke and Your Patients' Health: The Air Quality Index*. https://www.epa.gov/wildfire-smoke-course/wildfire-smoke-and-your-patients-health-air-quality-index

[22] Deep, B., Mathur, I., & Joshi, N. (2020). Coalescing IoT and Wi-Fi technologies for an optimized approach in urban route planning. Environmental Science and Pollution Research, *27*(27), 34434-34441.

[23] Cardinal, D. (2018, November 21). *How Air Quality and the AQI Are Measured - ExtremeTech*. Extreme Tech. https://www.extremetech.com/electronics/280956-how-air-quality-and-the-aqi-are-measured

[24] Meyer, M. D., & Elrahman, O. A. B. T.-T. and P. H. (Eds.). (2019). *Chapter 4 - Air and water pollution: An important nexus of transportation and health* (pp. 65-106). Elsevier. https://doi.org/https://doi.org/10.1016/B978-0-12-816774-8.00004-9

[25] Lindsey, R. (2018). Climate change: atmospheric carbon dioxide. *National Oceanographic and Atmospheric Administration, News & Features. August*.

[26] Carinci, F., Benedetti, M. M., Klazinga, N. S., & Uccioli, L. (2016). Lower extremity amputation rates in people with diabetes as an indicator of health systems performance. A critical appraisal of the data collection 2000-2011 by the Organization for Economic Cooperation and Development (OECD). Acta Diabetologica, *53*(5), 825-832.

[27] WHO. (2020). *Air pollution*. https://www.who.int/health-topics/air-pollution#tab=tab_1

[28] Hanson, M., Gluckman, P., & Bustreo, F. (2016). Obesity and the health of future generations. The Lancet Diabetes & Endocrinology, *4*(12), 966-967.

[29] Kioumourtzoglou, M.-A., Schwartz, J. D., Weisskopf, M. G., Melly, S. J., Wang, Y., Dominici, F., & Zanobetti, A. (2016). Long-term PM2.5 Exposure and Neurological Hospital Admissions in the Northeastern United States. Environmental Health Perspectives, *124*(1), 23-29. https://doi.org/10.1289/ehp.1408973

[30] Mandell, J. T., Idarraga, M., Kumar, N., & Galor, A. (2020). Impact of Air Pollution and Weather on Dry Eye. Journal of Clinical Medicine, *9*(11), 3740.

[31] Huang, Y.-C. T., & Ghio, A. J. (2006). Vascular effects of ambient pollutant particles and metals. Current Vascular Pharmacology, *4*(3), 199-203. https://doi.org/10.2174/157016106777698351

[32] Künzli, N., & Tager, I. B. (2005). Air pollution: from lung to heart. Swiss Medical Weekly, *135*(47-48), 697-702.

[33] Suzuki, T., Hidaka, T., Kumagai, Y., & Yamamoto, M. (2020). Environmental pollutants and

the immune response. Nature Immunology, 1-10.

[34] Beelen, R., Raaschou-Nielsen, O., Stafoggia, M., Andersen, Z. J., Weinmayr, G., Hoffmann, B., Wolf, K., Samoli, E., Fischer, P., & Nieuwenhuijsen, M. (2014). Effects of long-term exposure to air pollution on natural-cause mortality: an analysis of 22 European cohorts within the multicentre ESCAPE project. The Lancet, *383*(9919), 785-795.

[35] Balmes, J. R., Fine, J. M., & Sheppard, D. (1987). Symptomatic Bronchoconstriction after Short-Term Inhalation of Sulfur Dioxide1, 2. Am Rev Respir Dis, *136*, 1117-1121.

[36] Kagawa, J. (1985). Evaluation of biological significance of nitrogen oxides exposure. The Tokai Journal of Experimental and Clinical Medicine, *10*(4), 348-353.

[37] Shahid, M., Natasha, Dumat, C., Niazi, N. K., Xiong, T. T., Farooq, A. B. U., & Khalid, S. (2020). *Ecotoxicology of Heavy Metal(loid)-Enriched Particulate Matter: Foliar Accumulation by Plants and Health Impacts* (pp. 1-49). Springer New York. https://doi.org/10.1007/398_2019_38

[38] Ghio, A. J., & Huang, Y.-C. T. (2004). exposure to concentrated ambient particles (CAPs): a review. Inhalation Toxicology, *16*(1), 53-59. https://doi.org/10.1080/08958370490258390

[39] Uysal, N., & Schapira, R. M. (2003). Effects of ozone on lung function and lung diseases. Current Opinion in Pulmonary Medicine, *9*(2), 144-150. https://doi.org/10.1097/00063198-200303000-00009

[40] Chauhan, A. J., Krishna, M. T., Frew, A. J., & Holgate, S. T. (1998). Exposure to nitrogen dioxide (NO2) and respiratory disease risk. Reviews on Environmental Health, *13*(1-2), 73-90.

[41] Rastogi, S. K., Gupta, B. N., Husain, T., Chandra, H., Mathur, N., Pangtey, B. S., Chandra, S. V, & Garg, N. (1991). A cross-sectional study of pulmonary function among workers exposed to multimetals in the glass bangle industry. American Journal of Industrial Medicine, *20*(3), 391-399. https://doi.org/10.1002/ajim.4700200311

[42] Tager, I. B., Balmes, J., Lurmann, F., Ngo, L., Alcorn, S., & Künzli, N. (2005). Chronic exposure to ambient ozone and lung function in young adults. *Epidemiology (Cambridge, Mass.)*, *16*(6), 751-759. https://doi.org/10.1097/01.ede.0000183166.68809.b0

[43] Kuo, C.-Y., Wong, R.-H., Lin, J.-Y., Lai, J.-C., & Lee, H. (2006). Accumulation of chromium and nickel metals in lung tumors from lung cancer patients in Taiwan. *Journal of Toxicology and Environmental Health. Part A*, *69*(14), 1337-1344. https://doi.org/10.1080/15287390500360398

[44] Nawrot, T., Plusquin, M., Hogervorst, J., Roels, H. A., Celis, H., Thijs, L., Vangronsveld, J., Van Hecke, E., & Staessen, J. A. (2006). Environmental exposure to cadmium and risk of cancer: a prospective population-based study. The Lancet. Oncology, *7*(2), 119-126. https://doi.org/10.1016/S1470-2045(06)70545-9

[45] Halkos, G., & Argyropoulou, G. (2020). Pollution and Health Effects: A Nonparametric Approach. Computational Economics, 1-24.

[46] IQAIR. (2019). *Empowering the World to Breathe Cleaner Air | IQAir*. https://www.iqair.com/us/

[47] Altemose, B., Robson, M. G., Kipen, H. M., Strickland, P. O., Meng, Q., Gong, J., Huang, W., Wang, G., Rich,

D. Q., & Zhu, T. (2017). Association of air pollution sources and aldehydes with biomarkers of blood coagulation, pulmonary inflammation, and systemic oxidative stress. Journal of Exposure Science & Environmental Epidemiology, *27*(3), 244-250.

[48] Bhandari, B. S., & Bijlwan, K. (2019). Effects of Atmospheric Pollutants on Biodiversity. In *Global Perspectives on Air Pollution Prevention and Control System Design* (pp. 142-173). IGI Global.

[49] Ewan, K. B., & Pamphlett, R. (1996). Increased inorganic mercury in spinal motor neurons following chelating agents. Neurotoxicology, *17*(2), 343-349.

[50] Forde, M. S., Dewailly, E., Robertson, L., Sidi, E. A. L., Côté, S., Dumas, P., & Ayotte, P. (2014). Prenatal exposure to persistent organic pollutants and polybrominated diphenyl ethers in 10 Caribbean countries. Environmental Research, *133*, 211-219.

[51] McCarthy, N. (2020). *How does air pollution affect the economy? | World Economic Forum*. World Economic Forum. https://www.weforum.org/agenda/2020/02/the-economic-burden-of-air-pollution

[52] Myllyvirta, L. (2020). Quantifying the economic costs of air pollution from fossil fuels. *Centre for Research on Energy and Clean Air*.

[53] MacCarthy Niall. (2020). *Chart: The Economic Burden Of Air Pollution | Statista*. Statista. https://www.statista.com/chart/20804/costs-of-air-pollution-from-fossil-fuels/

[54] Lave, L. B., & Seskin, E. P. (2013). *Air pollution and human health*. Routledge.

[55] Ali, S. H., & Puppim de Oliveira, J. A. (2018). Pollution and economic development: an empirical research review. Environmental Research Letters, *13*(12), 123003. https://doi.org/10.1088/1748-9326/aaeea7

[56] OECD. (2016). *Economic Consequences of Outdoor Air Pollution*. Organisation for Economic Co-operation and Development.

[57] Qiu, Q., Wang, Y., Qiao, S., Liu, R., Bian, Z., Yao, T., & Nguyen, T. S. (2020). Does air pollution affect consumer online purchasing behavior? The effect of environmental psychology and evidence from China. *Journal of Cleaner Production*, 120795.

[58] Karthik, T., & Gopalakrishnan, D. (2014). Environmental analysis of textile value chain: an overview. In *Roadmap to Sustainable Textiles and Clothing* (pp. 153-188). Springer.

[59] Haseeb, M., Haouas, I., Nasih, M., Mihardjo, L. W. W., & Jermsittiparsert, K. (2020). Asymmetric impact of textile and clothing manufacturing on carbon-dioxide emissions: Evidence from top Asian economies. Energy, *196*, 117094. https://doi.org/https://doi.org/10.1016/j.energy.2020.117094

[60] Heydebreck, F., Tang, J., Xie, Z., & Ebinghaus, R. (2016). Emissions of Per- and Polyfluoroalkyl Substances in a Textile Manufacturing Plant in China and Their Relevance for Workers' Exposure. Environmental Science & Technology, *50*(19), 10386-10396. https://doi.org/10.1021/acs.est.6b03213

[61] Bucklow, J., Perry, P., & Ritch, E. (2017). The influence of eco-labelling on ethical consumption of organic cotton. In *Sustainability in fashion* (pp. 55-80). Springer.

[62] Parvin, F., Islam, S., Akm, S. I., Urmy, Z., & Ahmed, S. (2020). A Study on the Solutions of Environment Pollutions and Workers Health Problems Caused by Textile Manufacturing

Operations. Biomedical Journal of Scientific & Technical Research, *28*(4), 21831-21844.

[63] *Environmental Standards & Rules for Leather & Textile Industry - Fibre2Fashion*. (2014). Fibre2Fashion. https://www.fibre2fashion.com/industry-article/7287/environmental-standards-for-reducing-pollution-from-textile-and-leather-industry

[64] Hemachandra, D. W. K. (2015). *Adoption of voluntary environmental practices: evidence from the textile and apparel industry in Sri Lanka*.

[65] Helan, R. (2020). *India has 21 of the world's 30 cities with the worst air pollution - CNN*. CNN. https://edition.cnn.com/2020/02/25/health/most-polluted-cities-india-pakistan-intl-hnk/index.html

[66] Broom, D. (2020). *India dominates the list of the world's most polluted cities | World Economic Forum*. World Economic Forum. https://www.weforum.org/agenda/2020/03/6-of-the-world-s-10-most-polluted-cities-are-in-india/

[67] ICCT. 2020 (n.d.). *China: Light-duty: Emissions | Transport Policy*. Tnansportpolicy.Net. Retrieved November 26, 2020, from https://www.transportpolicy.net/standard/china-light-duty-emissions/

[68] *Global Trends in Renewable Energy Investment 2018 | Capacity4dev*. (n.d.). Retrieved November 26, 2020, from https://europa.eu/capacity4dev/unep/documents/global-trends-renewable-energy-investment-2018

Chapter 2

Implication of Secondary Atmospheric Pollutants in the Air Quality: A Case-Study for Ozone

David Galán Madruga

Abstract

Air quality and Public Health are concepts linked to each other. Within the frame of Public Health, a wide range of external factors, derived from rising wastes towards all environmental compartments, may generate harmful effects on human health. In particular, the release of polluting compounds into the ambient air coming from emission sources is a paramount concern, given that atmospheric pollution is considered the most significant environmental risk for human beings. In this context, while this chapter to provide an overview of the most critical air pollutants that can depict air quality status in terms of exposure, potential effects, emission sources, and types of pollutants, the principal purpose is focused on secondary atmospheric pollutants, emphasizing to tropospheric ozone as a significant pollutant within this group. In this sense, aspects such as the atmospheric ozone chemistry responsible for its formation and its spatial distribution into vast territories, including urban, suburban, and rural environments, were conveniently explained. Based on displayed evidence, primaries air pollutants, mainly nitrogen oxides, volatile organic compounds, and carbon monoxide, are responsible for the tropospheric ozone's formation; therefore, reducing their levels could be translated into a decrease of ozone concentrations at the ground-level. Attending to the ozone distribution, the revealed findings lead to the next concentration gradient: higher ozone levels in rural, followed by suburban and urban sites, respectively. Finally, it can be concluded that the importance of tropospheric ozone within air quality lies in the possibility of producing harmful effects on human health and generating climate changes, either directly or indirectly.

Keywords: air quality, atmospheric pollution, emission sources, secondary polluting compounds, tropospheric ozone's chemistry, and importance of ozone within air quality frame

1. Introduction

The human being's health is regulated by different agents, such as adequate alimentation, appropriate sanitation and hygiene, safe workplaces, and sustainable cities, among others. In this sense, environmental health points to be one of the most relevant factors influencing human beings' well-being, given the existing direct relationship between human beings and the surrounding environment. Therefore, both variables (environment and human being) should keep an equilibrium.

Nevertheless, within that binomial, human activities can negatively affect the whole environment generating environmental variations at the contamination level, or climate change, at the global level [1–3]. As a consequence, a deteriorated environment can seriously affect the human being's health.

In this sense, the population growth featured over the last decades has been translated into a fast industrial grown, an increment of transportation networks and energy demands, generating a negative effect on the different environmental departments due to rising wastes [4]. In particular, the rise of polluting emissions into the atmosphere leads to a decrease in air quality, thereby becoming a public health issue. In fact, atmospheric pollution is considered the major environmental risk to human being's health worldwide [5].

According to Landy (see [6]), atmospheric pollution is defined as the presence in the air of one or more contaminants in such a concentration and of such duration as causing a nuisance or being injurious to human life, animal life, or vegetation. Therefore, atmospheric pollution and air quality are intimately linked, so a lower level of atmospheric pollution carries a higher air quality and vice versa.

Atmospheric pollution covers a wide variety of pollutants, such as gaseous compounds (organic and inorganic chemicals) and airborne particulate matter. The pollutants used as air quality indicators are defined as "criteria air pollutants" [see https://www.environment.gov.au/protection/air-quality/air-pollutants, accessed October 28, 2020], and they are described as coming. So, polluting species as nitrogen oxides (NO_x), sulfur oxides (SO_2), carbon monoxide (CO), ozone (O_3), particulate matter (PM), and lead (Pb) are air pollutants commonly found in urban environments. In the case of nitrogen oxides, the most paramount compounds at the atmospheric level point to nitrogen monoxide (NO) and nitrogen dioxide (NO_2). In ambient air, NO is a highly unstable molecule rapidly oxidizing in the presence of oxygen, yielding nitrogen dioxide; therefore, they sustain a direct chemistry relationship. For this motive, NO is also considered a free radical [7]. NO_2 is a gaseous reddish-brown compound, toxic and irritating. It has a characteristic pungent odor and is a strong oxidant, and reacts with water to produce nitric acid (HNO_3). While at the high ambient NO concentrations, this oxidation reaction is based on the oxygen (O_2) presence [8], at the low levels, the molecular O_3 is responsible for oxidizing, within minutes, NO molecule [9]. SO_2 is a dense and colorless gas with a suffocating odor that readily soluble in water. Its oxidation generates the formation of sulfuric acid (H_2SO_4) by reacting with the surface of particles in the presence of metallic catalysts [10]. CO is a colorless, odorless, and tasteless gas, poorly soluble in water, and it has a slightly lower density than air [11]. O_3 is one of the most potent oxidizing agents due to its high redox potential. It is an essential constituent of the upper atmosphere [12]. Relative to the PM, it is needed to indicate that these pollutants are characterized in function on its equivalent aerodynamic diameter (EAD). It is a method for classifying PM [13]. At the atmospheric level, the most studied particles aimed at $PM_{2.5}$ (EAD of 2.5 μm or less) and PM_{10} particles (EAD $\leq$ 10 μm). As a general concept, atmospheric particulate matter consists of a complex mixture formed by solid and liquid organic and inorganic origin substances suspended in the air. Therefore, the particle composition can differ in distinct variables, such as emission sources and meteorological features affecting a specific location. Finally, Pb is a toxic metal, and it is associated with particulate matter in ambient air [14].

At the atmospheric level, a broad and important group of polluting agents are included within the volatile organic compounds' family (VOC's). Benzene (C_6H_6) is considered the major COV's with particular attention to urban environments. It is a highly volatile compound, and it is degraded rapidly in the upper atmosphere. Given its solubility in water, the atmospheric benzene can be deposited on surface waters and soil due to rain action, removing a minor amount of air benzene [https://www.who.int/ipcs/features/benzene.pdf, accessed November 9, 2020].

The ambient air's target pollutant levels will depend on geographic area, topographic, emission sources, and meteorological variables. Air pollutants' occurrence is directly related to release processes from emission sources and their formation through chemical reactions between several pollutants under specific meteorological conditions.

Firstly, this work pretends to introduce an overview concerning air quality in terms of exposure to air pollutants, types of contaminants, and emission sources. Within this frame, an essential aspect of the air quality is supported by the secondary atmospheric pollutants. In this sense, on the other hand, this work aims to display the importance of this type of compounds, focused mainly on ozone.

2. Exposure to air pollutants on human being and potential effects on health

Numerous scientific studies have revealed a direct relationship between exposure to poor air quality and the appearance of harmful effects on human health [15–17]. Broadly, long and short-term exposures to toxicant compounds suspended into the air may yield different toxicological impacts on humans, such as cardiovascular and respiratory diseases, eyes and skin irritation, and chronic illness, among others.

Exposure to air pollutants on the human being depends on both the target air pollutant's concentration, usually expressed in $\mu g/m^3$ and the exposure time to that polluting level with a unit in minutes. There are different air pollutant exposure models (APEX). The scientific community has widely described APEX [18–20]. They are based on probabilistic analysis simulating different exposure environments using several study variables, such as population data, activities developed by people, traffic density, meteorological features, emission air pollutant data, potential fixed and mobile emission sources [https://www.epa.gov/sites/production/files/2019-11/documents/apex5_introduction_document.pdf, accessed November 13, 2020].

Given the considerable number of polluting compounds existing in the atmosphere, a brief review relative to potential effects of criteria air pollutants (further COV's) on human being's health is addressed.

2.1 Nitrogen oxides

Exposure to nitrogen oxides may raise the risk of respiratory infections [21], such as burning eyes, sore throat, or cough [22]. Exposure to levels ranged between 1300 and 3800 $\mu g/m^3$ (0.7–2.0 ppm, parts of million) for 10 min may increase the inspiratory and expiratory flow resistance [8].

2.2 Sulfur oxide

Sulfur dioxide produces an extensive glossary of harmful effects on human health. The most likely results point to wheezing, shortness of breath, and chest tightness, which are intensified during physical activity development. According to the American Environmental Pollution Agency [23], an increase in respiratory symptoms and a lung ability reduction due to durable exposures at high sulfur dioxide levels may be produced. Similarly, it has been linked to cardiovascular disease [24].

2.3 Carbon monoxide

The symptoms derived from carbon monoxide exposure are not always obvious. Once inhaled, CO readily reacts with existing hemoglobin in the blood, which is the protein responsible for the transport of O_2 in blood [25], reducing the blood's oxygen-carrying capacity [26]. Carbon monoxide has a greater affinity for hemoglobin than oxygen (>200 times, approximately) [27]. The amount of hemoglobin that has been reacted with CO is named carboxyhemoglobin [28].

At low concentrations, the most common symptom results in a headache. However, other effects, like dizziness, feeling and being sick, tiredness and confusion, stomach pain, shortness of breath, and difficulty breathing, are willing to appear [29]. Finally, the inhalation of carbon monoxide at high concentrations is lethal [30].

2.4 Ozone

In the function of the exposure concentration, the effects occasioned by ozone inhalation can be classified as acute and chronic. The first group embraces inflammatory mediators' alterations, morbidity, and mortality, while the second points to lung function reduction, atherosclerosis and asthma development, and reduction in life expectancy [31].

2.5 Airborne particulate matter

The major effects of atmospheric particles on human being's health appear in the cardiovascular system due to the mechanisms of systemic inflammation, direct and indirect coagulation activation, and direct translocation into the systemic circulation [32]. Similarly, scientific evidence shows that ozone exposure exacerbates respiratory diseases. The inhalation of particulate matter produces oxidative stress and inflammation, leading to pulmonary anatomic and physiologic remodeling [33].

2.6 Lead

Exposure to Pb may affect adults, children, and unborn babies' health. It is stored in bones and teeth, even damaging different body parts like the liver, kidneys, and brain. Pb circulates in blood once into the body. The levels of Pb in blood express micrograms of Pb per deciliter of blood [34]. Pb's intoxication may drive to the next symptoms: abdominal pain, anemia, constipation, headaches, irritability, loss of concentration and memory, and sleep disorders [35].

2.7 Volatile organic compounds

Broadly, inhaled volatile organic compounds are associated with oxidative stress and decreased lung function [36]. In the case of benzene, since 1987, it has been included in group 1 of the IARC carcinogenic classification system [37].

Through the International Programme on Chemical Safety, World Health Organization has developed a series of documents concerning "Environmental Health Criteria, EHC" (more than 200 issues). While these EHC encompass for each studied air pollutant different aspects relative to emission sources, atmospheric levels, among others, any reader interested in delving into subjects regarding the effects of air pollutants on human health can access the next link: https://www.who.int/ipcs/publications/ehc/ehc_numerical/en/, accessed November 13, 2020.

Finally, it is relevant to highlight Public Administrations's growing interest in controlling and assessing toxicant compounds in ambient air to protect human being's health. Within this frame, the European Union has developed Air Quality Standards for setting limit or target values for several air pollutants. Directive 2008/50/EC (see [38]) lays down the following limit values: 40 μg/m^3 for NO_2 and PM_{10}, 5 μg/m^3 for C_6H_6, 0,5 μg/m^3 for Pb as calendar year averages, 10 mg/m^3 for CO as maximum daily eight-hour mean and 125 μg/m^3 as one-day average (not to be exceeded more than three times a calendar year).

3. Emission sources of air pollutants

3.1 Natural sources

Firstly, natural emission sources of air pollutants do not depend on human activity. They mainly characterize the occurrence of polluting agents into the atmosphere at the local scope. Among others, it is worth mentioning as natural origins of air pollutants, forest fires, dust storms, volcanos, emissions from the water surface, microbial activity, and anaerobic degradation of organic material in terrestrial environments.

Natural sources' contribution to air pollutants' presence in ambient air is lower quantitatively than that produced by anthropogenic sources.

3.2 Anthropogenic sources

Anthropogenic emission sources result from human activity. Therefore, large urban areas support highly-polluting emissions into the ambient air due to many emission focus (road traffic, heating house, and industry, mainly), which generates relevant health issues.

Dominant emission air pollutant sources derived from human activity are:

- Fuel combustion from motor vehicles
- Power generation
- Industrial activities
- Municipal and agricultural waste sites as well as its incineration/burning
- Residential cooking/heating, and lighting with polluting fuels (https://www.who.int/airpollution/ambient/pollutants/en/, accessed November 16, 2020)

4. Types of pollutants

At the scientific level, there are several categories of classifying air pollutants. So, they can be ranked in terms of the type of emission source (primary or secondary pollutant), chemical composition (gaseous agents or particulate matter), place of release into the atmosphere (indoor or outdoor site) [39]. This section will encompass their possible origin in emission sources' function within this work's common thread.

4.1 Primary pollutants

Primary polluting compounds involve those pollutants that are directly emitted into the atmosphere. Their origin can be natural (for example, volcanos or grassland fires) and anthropogenic (industrial and vehicular emissions).

4.2 Secondary pollutants

Secondary air pollutants are not generated from emission sources, but they need at least two primary pollutants to react with each other to yield a secondary agent [40].

Based on the previous description, NO_x, SO_2, CO, PM are examples of primary air pollutants, whereas O_3 and peroxyacylnitrates (such as peroxyacetylnitrate, peroxy-propionyl nitrate, and peroxybenzoylnitrate) are secondary pollutants.

5. Ozone

5.1 Description

The ozone is a powerful oxidant agent capable of oxidizing organic compounds and a precursor to yield hydroxyl radicals [41]. Only fluorine, oxygen fluoride, and atomic oxygen have a higher redox potential than O_3. It is formed by three oxygen atoms and is a greenhouse gas [42]. While ozone is a constituent of the Earth's upper atmosphere (stratosphere) [43], exceeding 1000 parts per billion (ppb) concentrations [44], it also occurs at ground-level (troposphere) [45]. Ground-level O_3 concentrations rise with the height from 20 to 200 ppb depending on proximity to primary air pollutants emission sources [46].

On the one hand, O_3 is beneficial when it is found in the atmosphere's upper layer, given that it forms a protective coating for human beings against ultraviolet solar radiation [47]. Nevertheless, high levels of tropospheric O_3 (at the ground-level) may produce human health's harmful effects due to its high oxidant capacity [48], which appoints this compound as the most significant photochemical pollutant [49].

5.2 Ozone atmospheric chemical: formation at the tropospheric level

Given that the ozone is cataloged as a secondary air pollutant, its occurrence into the ambient air is due to chemical reactions among primary polluting compounds emitted directly from emission sources. Therefore, its tropospheric formation is associated with anthropogenic and biogenic emissions [50]. Reactions ruling ground-level ozone generation have been vehemence studied over past decades [51].

The essential primaries pollutants involved in yielding tropospheric ozone are NO_x (NO and NO_2), COVs and CO. Solar radiation is a primordial agent for ground-level ozone formation.

In the case of NO_x, once NO is released from emission sources into the atmosphere, this is oxidized to generate NO_2. The formation rate of this reaction is highly dependent on the ambient NO concentration. While high NO levels lead to a fast conversion to NO_2, low levels slow down its production rate.

Although the oxidative NO reaction's predominant oxidant agent is O_3, molecular oxygen can also ease its conversion to NO_2.

A fraction of the generated NO_2 undergoes conversion to NO during diurnal hours due to a photolysis process. This re-conversion is translated into yielding tropospheric O_3, according to Eqs. (1) and (2) [52].

$$\begin{array}{c} 290\,\text{nm} < \text{solar radiation} < 430\,\text{nm} \\ \swarrow \\ NO_2 \rightarrow NO + O*(\text{atomic oxygen}) \end{array} \tag{1}$$

$$O* + O_2\,(\text{molecular oxygen}) \rightarrow O_3 \tag{2}$$

A minimum ambient air NO_2 concentration (0.02–0.03 ppb) is required for generating O_3.

Relative to VOCs, the oxidation of these polluting compounds yields ground-level O_3. At the atmospheric level, this oxidative reaction is carried out in the presence of NO_x and sunlight [53]. The chemistry reactions responsible for the tropospheric O_3 generation from VOC's are shown in Equations from 3 to 6 [54].

$$VOCs + OH + O_2 \rightarrow RO_2 + H_2O \tag{3}$$

$$RO_2 + NO + O_2 \rightarrow NO_2 + H_2O \tag{4}$$

$$H_2O + NO \rightarrow NO_2 + OH \tag{5}$$

$$NO_2 + \text{sunlight} + O_2 \rightarrow NO + O_3 \tag{6}$$

Therefore, the VOCs oxidation is regulated by hydroxyl radicals (OH), oxygen, and NO_x.

Finally, the oxidative CO process drives the formation of hydroperoxyl radical (HO_2), which reacts with atmospheric NO for yielding NO_2, according to Eqs. (7) and (8) [55]. The generation of ground-level O_3 from NO_2 oxidation has already been explained previously.

$$CO + OH \rightarrow CO_2 + HO_2 \tag{7}$$

$$HO_2 + NO \rightarrow NO_2 + OH \tag{8}$$

At the European level, the highest tropospheric ozone is located on the Mediterranean coast (see **Figure 1**). According to the chemical formation of ground-level O_3, preventive measures driving toward abatement of NO_x, VOCs, and CO concentrations in ambient air would assist control of tropospheric O_3 levels.

It is relevant to highlight that the participation of ground-level O_3 in the atmospheric chemistry generates free radicals, such as hydroxyl radical (OH^*) [56], which governs the atmospheric lifetime of many species and their potential to contribute to climate change.

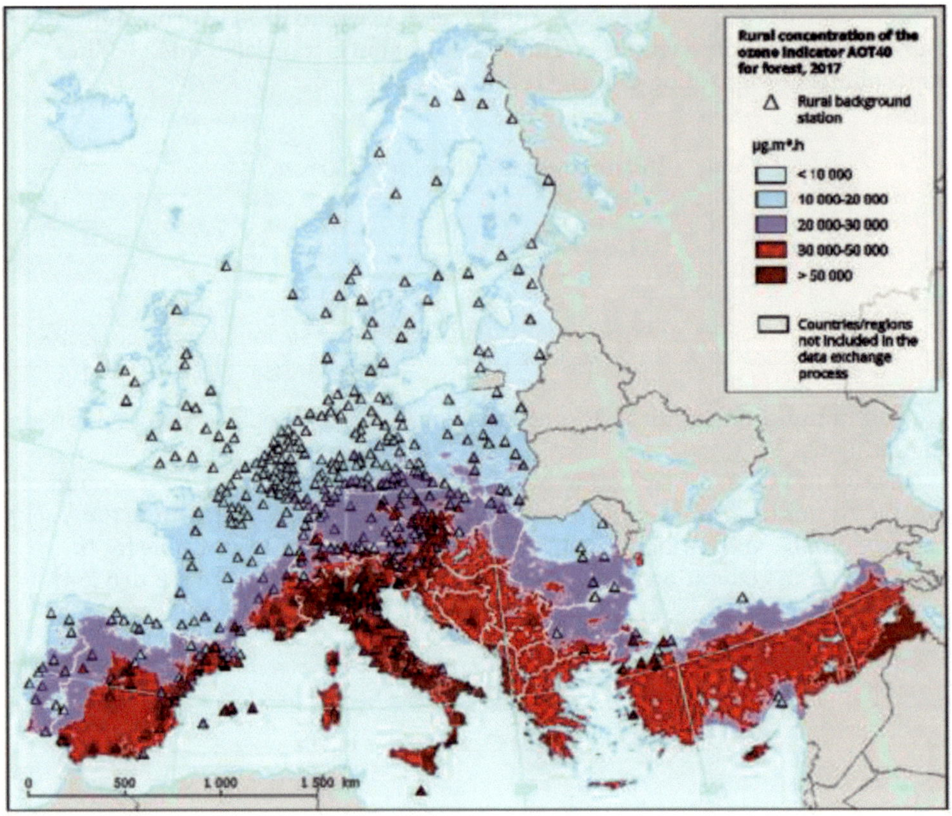

Figure 1.
Accumulated ozone exposure values for forest — over a threshold of 40 parts per billion (AOT40f) — for 2017, as calculated for the fusion maps and as measured at rural background stations. Source: European Environment Agency (Available at: https://www.eea.europa.eu/data-and-maps/figures/rural-concentration-of-the-ozone-5, accessed 18 November 2020).

5.3 Measurement techniques

Currently, there are several methodologies for measuring tropospheric O_3. Broadly, they can be classified into two clusters: (i) Those methods needing a sampling process and an ulterior analysis into the laboratory, and (ii) those methods measuring in real-time, collecting and determining the tropospheric O_3 levels in situ (at the target sampling point). Whereas in the first case, they are defined as passive methodology, in the second case, active methods.

5.3.1 Passive methodology

Passive sampling methods are valuable instruments for air pollution monitoring. They support ground-level O_3 measures' simultaneous performance at numerous sampling points, which results in remarkable relevance. Their implementation on an extensive study surface would picture spatial distribution concerning O_3 levels within that domain. Characteristics such as small, simple, lightweight instrumentation, low-cost, and minimum maintenance give it advantageous features against active methodology.

Fick's First Law, which postulates a mass transfer phenomenon through a layer of gas or a membrane, can explain the sampling process. It is based on a concentration gradient process, relating the flow of target gas that diffuses from regions of high concentration to regions of low concentration with the exposure time and the area of the passive sampler [57].

The gas collection is based on ozone's chemical reactions with a specific reagent for generating a final product, which is determined in the laboratory using appropriate analytical techniques [58].

Within this methodology, two types of samplers are available commercially: Radiello and Ogawa sampler, respectively. Radiello type sampler sustains a radial diffusion (the diffusive process occurs in all space directions), whereas Ogawa only supports an axial diffusion.

Suppose any reader needs more information about Radiello or Ogawa sampler. In this case, this one can be satisfied in the following links: https://www.restek.com/pdfs/radiello-manual.pdf for Radiello sampler and http://ogawausa.com/wp-content/uploads/2014/04/proozone.pdf for Ogawa sampler (both they were accessed November 19, 2020).

The spectrophotometric analysis is usually used for determining the formation of products generated by collecting ozone using passive samplers. This technique is based on colorimetric processes.

5.3.2 Active methodology

The traditional methods for monitoring ground-level ozone have been the automatic methods in continuous. Those methods are based on the direct measure of a pollutant in the air matrix. In the case particular of O_3, there are two specific methods:

- Chemiluminescence in the gas phase. It is based on photons' detection produced in the exothermic reaction between O_3 and ethene or nitrogen oxide [59].

  The chemiluminescent light emission's intensity (from 350 to 550 nm) is proportional to the ozone concentration in ambient air. This method is appropriate for O_3 measures into the concentration range from 0.001 to 100 ppm, obtaining a linear response. It does not have known interferences, and its response time is 1 sec, reaching up to 2% accuracy at 50 ppb ozone [60].

- Ultraviolet photometric. It is based on absorption that the molecular O_3 presents in the spectrum's ultraviolet region, whose maximum is situated highly close to the mercury emission line at 254 nm [61]. The instruments based on that technique evaluate O_3 levels from the existing relationship between the sequentially transmitted light intensities at 253.7 nm. The ultraviolet photometric is a more stable method than the chemiluminescence. However, it has a response time significantly higher (30 sec). Similarly, it can present interferences with carbonyl or aromatic compounds that absorb in the same UV region.

5.4 Ozone levels in the ambient air

The tropospheric O_3 concentrations are usually expressed at ppb or μg/m^3 (mass per unit volume). The conversion factor between both units depends on pollutant type (in this case, molecular O_3), ambient air temperature, and pressure, according to Eq. (9).

$$\mu g / m^3 = ppb * \frac{\text{molecular weight}}{\text{molecular volume (liters)}} \tag{9}$$

Where

$$\text{molecular volume} = 22.4\,\text{L} * \frac{\text{absolute temperature}(\text{k})}{273.15\,\text{k}} * \frac{1013.25\,\text{hPa}}{\text{atmospheric pressure}(\text{hPa})}$$

For tropospheric O_3, in order to transform ppb to μg/m^3, a conversion factor of 2.142, 2.066, and 1.962 should be used (conversion make to 0, 10, and 25°C, respectively) (http://www.apis.ac.uk/unit-conversion, accessed November 23, 2020).

A relevant research study used data monitored by several countries' air quality monitoring networks for reporting tropospheric O_3 patterns [62]. Those O_3 data grouped 808 urban and 300 rural fixed stations. In the urban environments, the reached findings reported annual mean values of daily O_3 concentrations ranged from 19.5 ppb in the United Kingdom to 27.2 ppb in the United States during the studied period, while in South Europe Region (Italy, France, and Spain), the levels were next to 24 ppb. The lowest O_3 concentrations occurred on weekdays, whereas

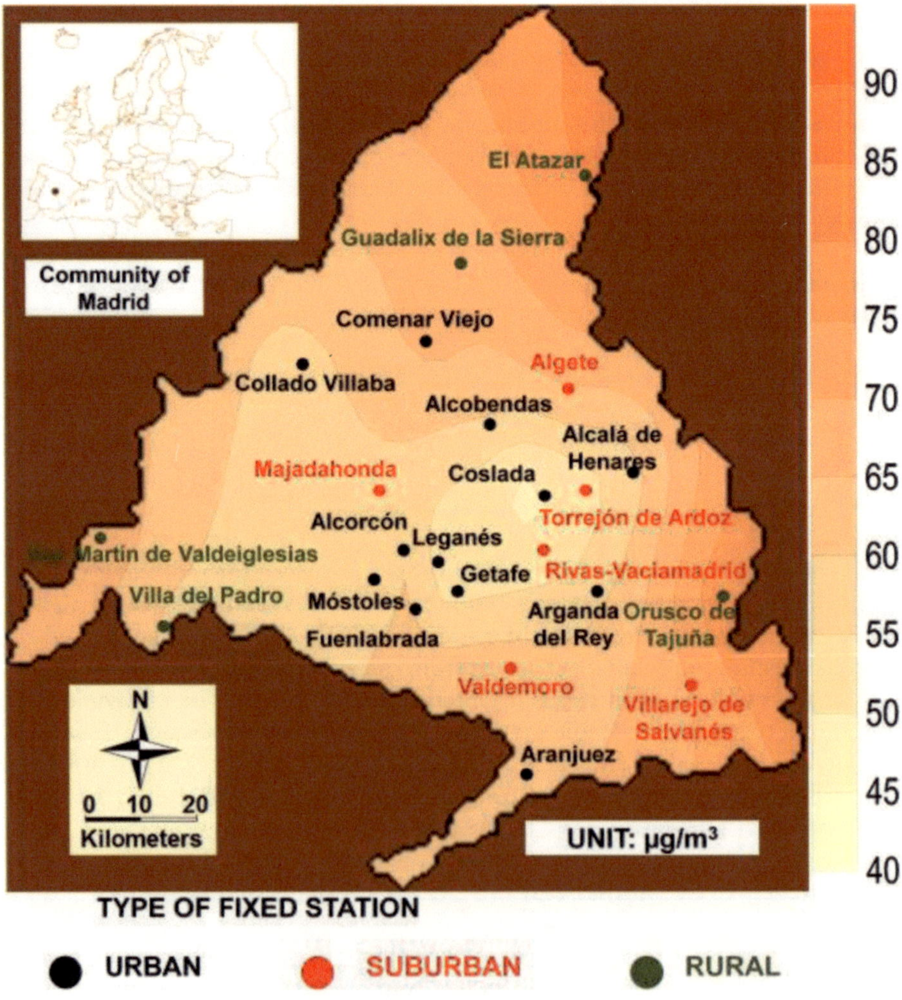

Figure 2.
Spatial distribution of the annual average O_3 concentrations within the Community of Madrid (2018). Source: Own elaboration using O_3 data monitored in 2018 and obtained from Community of Madrid's open-access database (https://www.comunidad.madrid/, accessed 20 November 2020).

those highest ones were found during weekends. On Monday, the daily mean O_3 concentrations were 1.4–3.8% higher than on other weekdays.

In rural environments, higher tropospheric O_3 concentrations than at urban sites were exhibited. The annual daily mean O_3 concentrations ranged between 28.5 ppb in the United Kingdom to 36.8 ppb in Japan over the investigated time. The O_3 concentrations between weekdays and weekends showed high stability (differences lower than 1 ppb).

While the highest levels of primary air pollutants are found in urban areas, given that their dominant emission sources are within these zones, the highest O_3 concentrations are pointed in semi-urban and rural environments, according to its tropospheric chemistry. The major reason sustaining this behavior is that tropospheric O_3 acts secondary pollutant, which needs a particular time for its formation in ambient air, displaced an absolute distance during that time concerning the prevalent emission primary source.

Ground-level O_3 data monitoring at 23 fixed stations from the Community of Madrid's air quality monitoring network during 2018 was used for picturing concentration gradient along all studied surface in order to knowledge the spatial distribution of tropospheric ozone into a typical South Europe Region. Community of Madrid is located in the center of Iberian Peninsula, it has a population of over 6,500,000 inhabitants, and it consists of 179 municipalities. The iso-concentration map of annual average O_3 levels in the Community of Madrid in 2018 was built by using a Geographic Information System. The concentrations at the not measured monitoring points were extrapolated from measurement points using the Kriging method as a geostatistical estimation tool [63].

Figure 2 shows spatial variation within the Community of Madrid's territory. As can be seen, higher annual O_3 concentrations at rural sites than suburban and urban areas were averaged during 2018, reaching a value of 70.8, 60.5, and 57.5 $\mu g/m^3$ at rural, suburban, and urban sites, respectively; therefore, higher levels were represented aways from the emission sources for primary air pollutants, such as NO, NO_2, CO, C_6H_6, SO_2, and particulate matter.

It is necessary to highlight that the annual O_3 typical average in cities reaches values between 20 and 30 ppb [64]. In the previous case study, the annual average concentration of O_3 in urban areas fell into the indicated range (29.31 ppb, it expressed to 25°C temperature and 1013.25 hPa pressure).

5.5 Current legislation

Over the last decades, the Public Administrations have shown a growing interest in environmental issues, particularly toward the air quality, rising the air pollutants

Objective	Averaging time	Target Value
Human health's protection	Maximum daily eight-hour mean	120 µg/m3 not to be exceeded on more than 25 days per civil year averaged over three years
Vegetation's protection	From May to July	AOT40 (calculated from 1 h values) 18 000 µg/m3 • h averaged over five years

AOT40 indicates the sum of the difference between hourly concentrations greater than 80 $\mu g/m^3$ and 80 $\mu g/m^3$ over a given period using one-hour values monitored from 8.00 a.p. to 20.00 p.m.

Table 1.
Target values relative to ground-level O_3.

control by implementing Air Quality Standards to reduce human beings' exposure to atmospheric polluting.

In this sense, European Union developed Directive 2008/50/CE [38] for setting air quality objectives in order to reduce harmful effects on human health and the environment. In the case of tropospheric O_3, the Directive lays down data quality objectives for measurements ambient air ground-level O_3 (it sets a measurement uncertainty of 15% for fixed measurements, with a minimum capture of data of 90% (during summer) and 75% (winter), as well as target values (see **Table 1**).

In this context, the Member States establish air quality monitoring networks (AQMNs) in their territories for testing compliance with those air quality objectives. In order to constitute an AQMN, the Directive 2008/50/EC sets criteria concerning classification, location, and a minimum number of sampling points. Nevertheless, aspects relative to each fixed station's representativeness within AQMN in a concrete area are not evaluated, which can be considered a limitation of the Air Quality Standard. As a consequence, it does not exist a harmonized methodology allowing assess the fixed stations' representativeness. Therefore, this issue should be addressed in terms of air quality management.

6. Consultation sites

In order to further deeper in Sections exhibiting formation and levels of ozone, the next links are displayed:

Chemistry formation of ozone.

https://earthobservatory.nasa.gov/features/ChemistrySunlight/chemistry_sunlight3.php (accessed November 12, 2020).

http://ozone.meteo.be/meteo/view/en/1547746-Formation+of+ozone.html (accessed November 23, 2020).

Levels of tropospheric ozone.

https://pubmed.ncbi.nlm.nih.gov/ (accessed October 29, 2020).

https://www.sciencedirect.com/ (accessed November 3, 2020).

In the two last databases, the Keywords for searching scientific documents involving O_3 levels are: (O_3 or ozone) & (ground-level or tropospheric) & (ambient air).

7. Importance of presence of ground-level ozone in air quality

Based on the previously reported information, tropospheric ozone is a highly reactive air pollutant. For this reason, it plays a crucial role in the ground-level atmospheric chemistry, given that its participation generates free radicals. At the health level, these last compounds are responsible for forming other ones that can induce potential human health's adverse effects, for example, pollutants driving acid rain or those implicated in deposition processes.

Therefore, tropospheric ozone is responsible for the emergence of harmful effects on human health in direct (by itself) and indirect ways (through the generation of other toxic compounds).

On the other hand, at the climate level, the tropospheric O_3 is a greenhouse gas that may produce climate changes by itself or by forming other air pollutants.

Given that its occurrence in ambient air depends on both the presence of other primaries atmospheric pollutants and meteorological conditions, Public Health's preventive measures focused on human health's protection and climate change should be implemented in terms of assessing air quality and executing air quality plans in order to reduce primaries atmospheric pollutant levels.

8. Conclusions

Air quality and human health are concepts associated with each other, given that good air quality influences adequate life quality in terms of health and well-being, and vice-versa.

Nevertheless, an increase of polluting emissions into the atmosphere due to growing industrial activities and rising private and public transport networks, among others, threatens clean air quality.

In this sense, responsibility administrations set criteria for evaluating the air quality in order to reduce human exposure to airborne pollutants. Within this context, the European Union lays down Air Quality Standards involving air quality objectives, such as limit and target values. According to these guidelines, Member States sustain air quality monitoring networks into their territories to assess the air quality and test the compliance with those air quality objectives.

The presence and pollutants' chemical composition scattered into the ambient air is tremendously heterogeneous, which difficulties the air quality assessment. Among pollutants compendium presents in the air, primaries and secondary compounds are found.

Knowledge of secondary's air pollutants is fundamental to elucidate their importance within the air quality frame. In particular, the ozone encompasses relevant aspects, given its high reactivity, in terms of (i) health and (ii) climate.

Tropospheric ozone plays a critical role within both items, either directly or indirectly. It is able to induce adverse effects on humans at the health level, as changes in climate. Both effects could be generated by ground-level ozone itself or by participating in the formation of other air pollutants with the capability of producing these effects.

It is generated from primary air pollutants, for what is displaced to zones where there are not prevailing emission sources for primary air pollutants. In this sense, the highest tropospheric O_3 levels are found in rural environments instead of suburban and urban zones (typical of primary air pollutants).

As a remark conclusion, attending to the ambient O_3 formation, control measures on emission sources of primaries air pollutants should be implemented in order to abate the potential tropospheric O_3 effects on health and climate.

Conflict of interest

The author declares no conflict of interest. This work does not have commercial purposes, only scientific ones.

Other declarations

This work did not receive any specific grant from funding agencies in the public, commercial, or not-for-profit sectors.

Author details

David Galán Madruga
Atmospheric Pollution Area, National Center for Environment Health, Carlos III Health Institute, Madrid, Spain

*Address all correspondence to: david.galan@isciii.es

References

[1] Sharif A, Godil DI, Xu B, Sinha A, Khan SAR, et al. Revisiting the role of tourism and globalization in environmental degradation in China: Fresh insights from the quantile ARDL approach. Journal of Cleaner Production. 2020;1:122906. DOI: 10.1016/j.jclepro.2020.122906

[2] Zhang Y, Khan SAR, Kumar A, Golpîra H, Sharif A. Is tourism really affected by logistical operations and environmental degradation? An empirical study from the perspective of Thailand. Journal of Cleaner Production. 2019;227:158-166. DOI: 10.1016/j.jclepro.2019.04.164

[3] Khan SAR, Sharif A, Golpîra H, Kumar A. A Green Ideology in Asian Emerging Economies: From Environmental Policy and Sustainable Development. Sustainable Development. 2019;27(2)

[4] Sharif A, Afshan S, Chrea S, Amel A, Khan SAR. The role of tourism, transportation and globalization in testing environmental Kuznets curve in Malaysia: new insights from quantile ARDL approach. Environmental Science and Pollution Research. 2020;27: 25494-25509. DOI: 10.1007/s11356-020-08782-5

[5] Madruga DG, Úbeda RM, Terroba JM, dos Santos SG, García-Cambero JP. Particle-associated Polycyclic Aromatic Hydrocarbons in a Representative Urban Location (Indoor-outdoor) from South Europe: Assessment of Potential Sources and Cancer Risk to Humans. Indoor Air. 2019;29(5):817-827. DOI: 10.1111/ina.12581

[6] Landy M. Environmental Impact Statement Glossary. A Reference Source for EIS Writers, Revuewers, and Citizens., IFI/Plenum Publishing Corporation, New York, 1979

[7] Ford PC, Miranda KM. The solution chemistry of nitric oxide and other reactive nitrogen species. Nitric Oxide. 2020;103:31-46. DOI: 10.1016/j.niox.2020.07.004

[8] Bagg J. The formation and control of oxides of nitrogen in air pollution, In: Strauss, W. ed. Air Pollution Control, Part I, NY, Wiley-Interscience. 1971;35-94

[9] Lin C, Feng X, Heal MR. Temporal persistence of intra-urban spatial contrasts in ambient NO_2, O_3 and O_x in Edinburgh, UK. Atmospheric Pollution Research. 2016;7:734-741. DOI: 10.1016/j.apr.2016.03.008

[10] Roy P, Sardar A. SO_2 Emission Control and Finding a Way Out to Produce Sulphuric Acid from Industrial SO_2 Emission. Journal of Chemical Engineering & Process Technology. 2015;6(2): 1000230. DOI: 10.4172/2157-7048.1000230

[11] WHO. Air Quality Guidelines for Europe. Second Edition. 2000. World Health Organization. Regional Office for Europe. Copenhagen. WHO Regional Publications, European Series N° 91. Page 75

[12] Environmental Health Criteria 4. Oxides of nitrogen. World Heath Organization1997. ISBN 92 4 154064 8. (Available at: http://www.inchem.org/documents/ehc/ehc/ehc004.htm, accessed November 6, 2020)

[13] Farahat A, El-Askary H, Dogan AU. Aerosols size distribution characteristics and role of precipitation during dust storm formation over Saudi Arabia. Aerosol and Air Quality Research. 2016;16(10): 2523-2534. DOI: 10.4209/aaqr.2015.11.0656

[14] Mari M, Martí N, Schuhmacher M, Domingo JL. Monitoring PCDD/

Fs, PCBs and metals in the ambient air of an industrial area of Catalonia, Spain. Chemosphere. 2008;73(6):990-998. DOI: 10.1016/j.chemosphere.2008.06.025

[15] Kim D, Chen Z, Zhou LF, Huang SX. Air pollutants and early origins of respiratory diseases. Chronic Diseases and Translational Medicine. 2018;4:75-94. DOI: 10.1016/j.cdtm.2018.03.003

[16] Zhu F, Ding R, Lei R, Cheng H, Liu J, et al. The short-term effects of air pollution on respiratory diseases and lung cancer mortality in Hefei: A time-series analysis. Respiratory Medicine. 2019;146:57-65. DOI: 10.1016/j.rmed.2018.11.019

[17] London Jr NR, Lina I, Ramanathan Jr M. Aeroallergens, air pollutants, and chronic rhinitis and rhinosinusitis. World Journal of Otorhinolaryngology-Head and Neck Surgery. 2018;4:209-215. DOI: 10.1016/j.wjorl.2018.08.006

[18] Jerret M, Arain A, Kanaroglou P, Beckerman B, Potoglou D, et al. A review and evaluation of intraurban air pollution exposure models. Journal of Exposure Analysis and Environmental Epidemiology. 2005;15:185-204

[19] Klepeis NE. Modeling Human Exposure to Air Pollution. 2006. In: Human Exposure Analysis, ed., Ott et al., CRC Press

[20] Beevers SD, Kitwiroon N, Williams ML, Kelly FJ, Anderson HR, Carslaw DC. Air pollution dispersion models for human exposure predictions in London. 2013. Journal of Exposure Science and Environmental Epidemiology. 23, 647-653

[21] Chen TM, Gokhale J, Shofer S, Kuschner WG. Outdoor air pollution: Nitrogen dioxide, sulfur dioxide, and carbon monoxide health effects. The American Journal of the Medicine Sciences. 2007;333:249-256. DOI: 10.1097/MAJ.0b013e31803b900f

[22] McKay CA. Toxin-induced respiratory distress. Emerging Medicine Clinica North America. 2014;32(1):127-147

[23] U.S. EPA. Integrated Science Assessment for Sulfur Oxides - Health Criteria. September 2008. EPA/600/R-08/047F

[24] Amsalu E, Guo Y, Li H, Wang T, Liu Y, et al. Short-term effect of ambient sulfur dioxide (SO2) on cause-specific cardiovascular hospital admission in Beijing, China: A time series study. 2019;208:74-81. DOI: 10.1016/j.atmosenv.2019.03.015

[25] Blanco A, Blanco G. Medical Biochemistry. 2017. Chapter 3 Proteins:21-71. DOI: 10.1016/B978-0-12-803550-4.00003-3

[26] Blumenthal I. Carbon monoxide poisoning. Journal of Royal Society of Medicine. 2001;94:270-272. DOI: 10.1177/014107680109400604

[27] Ganong WF. Review of Medical Physiology. Norwalk Ct: Appleton & Lange. 1995

[28] Neto CA, Yanagihara JI, Turri F. A carbon monoxide transport model of the human respiratory system applied to urban atmosphere exposure analysis. Journal of the Brazilian Society of Mechanical Sciences and Engineering. 2008;30(3):253-260. DOI: 10.1590/S1678-58782008000300011

[29] Department of Health, London. Carbon Monoxide: The Forgotten Killer. Letter from the Chief Medical Officer, PL/CMO/98/5. London: DoH. 1998

[30] Weaver LK. Carbon monoxide poisoning. Critical Care Clinic. 1999;15(2):297-317. DOI: 10.1016/S0749-0704(05)70056-7

[31] WHO. Air Quality Guidelines. Global Update. Particulate matter, ozone, nitrogen dioxide and sulfur dioxide. 2005. World Health Organization. Regional Office for Europe. Copenhagen.

[32] Anderson JO, Thundiyil JG, Stolbach A. Clearing the Air: A Review of the Effects of Particulate Matter Air Pollution on Human Health. Journal of Medical Toxicology. 2012;8:166-175. DOI: 10.1007/s13181-011-0203-1

[33] Kyung SY, Jeong SH. Particulate-Matter Related Respiratory Diseases. Tuberculosis and Respiratory Diseases. 2020;83:116-121. DOI: 10.4046/trd.2019.0025

[34] Azam AG, Zanjani BR, Mood MB. Effects of air pollution on human health and practical measures for prevention in Iran. Journal of Research in Medical Sciences. 2016;21:65. DOI: doi: 10.4103/1735-1995.189646

[35] Tiwari S, Tripathy IP, Tiwari HL. Effects of lead on environment. International Journal of Emerging Research in Management & Technology. 2013;2(6):1-5

[36] Kwon JW, Park HW, Kim WJ, Kim MG, Lee SJ. Exposure to volatile organic compounds and airway inflammation. Environmental Health. 2018;17:65. DOI: 10.1186/s12940-018-0410-1

[37] IARC 2012. A Review of Human Carcinogens: Chemical Agents and Related Occupations, IARC Monographs on the Evaluation of Carcinogenic Risks to Humans, 100F. World Health Organization, pp. 249-294. International Agency for Research on Cancer. (Available at: https://monographs.iarc.fr/wp-content/uploads/2018/06/mono100F.pdf, accessed November 11, 2020)

[38] Directive 2008/50/EC of the European Parliament and of the Council on 21 May 2008 on ambient air quality and cleaner air for Europe. (Available at: https://eur-lex.europa.eu/legal-content/EN/TXT/PDF/?uri=CELEX:32008L0050&from=en, accessed November 20, 2020)

[39] Bernstein JA, Alexis N, Barnes C, Bernstein IL, Bernstein JA, et al. Health effects of air pollution. Journal of Allergy and Clinical Immunology. 2004;114(5):1116-1123. DOI: 10.1016/j.jaci.2004.08.030

[40] Kadam V. Thesis "Multifunctional Air Filtration for Respiratory Protection using Electrospun Nanofibre Membrane". November 2018. RMIT University

[41] Kim MS, Cha D, Lee KM, Lee HJ, Kim T. Modeling of ozone decomposition, oxidant exposures, and the abatement of micropollutants during ozonation processes. Water Research. 2020;169:115230. DOI: 10.1016/j.watres.2019.115230

[42] IPCC/TEAP Special Report: Safeguarding the Ozone Layer and the Global Climate System. Page 86. (Available at: https://www.ipcc.ch/site/assets/uploads/2018/03/sroc_full-1.pdf, accessed November 21, 2020)

[43] Zhao K, Hu C, Yuan Z, Xu D, Zhang S, et al. A modeling study of the impact of stratospheric intrusion on ozone enhancement in the lower troposphere over the Hong Kong regions, China. Atmospheric Research. 2021. 247:105158. DOI: 10.1016/j.atmosres.2020.105158

[44] McPeters, R.D., Labow, G.J., Logan, J.A. Ozone climatological profiles for satellite retrieval algorithms. Journal of Geophysical Research. 2007;112:D05308. DOI: 10.1029/2005JD006823

[45] Liu P, Song H, Wang T, Wang F, Li X, et al. Effects of meteorological conditions and anthropogenic

precursors on ground-level ozone concentrations in Chinese cities. Environmental Pollution. 2020;262:114366. DOI: 10.1016/j.envpol.2020.114366

[46] Jury MR, Fontanez-Vázquez ILF. Ozone structure in Caribbean hurricanes. Heliyon. 2020;6:e05366. DOI: 10.1016/j.heliyon.2020.e05366

[47] Staehelin J, Petropavlovskikh I, De Mazière M, Godin-Beekmann S. The role and performance of ground-based networks in tracking the evolution of the ozone layer. Comptes Rendus Geoscience. 2018;350:354-367. DOI: 10.1016/j.crte.2018.08.007

[48] Valacchi G, Pecorelli A, Belmonte G, Pambianchi E, Cervellati F, et al. Protective Effects of Topical Vitamin C Compound Mixtures against Ozone-Induced Damage in Human Skin. Journal of Investigative Dermatology. 2017;137(6):1373-1375. DOI: 10.1016/j.jid.2017.01.034

[49] Caballero S, Galindo N, Pastor C, Varea M, Crespo J. Estimated tropospheric ozone levels on the southeast Spanish Mediterranean coast. Atmospheric Environment. 2007;41(13):2881-2886. DOI: 10.1016/j.atmosenv.2006.12.047

[50] Bromberg PA. Mechanisms of the acute effects of inhaled ozone in humans. Biochimica et Biophysica Acta (BBA)-General Subjects. 2016;1860(12):2771-2781. DOI: 10.1016/j.bbagen.2016.07.015

[51] Sillman S. The relation between ozone, NOx and hydrocarbons in urban and polluted rural environments. Atmospehric Environment. 1999;33:1821-1845. DOI: 10.1016/s1352-2310(98)00345-8

[52] Hargreaves KL, Fowler D, Storeton-West RL. The exchange of nitric oxide, nitrogen dioxide and ozone between pasture and the atmosphere. Environmental Pollution. 1992;75(1):53-59. DOI: 10.1016/0269-7491(92)90056-G

[53] Gao Y, Zhang J, Yan F, Leung LR, Luo K, et al. Nonlinear effect of compound extreme weather events on ozone formation over the United States. Weather and Climate Extremes. 2020;30:100285. DOI: 10.1016/j.wace.2020.100285

[54] European Environment Agency. Tropospheric ozone: background information. In: Tropospheric Ozone in EU - The Consolidated Report. 2008 (Available at: https://www.eea.europa.eu/publications/TOP08-98, accessed November 18, 2020)

[55] Amnuaylojaroen T, Macatangay RC, Khodmanee S. Modeling the effect of VOCs from biomass burning emissions on ozone pollution in upper Southeast Asia. Heliyon. 2019;5:e02661. DOI: 10.1016/j.heliyon.2019.e02661

[56] Elshorbany Y, Barnes I, Becker KH, Kleffman J, Wiesen P. Sources and Cycling of Tropospheric Hydroxyl Radicals – An Overview. Physical Chemistry. 2010;224:967-987. DOI: 10.1524.zpch.2010.6136

[57] Galán Madruga D, Fernández Patier R, Sintes Puertas MA, Romero García MD, Cristóbal López A. Characterization and Local Emission Sources for Ammonia in an Urban Environment. Bulletin of Environmental Contamination and Toxicology. 2018;100:593-599. DOI: 10.1007/s00128-018-2296-6

[58] Cerrato-Alvarez M, Frutos-Puerto S, Miró-Rodríguez C, Pinilla-Gil E. Measurement of tropospheric ozone by digital image analysis of indigotrisulfonate-impregnated passive sampling pads using a smartphone camera. Microchemical Journal. 2020;154:104535. DOI: 10.1016/j.microc.2019.104535

[59] Zahn A, Weppner J, Widmann H, Schlote-Holubek K, Burger B et al. A fast and precise chemiluminescence ozone detector for eddy flux and airborne application. Atmospheric Measurement Techniques. 2012;5:363-375. DOI: 10.5194/amt-5-363-2012

[60] CIEMAT. Química analítica de los Contaminantes Medioambientales. 2002. Serie ponencias. Ed. Ciemat, Madrid, España

[61] Sexton FW, McElroy FF, Michie RM, Thompson VL, Bowen JA. Performance test results and comparative data for designated reference and equivalent methods for ozone. EPA-600/4-83-003. U.S. Environmental Protection Agency. 1981. Research Triangle Park, NC 27711

[62] Sicard P, Paoletti E, Agathokleous E, araminiené V, Proietti C, et al. Ozone weekend effect in cities: Deep insights for urban air pollution control. Environmental Research. 2020;191:110193. DOI: 10.1016/j.envres.2020.110193

[63] Galán-Madruga D. A methodological framework for improving air quality monitoring network layout. Applications to environment management. Journal of Environmental Sciences. 2021;102:138-147. DOI: 10.1016/j.jes.2020.09.009

[64] Jaffe DA, Cooper OR, Fiore A, Henderson BH, Tonnesen GS, et al. Scientific assessment of background ozone over the U.S.: Implications for air quality management. Elementa (Wash D C). 2018;6:56

Chapter 3

Particle and Carbon Monoxide Atmospheric Pollution in the City of Tepic, Nayarit, Mexico

Mario García, Héctor Ulloa, Omar García, Hermes Ramírez, Aida Fajardo, Claudia Saldaña, Sarah Messina and Yamilet Rodríguez

Abstract

Actively caring for the environment is an issue that prevails in the international debate, and our country takes part on this argument. One aspect of environmental deterioration is, with no doubt, atmospheric pollution; a constant in modern societies, which, in the attempt to find growth and development, impact the natural and urban environment they inhabit. A distinctive feature of commercial and economic exchange are the strategic cities, so-called capitals, in addition to the territories where coastal tourism predominates as an engine of regional human development. In this balance, which is far from being sustainable and fair, an exponential consumer market dominates and generates the progressive increase in the use of fossil fuels. The former being emitted into the atmosphere, in such a way that they alter chemical composition and cause harmful air quality. In addition, the territory is impacted, intoxicating the soil and water, which are the final deposit. The objective of this work is to determine the temporal behavior of atmospheric pollutants in the city of Tepic, Nayarit, Mexico and to identify the dominant pollution indicators considering the international and national context. Through the statistical analysis of the databases of particles smaller than 2.5 micrometers, particles smaller than 10 micrometers and carbon monoxide, the behaviors of these pollutants in the study area were obtained. Among the most significant results, particles smaller than 2.5 micrometers showed maximum levels outside the norm almost all year round, with values reaching 170 micrograms per cubic meter. On the other hand, the particles smaller than 10 micrometers presented satisfactory levels in their average behavior; however, the maximum concentrations remained outside the official Mexican standard. Regarding the analysis of CO, a behavior within the maximum permissible limits of protection for the population was reported; Nevertheless, by favoring the formation of tropospheric ozone, its contribution is significant, especially when the atmosphere is highly photo-reactive. This research can be used as a timely tool for mitigation of climate change, where the results contribute to the review and rethinking of public management of the environment towards sustainable development.

Keywords: air pollution, particles, carbon monoxide, Tepic, Mexico

 IntechOpen

1. Introduction

The emission of pollutants into the atmosphere has caused damage to the environment, destroying and large areas of flora and fauna, necessary for the natural balance of the planet. The use of fossil fuels raises air pollution to harmful levels for humans, favors the accumulation of greenhouse gases and destroys the ozone layer, in addition to the acid rain phenomenon that has affected Europe, Asia and North America; in fact, it is considered a significant problem in China so far in the 21st century [1–4]. The most relevant impact occurs in urban areas and industrial complexes; however, adverse effects have been observed even in distant areas [5].

Although air pollution is a cross-cutting issue that needs to be addressed in a comprehensive manner, the media discourses, being transformed into immediate solutions not for prevention and control in the long term but for mitigation, stand out. Consequently, the centralization of goods and services in the territory negatively affects the well-being of the population [6–11]. In this context, Khan et al. [12], reported that the use of renewable energy in logistics operations improves the environmental and economic state and reduces spending on public health.

The role of air quality indexes is pivotal in controlling emissions from both mobile and stationary sources. No traffic congestion problems would be observed if the use of means of transport were planned; Transportation is inefficient due to the metropolis great extension and functional segregation [6, 13]. While pollution in developed countries is due to industrialization, in most countries of Latin America and the Caribbean, it is influenced by motor vehicles [14–18].

Climate, meteorological and topographic conditions influence the pollutant concentration, transport and dispersion, as well as the magnitude of the impact on the population and its habitat [14–18]. The statistical analysis of meteorological and pollutant data makes possible to identify a relationship between local emission sources and to have a better understanding of the spatio-temporal trends of air quality indexes; whose variation depends on the distance between sources and traffic volumes [14, 19]. In addition, cities located in depressions are associated with the formation of thermal inversions, an optimal condition for the assembling of pollutants, at harmful rates, in the lower layer of the atmosphere [20]. These changes avoid the dispersion of pollutants, especially in winter.

Several studies report that the association of pollutant-meteorological variables impact human health. Therefore, incorporating atmospheric monitoring stations in cities for their evaluation and control is fully justified [19]. Another study by Tiwari et al. [21], reported that rain, wind speed, and surface temperature are the factors that control the evolution of PM10 and PM2.5 particles over Guwahati, India. Pollution levels are comparable with those obtained in other megacities and explain the effects on health in recent years.

Mexico is no stranger to environmental problems; in fact, air pollution in cities has impacted the life quality of society, as well as its health and the environment. The metropolitan area of the Valley of Mexico (**Figure 1**), Guadalajara and Monterrey are included among the metropolises that are characterized by suffering from environmental issues [22–24]. The objective of this work is to determine the spatial–temporal behavior of atmospheric pollutants in the city of Tepic, Nayarit, Mexico and define, based on the available information, the dominant pollution indicators in the area. The results can contribute to the improvement of atmospheric contingency plans, in addition to being an alternative for public management of environment aspects.

Figure 1.
Pollution in CDMX under thermal inversion scenarios. Own source (01/12/2019).

2. Case study and methodology

2.1 Description of the study area: Tepic City, Nayarit, Mexico

The state of Nayarit is located in the central western region of Mexico and has an approximate area of 27,857 km^2 that covers 1.4% of the national territory (**Figure 2**). Its geographic coordinates correspond to 23° 05′-20° 36′ north latitude and 103° 43′-105° 46′ west longitude [25–27]. It has an approximate population of 1,181,050 inhabitants [28] and it is composed of 20 municipalities, the city of Tepic being the state capital. Within its growth and development potential, tertiary and service economic activities predominate, as a consequence of its tourist and commercial explosion.

From its side, the municipality of Tepic is located in the center of the state with a land area of 1,983.3 km^2 which represents 7.25% of the state of Nayarit. It has an approximate population of 413,608 inhabitants, highlighting the towns of Tepic, Francisco I. Madero, Bellavista and Camichín de Jauja. Its geographic coordinates are 21° 51′-21° 24′ north latitude and 104° 34′-105° 05′ west longitude [26, 27, 29].

Regarding its orography, 72.5% of the land relief of the municipality corresponds to mountains, the rest is represented by hills, plains and small valleys. The mountain ranges that cross the municipality are the Neovolcanic Axis and the Sierra Madre Occidental. Its main elevations correspond to the volcanoes of Sangangüey, San Juan, Las Navajas and El Rincón hill, with altitudes of 2340, 2180, 1680 and 1600 meters above sea level, respectively.

Two main types of climate predominate; subhumid warm climate with rain in summer that affects 66.06% of the municipal geography and the subhumid semi-warm climate with rain in summer, which benefits the remaining 33.94%. Rain concentration of 91.05% is observed between the months of July and October. Average annual rainfall is 1,121 mm. In this order, **Table 1** shows monthly values of temperatures (maximum and minimum); the average temperature oscillates between 21.1° C, with north prevailing winds, with an average speed of 8 km/h [26, 27, 31].

Figure 2.
Geographical location of Tepic, Nayarit, Mexico. Source: Google images.

The city of Tepic has developed around a significant vehicular complex and extraordinary industrial activity dominated by the "El Molino" sugar mill with more than a century of operations. Both sources have increased the discharge of pollutants into the atmosphere, such as total suspended particles and carbon monoxide, among others. This sugar factory, due to its magnitude and economic impact, remains within the top 10 in performance at the national level and the most important in the state of Nayarit, with the countryside, suppliers, contractors, transporters and employees, as sectors that benefit the most. In this way, Khan et al. [32], reported that trade and transport infrastructure are positively correlated with per capita income, while logistics operations are negatively associated with social and environmental problems (climate change). In addition, air pollution in the territory contributes to the generation of acid rains, water pollution and a harmful impact on human health. In this scenario, the search for better living conditions has an impact on the habitat, exceeding air quality indexes to levels that are harmful to the population.

In relation to this, Hernández [33] showed that "El Molino" is a source of atmospheric pollution. It is noteworthy that, due to its location close to the city center, due to the burning of sugarcane and its transportation, the damage to health increases 2.6 times during the harvest season.

In coincidence, the work carried out in El Salvador by Castillo and Rivera [34], reported that the sugar sector, despite its high economic and nutritional value, is a source of pollution. Accordingly, with information from 2011 to 2013, Alatorre and Llanos [35] identified that the most significant peaks in CO concentrations occur during the morning (6:00–9:00 hours), associated with low temperatures (lower temperatures, higher concentration of particles in the air) and to vehicular and industrial influence. This is proven by the fact that the most significant values were recorded in the months of December to April. Furthermore, the synergy

Month	January	February	March	April	May	June	July	August	September	October	November	December
MaxTR	28	29	31	32	35	33	33	32	33	33	31	30
DMaxT	25	24	28	29	32	31	28	30	29	30	29	28
DMinT	16	15	15	17	16	22	22	22	23	21	14	12
MinTR	9	9	8	11	11	17	20	20	20	16	7	4

Maximum temperature recorded (MaxTR), Daily maximum temperature (DMaxT), Minimum temperature recorded (MinTR), Daily minimum temperature (DMinT). Temperature (°C). [30].

Table 1.
Tepic average climatic parameters.

of pollutants in Tepic can negatively impact the presence of thermal inversions (**Figure 3**). The geography and orography of the city allow this behavior to occur frequently, concentrating pollution for long periods and reaching harmful levels for the exposed population [10, 36]. Finally, other essential activities of the municipality are agriculture and livestock, followed by logging, which bring with them multifactorial problems related to air, soil and water pollution. The list is completed by an excessive generation of urban solid waste, the elimination of electronic devices (e-waste), in addition to the high stress to which sanitary landfills or open dumps are subjected [37].

2.2 Methodology

The presence of phenomena that have affected the territory over time, leads researchers, decision makers and society to search for instruments to analyze and understand the root causes of the problem. In order to apprehend reality and identify solid bases to act in the prevention and mitigation of the risk to which living beings are continuously exposed, particularly human beings. Determining the magnitude and influence of pollution indicators and rethinking control methods in cities provides an indispensable tool for the preservation of the society-nature balance. In this synergy, problems and areas of opportunity can be identified in the management of the territory.

In the first instance, information corresponding to the database of atmospheric pollutants and meteorological variables (two monitoring stations) was collected in collaboration with the Ministry of Rural Development and Environment (SEDERMA) of Nayarit. Data were selected, validated and processed with statistical methods. To choose the indicators that best describe the problem in the territory, the Official Mexican Standard (NOM) is used as reference for the criteria and their relationship with international standards (**Table 2**). The reliability and updating of data, and the spatio-temporal trends are considered as key information in the description-evaluation of the phenomenon.

2.3 Stochastic processes

Using a stochastic model, it is possible to know the future of processes that develop over time, predict the behavior of physical and atmospheric phenomena, know their evolution and culminate with decision-making considering the best solution. The time *series or temporal series* that record observations of a process as

Figure 3.
(a-b) Pollution in Tepic - thermal inversion (05/30/2016–2109,35 am) source: Own image.

Pollution indicator	World Health Organization (WHO)*		Environmental Protection Agency (EPA)*		Official Mexican Standards (NOM) - DOF (1994)	
	ppm	μg/m³	ppm	μg/m³	ppm	μg/m³
Ozone (O_3)	- 0.05 (8 h)	- 100	0.12 (1 h) 0.075 (8 h)	235 150	0.095 (1 h) 0.070 (max 1 year)	215 157
Carbon monoxide (CO)	9 (8 h) 26 (1 h)	10290 29725	9 (8 h) 35 (1 h)	10290 40000	11 (8 h)- (1/year)	12595
Nitrogen dioxide (NO_2)	0.106 (1 h) 0.023 (annual)	200 40	0.25 (1 h) 0.053 (annual)	470 100	0.21 (1 h)- (1/year) -	395 -
Sulfur dioxide (SO_2)	0.007 (24 h) 0.191 (10 min)	20 500	0.14 (24 h) 0.03 (annual)	365 80	0.11 (24 h)- (1/year) 0.03 (annual)	289 80
Breathable particulate material (PM_{10})**	—	50 (24 h) 20 (annual)	—	150 (24 h) 50 (annual)	—	75 (24 h) 40 (annual)
Fine particulate material ($PM_{2.5}$)**	—	25 (24 h) 10 (annual)	—	35 (24 h) 15 (annual)	—	45 (24 h) 12 (annual)

*[38].
**[39]. ppm: parts per million, μg/m³: microgram/cubic meter.

Table 2.
Air quality standards (average pollution indicators).

a function of time, are justified to predict the future based on past data, know and control the process that generates it and obtain a description of its characteristics, among others [40].

In the case of Tepic, a pollution analysis was performed using this series-based methodology using spectral analysis and correlation functions. The tool used was the Statistical Package for the Social Sciences [41]; the solution provides adequate control that provides certainty and decision opportunity. In addition, environmental indicators (**Table 3**) are instruments for monitoring through the systematic collection of data obtained and time series. Overall, the indicators, variables and indices synthetically reflect a social concern regarding the environment and are coherently inserted into the decision-making process.

In order to know the current state of air quality, SEDERMA operates the air quality monitoring network, which measures pollutants: carbon monoxide (CO), nitrogen oxides (NO_X), nitrogen dioxide (NO2), ozone (O3), sulfur dioxide (SO2) and particles smaller than 10 and 2.5 micrometers (PM10 and PM2.5). The evaluation period is 2015, with the official institution being the provider of the information. PM10, PM2.5 and CO are evaluated, which are subjected to a spatial and temporal analysis. Regarding the inclusion criteria, the hourly average concentration of the pollutants is evaluated. Data outside the consistency range, negatives, zeros, and undetected were removed. Using a map, the study area was georeferenced and the monitoring stations were located, one located to the east over the "Primary School: PRIM" and the other to the west over the "Technological Institute of Tepic: ITT" (**Figure 4**).

Variable	Definition	Indicator	Equivalences – MAQI (100)
Air pollutants (CA)	PM_{10} $PM_{2.5}$ CO	75 μg/m^3 average in 24 hours 40 μg/m^3 annual average 45 μg/m^3 average in 24 hours 12 μg/m^3 annual average 11 ppm average in 8 hours	Metropolitan air quality index (Maximum average concentrations of protection for the population: 100). [39]

Table 3.
Definition of variables and indicators.

Figure 4.
Pollutant monitoring stations (a: PRIM, B: ITT). Source: Self-elaborated.

3. Results

Within the variety of aspects that characterize the Earth's atmosphere, one of them is its dynamic character; It is not static, but significant changes are constantly taking place. Thus, due to its influence on humans, it seeks to understand in depth atmospheric phenomena and even predict their formation [42]. Pollution is influenced by the atmosphere state; if the pollutants in the air exceed their concentration in excess of that existing in a normal environment, the same atmosphere is responsible for transporting, dispersing and even transforming them.

3.1 Monthly average behavior of particles smaller than 2.5 micrometers in 2015

The official Mexican standard for particles smaller than 2.5 micrometers (PM2.5) establishes a maximum permissible limit (threshold) of 45 μg/m^3 as an average in a 24-hour interval (or 12 μg/m³ annual average). According to the results obtained for this pollutant (PM2.5), during the month of January, a series of inconsistencies prevail. The homogeneous behavior observed is the result of the absence of information, deficiencies in the calibration and data validation stage during that month. Based on the information provided, the maximum values range between 50 and 80 μg/m³, that is to say, unsatisfactory levels with respect to current regulations. Regarding February, it was observed that average concentrations are maintained between 25 and 45 μg/m³ while the maximum values move in a range of 45 and 55 μg/m³ respectively. In this order, the month of March shows a negative trend in the average data, while the maximums only in some moments of the day exceed the regulations. In the case of April, the trend in the average values is maintained, although the maximums show higher levels in relation to the previous month, reaching concentrations between 55 and 75 μg/m³ (**Figure 5**).

On the other hand, during the month of May fluctuations are observed that keep the concentrations within the environmental regulations, both in their average and maximum levels. In June, a similar behavior to the previous month is maintained, although with a slight positive trend is its maximum peaks, however, the decrease in its average concentrations during the second half of the month due to the progressive influence of the wet period. For July, two schedules of maximum concentrations predominate; one at 4:00 p.m. and another at 6:00 p.m., reaching 60 μg/m³. The results for the months of August, September and October remain within the standards established in the NOMs, both in their average and maximum concentrations. For this region of the country, the influence of the rainy season is noticeable, maintaining a relatively clean atmosphere. The most significant values do not exceed 25 μg/m³ (**Figure 6**).

Finally, the months of November and December begin to experience a positive trend typical of the transition from the wet period to the dry and cold period. Average concentrations are kept within environmental regulations; however, its maximum levels range from 50 to 170 μg/m³ (**Figure 7**), indices of air quality that are harmful to the exposed population.

The analysis of the year 2015 for PM2.5, reports maximum levels outside the norm practically throughout the year, with the exception of the wet period, where a marked decrease in concentrations is observed. Maximum average concentrations reached maximums of 170 μg/m³ (**Figure 8**).

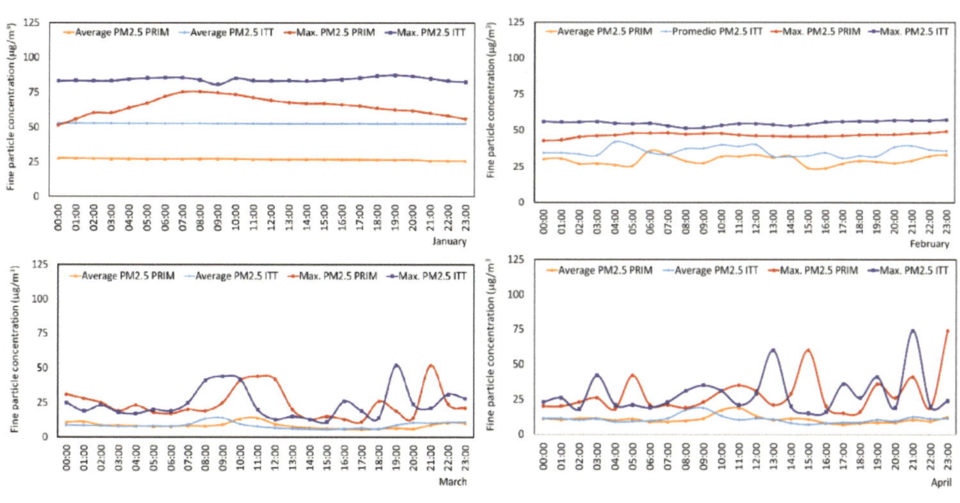

Figure 5.
(a–d) Monthly average behavior of $PM_{2.5}$ (24 h).

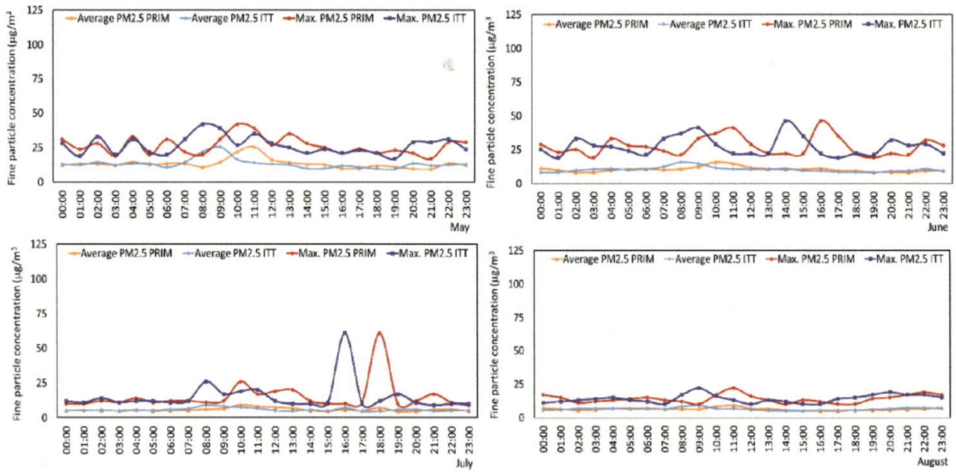

Figure 6.
(a–d) Monthly average behavior of $PM_{2.5}$ (24 h).

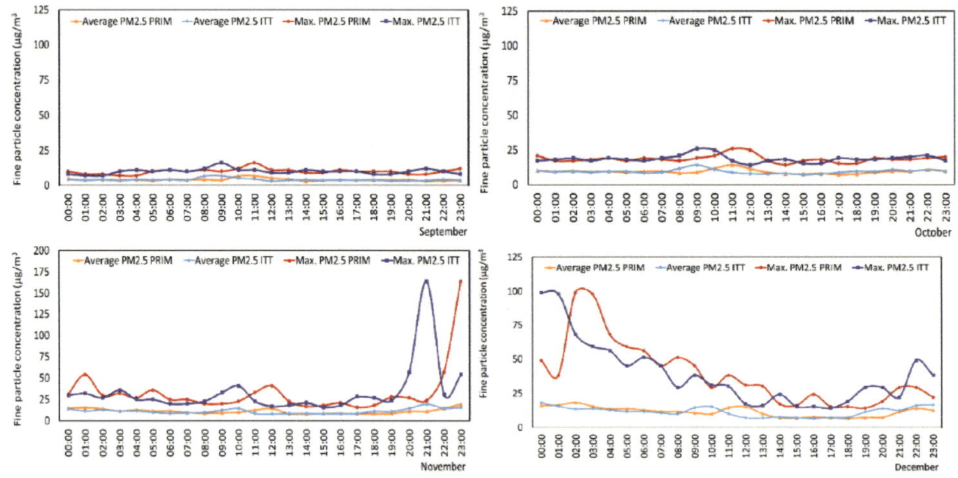

Figure 7.
(a–d) Monthly average behavior of $PM_{2.5}$ (24 h).

3.2 Monthly average behavior of particles smaller than 10 micrometers in 2015

In the case of particles smaller than 10 micrometers (PM10), current regulations define a maximum permissible limit of protection for population not being higher than 75 µg/m³ average in 24 hours (40 µg/m³ annual average). In a similar way to PM2.5, the average results on PM10 during the month of January report a series of inconsistencies, a consequence of the absence of reliable information, in addition to deficiencies in the calibration and validation stage of temporary data. Although maximum average concentrations are observed above the NOM (80 µg/m³), they are not considered relevant. In relation to February, concentrations have significant fluctuations between hourly records, being more important during 5:00 and 10:00 hours; Although its average levels do not exceed the norm for this pollutant, the maximum concentrations remain outside the standards throughout the period, with peaks reaching 140 µg/m³. During the months of March, April, May and June, concentrations within the NOM continue to be observed at their average levels, however, the maximum values for the two monitoring stations increased their maximum average

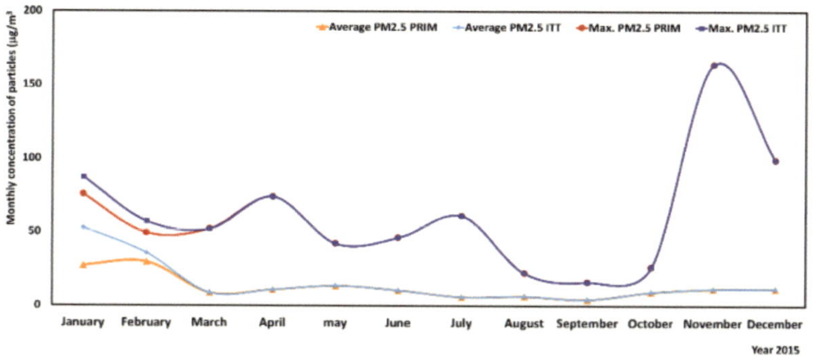

Figure 8.
Monthly average behavior of $PM_{2.5}$ for 2015.

concentrations of PM10. A common characteristic of these months is the progressive increase of pollutants starting at 7:00 a.m., remaining during the rest of the day, with maximums of up to 220 and 260 µg/m^3 (at 9:00 a.m. and 3:00 p.m.). Clearly, the maximum levels observed in the morning are related to the presence of thermal inversion, a natural phenomenon typical of the study area that contributes to the accumulation "in volume" of pollutants at air quality indices harmful to the population. On the other hand, the maximum levels in the afternoon are related to the intense productive activity of the sugar mill "el Molino" (stationary sources) and the vehicle fleet (**Figure 9**).

The continuity of the results allows us to observe that, at the end of June and the beginning of July, the negative trend is sufficiently noticeable due to the marked influence of the rainy season in the area; For this reason, the month of July perceives average concentrations within the NOM, while its maximum levels are located at the limit and slightly above 75 µg/m^3 (interval between 8:00 and 10:00 hours); this scenario remains without significant changes during the month of August. Meanwhile, September and October experience satisfactory average levels of air quality; Despite this, the maximum average concentrations define significant hourly fluctuations, highlighting the intervals from 4:00 a.m. to 10:00 a.m. for September and from 6:00 a.m. to 10:00 a.m. for October, even with a peak of extreme maximums at the beginning and end of the month (between seasons). These variations of maximum values move in a range from 75 to 180 µg/m^3. It is worth mentioning that these peaks define interactions coupled with a transition between the end of the wet period and the beginning of drier and colder periods (**Figure 10**).

Finally, the months of November and December observe critical conditions for the well-being of the population. **Figure 10** shows average concentrations of PM10 with peaks at the limit of the satisfactory threshold, while the maximum average values exceed the Mexican environmental regulations for this pollutant. PM10 levels vary between 75 and 200 µg/m^3, with higher volumes exceeding between 8: 00–12: 00 hours and 19: 00–22: 00 hours.

In addition, through the monthly analysis for 2015, satisfactory levels were observed in its average behavior; however, the average maximum concentrations remained outside the official norm; in fact, far above it. When making a comparison between the NOM and the WHO contamination standards for PM10, the maximums exceed 3.4 times the Mexican standard and 5.2 times the WHO indicators. Only the influence of the humid period in the interval from July to September, benefits the air quality in the city of Tepic, increasing again from October; the concentrations of 260 µg/m^3 during June show the magnitude of the environmental problems in the city (**Figure 11**).

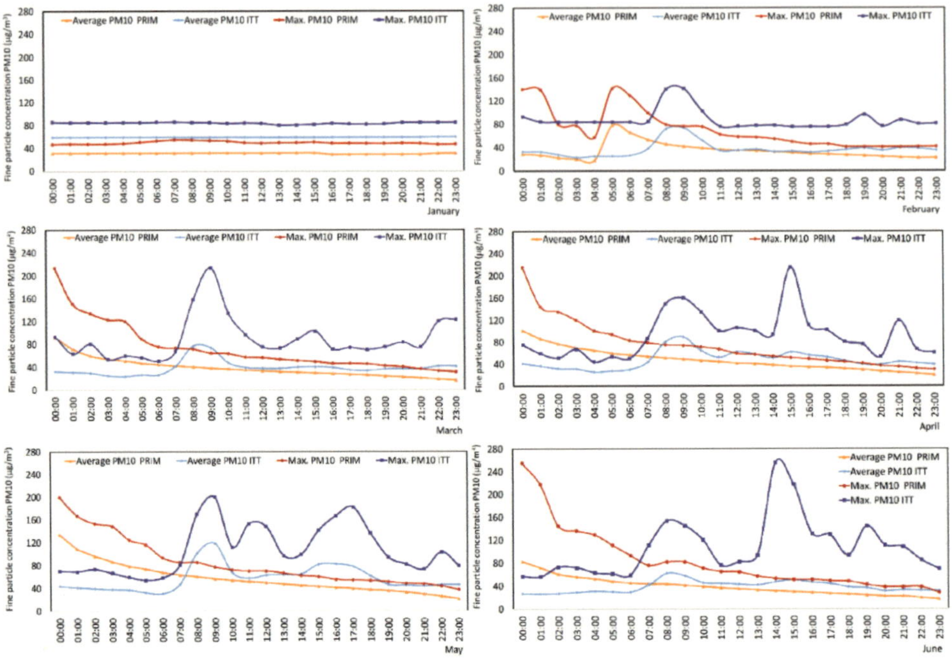

Figure 9.
(a–f) Maximum monthly average behavior of PM_{10} (24 h).

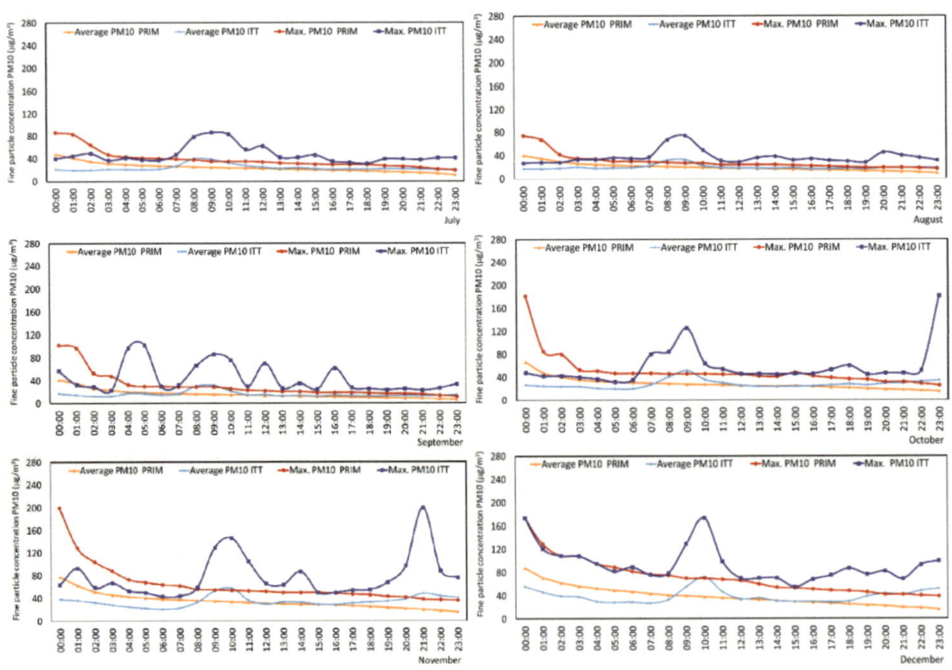

Figure 10.
(a-f) Maximum monthly average behavior of PM_{10} (24 h).

3.3 Average and maximum monthly CO behavior in 2015

Environmental researchers have verified that the tools used to determine the current state of a territory must be supported by sufficiently reliable data bases, in such a way that the results of the analysis allow the generation of strategic pollution control plans.

In this context, carbon monoxide (CO) is one of the pollutants that is emitted into the atmosphere most frequently; It is a primary gas, a product of the incomplete combustion of fossil fuels. However, its rapid ease of reacting with light (incident radiation), before raising its concentrations to out-of-standard rates, contributes to the generation of the secondary pollutant called ozone (O_3), which is a determining indicator in cities that experience smog photochemical. The NOM for pollutant CO establishes a concentration of 11 ppm as an average maximum limit of protection; this for 8 hours.

The aforementioned allows us to affirm that January, being a wet and cold month, presents maximum average concentrations outside the norm, especially during the first half of the day; while, as solar radiation increases, CO levels gradually decrease. For the month of February, only two small peaks with higher values are observed, but within the norms; this occurs in the morning and in the afternoon, –a typical behavior of the accumulation of pollutants due to the presence of thermal inversions and vehicular flow–. Regarding the months of March and April (**Figure 12**), the presence of two peaks of maximum values towards the PRIM station (center-west of the city) stands out, with values between 11 and 18 ppm (11, 00–14:00 and 11:00–15:00 hours, respectively).

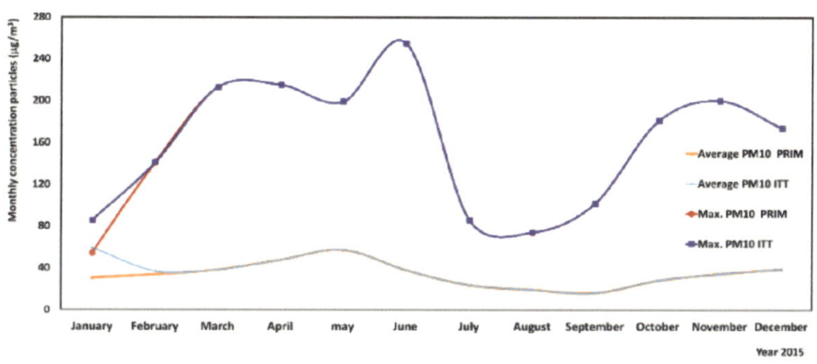

Figure 11.
Monthly average behavior of PM_{10} for 2015.

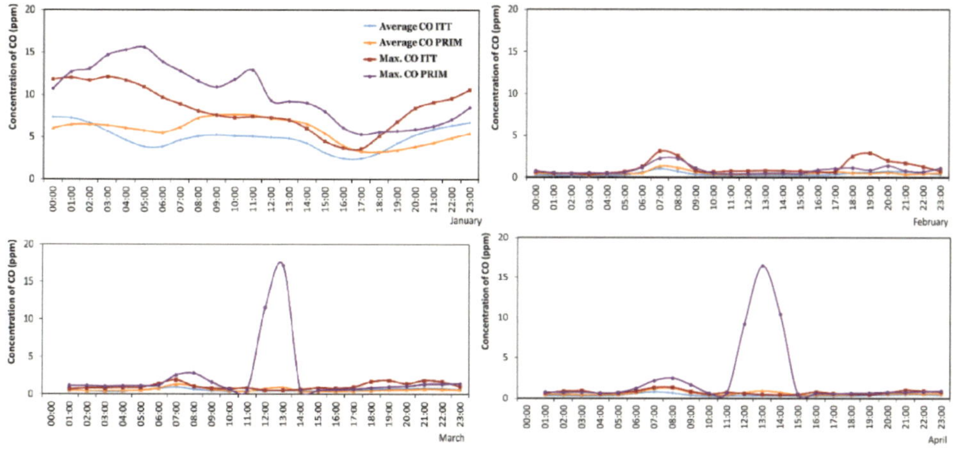

Figure 12.
(a–d) Average behavior and monthly maximum of CO (24 h).

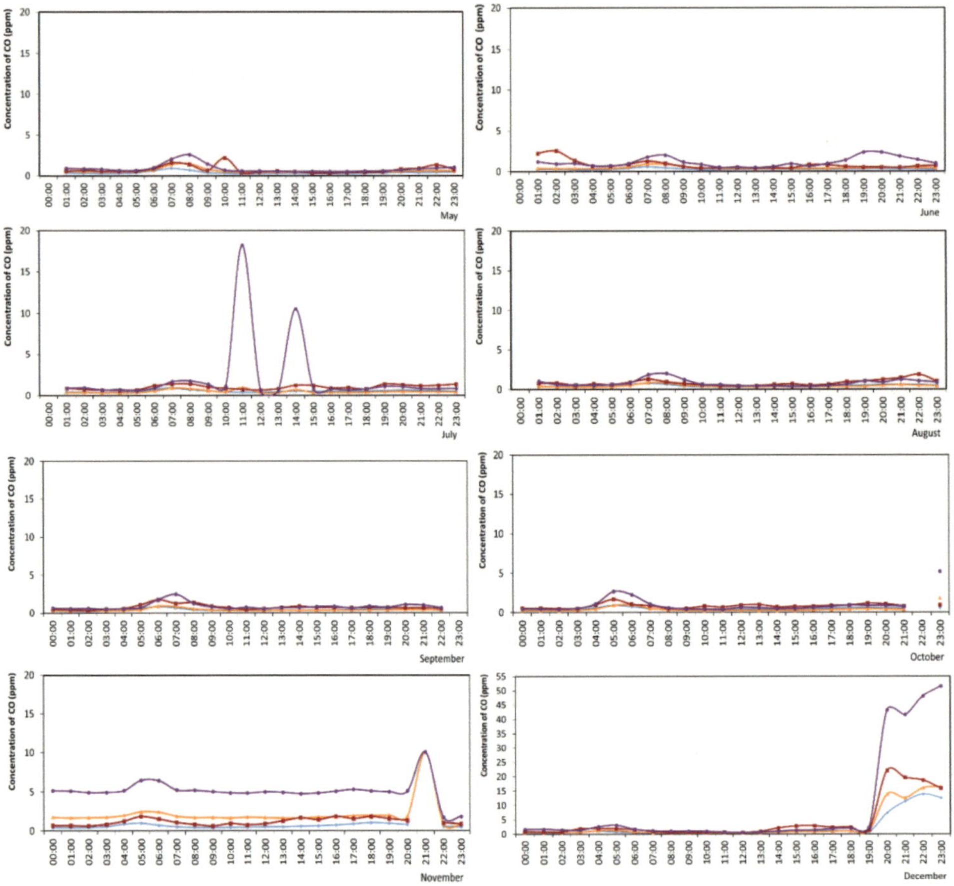

Figure 13.
(a–h) Average behavior and monthly maximum of CO (24 h).

On the other hand, in the interval from May to October, all the months present a behavior within the satisfactory standards, except July, which maintain concentration peaks outside the norm between 10:00 and 12:00 hours, with values 18 ppm. The month of November maintains its average levels in a satisfactory range, while the maximum averages are located at the limit in the evening hours (8:00 p.m. and 10:00 p.m.). Finally, December reports concentrations outside the NOM, especially at their maximum averages; this is observed at the end of the day, with values ranging between 11 and 52 ppm. This month contributes to a worrying scenario for the health of the exposed population (**Figure 13**).

Overall, CO keeps a behavior within the established regulations, however, by favoring the formation of tropospheric ozone, its contribution is significant especially when the atmosphere is highly photo-reactive.

4. Discussion and conclusions

The emission of pollutants into the atmosphere and local immission, in addition to the generation of waste in strategic cities, alter the habitat, the quality of life of the population and compromise sustainable urban development. The city of Tepic has prospered around a significant vehicle complex and extraordinary industrial activity, a scenario that produces air pollution. In Tepic a vehicular congestion prevails caused by inefficient transportation (public and private); it is a city that is gaining in extension and functional segregation. These factors contribute to the increase in the concentration of pollutants, which, added to the sugar mills (regional axis for development), represent a constant source of contamination and latent risk to the health of the population.

The sugar mill "el Molino" in its beginnings was an industry founded on the periphery of the city, however, the absence of uncontrolled real estate development plans, was propitiating a segregated growth in the surroundings, to such a degree that the synergy -city- industry- strengthened in deterioration of air quality, impacting on the most vulnerable age groups. In the work of Hernández [33], it was reported that damage to health increased 2.6 times in the harvest season, sugarcane burning and transportation, mainly the total suspended particles and CO; For its part, in the present study these levels rose to 3.7 times for PM2.5 and 3.4 times for PM10 with respect to the NOM (6.8 and 5.2 times for the WHO indicators).

It is observed that the diurnal patterns vary between seasonal periods (dry and wet period). The volumes and type of pollutants are significant for each place and region; in addition, the distances between sources and types of traffic are other aspects to consider.

Although it was not possible to determine the type of thermal inversion that occurs in the city, it is evident that it influences the concentration of pollutants, especially during the early hours and in the morning. Jang et al., [14]; Amarillo and Carreras [23] mention that cities located in depressions are associated with the formation of strong thermal inversions, an optimal condition for the accumulation of pollution in the lower atmosphere. In this context, the city of Tepic is located within a semi-closed valley, an ideal natural physical obstacle for pollutants to accumulate and prevent their dispersion, especially in the winter period. The geography and relief of the city contribute so that the anomaly of the temperature with the height, occurs frequently, concentrating the indices of air quality to levels not satisfactory for the population and, in the periods of greater cooling of the surface [43].

The results obtained show high air pollution at specific times, which can motivate local authorities to implement new criteria, create public policies, methodologies and innovate in mobility tasks, among others.

The current air quality monitoring stations are overwhelmed by the size of the city; It is suggested to incorporate three stations and distribute them in a homogeneous way in such a way that between the five they cover a greater coverage radius and a more comprehensive analysis. One element that can contribute to the evaluation of contamination by stationary sources is the assessment of the level of influence of the mill "El Molino" through the measurements "in situ" of the pollutants analyzed in this work (PM10, PM2.5, and CO, among others). The scaffolding must be laid for the configuration of an air quality monitoring network, supported by short and medium-term plans, where continuous monitoring is maintained on the databases obtained. This is useful in joint decision-making by the integrating axis Government and society [12]. It should be noted that various studies report an association between pollutants-meteorological variables and human health [32]; therefore, incorporating a mesh of five atmospheric monitoring stations in Tepic is fully justified [17], in addition, it will allow a comprehensive assessment of contamination and generate control mechanisms, as well as viable strategies for decision-making.

Tepic is a region that has the opportunity to progress under the objectives or characteristics pursued by sustainable development. Although it is true that there is an industrial productive sector that strengthens local economic development, it is possible to progress taking care of the local habitat, without compromising future generations and without risking the health of the vulnerable population [12].

The participation of the health sector and the Ministry of the Environment of Nayarit (SEMANAY) is significant in the sense of keeping track of cases of acute respiratory diseases in the different seasons of the year, and specifically, at the harvest season.

The study of pollutants in the air has great relevance to design strategies based on prediction, prevention, mitigation and resilience in the habitat. In agreement with Khan et al. [12], the present results can be a useful tool for the formulation of public policies and their implementation towards sustainable economic growth. Thus, the city of Tepic plays a key role in the state's economy and "el mill" is a central axis of the territory because the strategic lines that largely support the family economy converge there.

The continuous analysis of the CO pollutant for a longer period of time can favor the integral obtaining of plans and strategies for the control of emissions in the atmosphere of Tepic. Daytime patterns vary seasonally, weekly, and spatially depending on the distance between sources and volumes of traffic [14, 19].

Alatorre and Llanos [35] conducted a study on air pollution in the city of Tepic, reporting that there was no monitoring of air quality in the area; There was only one monitoring unit, but its operation was not optimal (constant interruptions in measurements). Emphasis is placed on the lack of dissemination of information, a situation that leads to high vulnerability in areas that lack it. That is, it is not disclosed or available to users. Through SEMANAY and with the information from 2011 to 2013, they identified that the most significant peaks in CO concentrations occur during the morning (6:00–9:00 am). They associated this with low temperatures (lower temperatures, higher concentration of particles in the air) and vehicular influence in this period of time (entry to schools and work). According to the National Institute of Ecology, the metropolitan area of Guadalajara with its 4.5 million inhabitants approximately, registers CO concentrations at an average of 10 ppm, while in Tepic (with an eighth of this population) levels have been observed by above 8 ppm. This is an indication that the proportion of contamination per inhabitant in Tepic is high.

It should be remembered that the sugar mill "El Molino" due to its historical impact and its economic importance represents an essential axis of local and regional development; However, despite being recognized as respectful of the environment and occupying the first place in the control of emissions of gases and solids released into the atmosphere, it is a constant source of air pollution and represents a risk to the health of the population [33]. Finally, although a period of one year is a short time for the analysis of air pollution, this diagnosis observes trends for PM10 and PM2.5; results that allow knowing the evolution of pollution in the short term and timely informing the population about its status in real time.

Author details

Mario García[1]*, Héctor Ulloa[1], Omar García[1], Hermes Ramírez[1], Aida Fajardo[1], Claudia Saldaña[2], Sarah Messina[2] and Yamilet Rodríguez[2]

1 Universidad de Guadalajara, México

2 Universidad Autónoma de Nayarit, México

*Address all correspondence to: megarcia@astro.iam.udg.mx

References

[1] Zhang, G., et al. Acid rain in Jiangsu province, eastern China: Tempo-spatial variations features and analysis, Atmospheric Pollution Research; 2017, http://dx.doi.org/10.1016/j.apr.2017.02.001.

[2] Dong, L., Dong, H., Fujita, T., Geng, Y., Fujii, M., 2015. Cost-effectiveness analysis of China's Sulfur dioxide control strategy at the regional level: regional disparity, inequity, and future challenges. J. Clean. Prod. 90, 345e359.

[3] Huang, L., Yang, J., Zhang, G. Chemistry and source identification of wet precipitation in a rural watershed of subtropical China. Chin. J. Geochem. 2012: 31 (4), 347e354.

[4] Chen, L., Heerink, N., van den Berg, M. Energy consumption in rural China: a household model for three villages in Jiangxi Province. Ecol. Econ. 2006: 58 (2), 407-420.

[5] Sans R. y Ribas J. Ingeniería ambiental. Contaminación y tratamientos. Ed. alfaomega marcombo. (1999). Pp. 148.

[6] Hajizadeh, Y., et al. Trends of BTEX in the central urban area of Iran: A preliminary study of photochemical ozone pollution and health risk assessment, Atmospheric Pollution Research (2017), http://dx.doi.org/10.1016/j.apr.2017.09.005.

[7] da Silva, D.B.N., Martins, E.M., Correa, S.M. Role of carbonyls and aromatics in the formation of tropospheric ozone in Rio de Janeiro, Brazil. Environ. Monit. Assess. 2016: 188, 1e13.

[8] Fard, R.F., Naddafi, K., Yunesian, M., Nodehi, R.N., Dehghani, M.H., Hassanvand, M.S. The assessment of health impacts and external costs of natural gas-fired power plant of Qom. Environ. Sci. Pollut. Res. 2016: 23, 20922e20936.

[9] Mario García, Héctor Ulloa, Hermes Ramírez, Miguel Fuentes, Silvia Arias, Martha Espinosa. Comportamiento de los vientos dominantes y su influencia en la contaminación atmosférica en la zona metropolitana de Guadalajara, Jalisco, México. Revista Iberoamericana de Ciencias. Vol. 1 No. 2. Pp:97-116. URL http://www.reibci.org/publicados/2014/julio/2200120.pdf.

[10] García Mario. Tesis Doctoral "Contaminación del aire en las Ciudades: La Zona Metropolitana de Guadalajara (2001-2010), hacia un modelo de prevención y mitigación. Jalisco, México. CONACYT/Universidad de Guadalajara. Pp. 321. 2013.

[11] Yu Zhang, Syed Abdul Rehman Khan, Anil Kumar, Hêriş Golpîra, Arshian Sharif. Is tourism really affected by logistical operations and environmental degradation? An empirical study from the perspective of Thailand. Journal of Cleaner Production. 2019, Vol. 227: 158-166. https://doi.org/10.1016/j.jclepro.2019.04.164.

[12] Syed Abdul Rehman Khan, Yu Zhang, Anil Kumar, Edmundas Zavadskas, Dalia Streimikiene. Measuring the impact of renewable energy, public healthexpenditure, logistics, and environmental performance on sustainable economic growth. Sustainable Development. 2020; Vol. 28: 833-843. https://doi.org/10.1002/sd.2034.

[13] Atkinson, R. Atmospheric chemistry of VOCs and NO x. Atmos. Environ. 2000: 34, 2063e2101

[14] Eunhwa Jang*, Woogon Do, Geehyeong Park, Minkyeong Kim, Eunchul Yoo. Spatial and temporal

variation of urban air pollutants and their concentrations in relation to meteorological conditions at four sites in Busan, South Korea. 2017: Vol.8, No.1, Pp: 89-100. https://doi.org/10.1016/j.apr.2016.07.009.

[15] Whiteman, C.D., Hoch, S.W., Horel, J.D., Charland, A. Relationship between particulate air pollution and meteorological variables in Utah's Salt Lake Valley. Atmos. Environ. 2014: 94, 742e753.

[16] Guttikunda, S.K., Gurjar, B.R. Role of meteorology in seasonality of air pollution in megacity Delhi, India. Environ. Monit. Assess. 2012: 184, 3199e3211.

[17] Colette, A., Granier, C., Hodnebrog, Ø., Jakobs, H., Maurizi, A., Nyiri, A., Bessagnet, B., D'Angiola, A., D'Isidoro, M., Gauss, M., Meleux, F., Memmesheimer, M., Mieville, A., Rouïl, L., Russo, F., Solberg, S., Stordal, F., Tampieri, F. Air quality trends in Europe over the past decade: a first multi-model assessment. Atmos. Chem. Phys. 2011: 11, 11657e11678.

[18] Simioni D. Contaminación atmosférica y conciencia ciudadana. *Comisión Económica para América Latina y el caribe (CEPAL).* ONU. Santiago de Chile. 2003. Pp. 278.

[19] Amarillo, Ana Carolina; Carreras, Hebe Alejandra. Quantifying the influence of meteorological variables on particle-bound PAHs in urban environments; Elsevier; Atmospheric Pollution Research; 7; 4; 2-2016; 597-602. https://doi.org/10.1016/j.apr.2016.02.006

[20] G Castejón Porcel, D Espín Sánchez, V Ruiz Álvarez, R García Marín, ... Runoff Water as A Resource in the Campo de Cartagena (Region of Murcia): Current Possibilities for Use and Benefits. Water 2018: 10 (4), Pp: 25.

[21] Tiwari, S., Dumka, U. C., Gautam, A. S., Kaskaoutis, D. G., Srivastava, A. K., Bisht, D. S., ... Solmon, F. Assessment of PM2.5 and PM10 over Guwahati in Brahmaputra River Valley: Temporal evolution, source apportionment and meteorological dependence. Atmospheric Pollution Research, 2017. Vol. 8, no 1, pág. 13-28. https://doi.org/10.1016/j.apr.2016.07.008.

[22] Instituto Nacional de Ecología y Cambio Climático (INECC). Informes de los estudios e investigaciones durante el periodo 2013-2018. Contaminación y Salud Ambiental. https://www.gob.mx/inecc. (2018) Consultation date: July, 2020.

[23] Amarillo, AC, Mateos, AC & Carreras, H. Distribución en origen de hidrocarburos aromáticos policíclicos ligados a PM 10 por factorización de matriz positiva en la ciudad de Córdoba, Argentina. Arch Environ Contam Toxicol 72, 380-390 (2017). https://doi.org/10.1007/s00244-017-0384-y

[24] Ramirez-Sanchez HU, Meulenert-PeñaAR,Garcia-GuadalupeME, Garcia-Concepción FO, Alcala-Gutierrez J and Ulloa-Godínez HH. The Influence of Air Pollutants on the Acute Respiratory Diseases in Children in the Urban Area of Guadalajara, Air Quality-Models and Applications. Dragana Popovic. InTech. (2011) http://www.intechopen.com/books/air-quality-models-and-applications. ISBN: 978-953-307-307-1.

[25] Instituto Nacional de Estadística, Geografía e Informática (INEGI). México en cifras/Nayarit. www.beta.inegi.org.mx/app/areasgeograficas/?ag=18. Consultation date: July,2018.

[26] Aguilar L. Cultura política y participación electoral en elecciones locales de la ciudad de Tepic, Nayarit. Universidad Autónoma de Nayarit.

Maestría en Desarrollo Económico Local. Tesis de Grado. 2010: Pp:268. http://www.uan.edu.mx/es/maestria-en-desarrollo-economico-local.

[27] Aguilar L. El miedo, la frustración y la indiferencia electoral y política en la ciudad de Tepic, Nayarit, 2011. Jalisco, México. Universidad de Guadalajara. Tesis de Grado. Pp:146. http://www.eumed.net/libros-gratis/2013a/1332/index.htm. 2012.

[28] Instituto Nacional de Estadística, Geografía e Informática (INEGI). Distribución de la población. Población total en viviendas particulares habitadas (número de personas), 2015. www.beta.inegi.org.mx/app/areasgeograficas/?ag=18. Consultation date: July 2018.

[29] Instituto Nacional de Estadística, Geografía e Informática (INEGI). Espacio y datos de México. División territorial. www.beta.inegi.org.mx/app/mapa/espacioydatos. Consultation date: July 2018.

[30] http://www.esacademic.com/dic.nsf/eswiki/1137532#sel=. Parámetros climáticos promedio de Tepic. Wunderground Weather, Tepic, Nayarit, México. Consultation date: July 2018.

[31] Enciclopedia de los municipios y delegaciones de México. Estado de Nayarit. Tepic/clima. www.inafed.gob.mx/work/enciclopedia/EMM18nayarit/. Consultation date: December, 2019.

[32] Syed Abdul Rehman Khan, Arshian Sharif, Hêriş Golpîra, Anil Kumar. A green ideology in Asian emerging economies: From environmental policy and sustainable development. Sustainable Development. 2019; Vol. 27: 1063-1075. https://doi.org/10.1002/sd.1958.

[33] Hernández Mojarro, E. E. (2008). Tesina. Contaminación ambiental causada por la quema de la caña de azúcar (Sacharum officinarum L.). Xalisco, Nayarit, México: El autor. Instituto Nacional de Estadística y Geografía. (2010). INEGI. Retrieved June, 2013, de www.inegi.gob.mx.

[34] Castillo A., Rivera D. Diagnóstico de las emisiones atmosféricas generadas por los ingenios azucareros de El Salvador. El Salvador, Centro América. *Tesis de licenciatura en Química y Farmacia de la Universidad de El Salvador*, 2004.

[35] Alatorre A. & Llanos M. Contaminación Atmosférica en Tepic, Nayarit, México. Educateconciencia, 2014: 4(4): 36-47. ISSN: 2007-6347. http://tecnocientifica.com.mx/educateconciencia/index.php/revistaeducate/article/view/297.

[36] Ramírez, H., Andrade, M., González, M., Celis, A. Contaminantes atmosféricos y su correlación con infecciones agudas de las vías respiratorias en niños de Guadalajara, Jalisco. Salud Pública de México. 2006: Vol. 48, Núm. 005. Pp. 385-394.

[37] Saldaña Durán, Claudia E.; Nájera González, Oyolsi. Identificación de sitios con potencial para la disposición final de residuos sólidos urbanos en el Municipio de Tepic, Nayarit, México. Revista Internacional de Contaminación Ambiental, V. 35, P. 69-77, 2019. http://Dx.Doi.Org/10.20937/Rica.2019.35.Esp02.07.

[38] Mihelcic J, Zimmerman J. Ingeniería Ambiental: Fundamentos, Sustentabilidad, Diseño. Alfaomega. México, 2012; pp. 720. ISBN: 978-607-707-317-8.

[39] Diario Oficial de la Federación (DOF) 2014. Norma Oficial Mexicana NOM-025-SSA1-2014, Salud ambiental. Valores límite permisibles para la concentración de partículas suspendidas

PM_{10} y $PM_{2.5}$ en el aire. Secretaria de Salud. http://siga.jalisco.gob.mx/aire/Normas.html.

[40] Justin A. Daniels. Advances in Environmental Research, Vol. 64. Nova Science Publishers, Inc. New York. Chapter 3: David Espín-Sánchez, Carmelo Conesa-García and Gregorio Castejón-Porcel. Temperature Inversions Due to Warm Air Advections at Low Levels: Significant Thermal Contrasts in the Vega Media of the Segura River (Southeast Spain). 2018: 139-177. ISSN: 2158-5717.Available from: https://www.researchgate.net/publication/.

[41] Statistical Package for the Social Sciences (SPSS), https://www.ibm.com/mx-es/analytics/spss-statistics-software. Consultation date: July, 2020.

[42] Strauss, W., Contaminación del aire: causas, efectos y soluciones. 2a Ed. México. Trillas. 2012: 208 p. ISBN: 978-607-17-0634-8.

[43] Global Forcast System GFS (2020). Numerical prediction models. Virtual soundings. https://www.tropicaltidbits.com/analysis/models/. Consultation date: December 2020.

Section 2

Air Pollution

Chapter 4

Characterization of Atmospheric Mercury in the High-Altitude Background Station and Coastal Urban City in South Asia

Manikanda Bharath Karuppasamy, Srinivasalu Seshachalam, Usha Natesan and Karthik Ramasamy

Abstract

This study is performed to evaluate the potential sources and seasonal variation of atmospheric mercury (Hg) emissions from regional sources and other influences in India. To achieve this, using the gold amalgam technique with an automated continuous mercury vapour analyzer (TekranTM 2537B). To assess the total gaseous mercury in high altitude mountain peak station at Kodaikanal & coastal/urban air in Chennai region, the impact of changing weather conditions is also evaluated. To compare the past and recent reports of mercury at different locations in the world. The average total gaseous mercury value in Chennai is 4.68 ng/m^3, which is higher as compared to Kodaikanal, where it is 1.53 ng/m^3. The association between TGM with meteorological parameters in ambient air such as temperature, relative humidity, rainfall intensity, the direction of wind and velocity of was studied. The TGM concentration in India are compared with other nations, the TGM levels are similar to the east and Southeast Asian countries, and also Europe, Sub-Saharan Africa and North America are the averages and maximum concentration generally smaller. This research will help to establish more effective management approaches to mitigate the impacts of atmospheric mercury on the rural and urban environment.

Keywords: ambient total gaseous mercury, meteorological parameters, high-altitude station, coastal urban city, global perspective

1. Introduction

The Atmosphere hosts almost all emissions from every source on the Earth's surface, freshwater bodies, oceanic surface and anthropogenic emissions. In the atmosphere, mercury occurs in the following three primary forms: The gaseous elemental mercury (GEM), reactive elemental mercury or divalent mercury (RGM) and particulate mercury (PHg) [1, 2]. There is a significant quantity of research which indicates that these elements in the environment, water and marine environments via a dynamic mixture of transport and transformation in natural and human (anthropogenic) [1, 3–5]. Mercury (Hg) stays as a natural substance with the biogeochemical cycle, which is involved in the Earth and is considered as a contaminant because of its long-range transport in the atmosphere [6, 7].

At World level, about 50 to 70% of total mercury discharge is through anthropogenic exercises, including petroleum product ignition, smelting of metals, burning of urban waste, the release of smoke from coal-burning power plants [8, 9]. In nature, mercury occurs in three unmistakable structures, GEM, RGM and PHg [10–13]. Among these three structures, RGM and PHg shift rapidly because of their characteristics such as high-water dissolvability and reactivity [14, 15]. The lifetime of GEM is 0.5 to 2 years, which is sufficient for its transportation worldwide level [16–18]. It is reported that East Asian Nations are a standout amongst the most critical patrons of worldwide anthropogenic mercury discharge [19–21]. Total gaseous mercury (Hg0) evasion is approximated to be 2900 mg/year (range 1900–4200 mg/yr) from the ocean [22, 23]. The ocean is therefore known to be the primary terrestrial Hg source worldwide, contrasted with approximately 2000 mg/yr from direct anthropogenic emissions. Hg usually occurs in geochemical reserves, but for several years human activities including mining and more recent burning of fossil fuels have increased the emission of Hg from the mineral source into the atmosphere [24, 25]. The background means the concentration of TGM in the northern hemisphere (1.3–1.6 ng m^{-3}), southern hemisphere (1.1–1.3 ng m^{-3}) and tropic regions (0.8–1.1 ng m^{-3}) respectively [26–28]. Various investigations have been completed worldwide on GEM mainly centred on the urban and rural locales, including mining and mechanical territories [29–34]. A thorough investigation of the air fluctuation, adding up to vaporous mercury and their relationship at the high-altitude station (Kodaikanal) of Southern India has been reported [35]. However, there is no complete investigation of the developed and developing urban regions of India and their contribution to TGM. This is the first research in India with a comparative and continuous observation of the temporal variations in TGM and its relationship to other meteorological parameters in urban and rural high-altitude stations. In general, the variation in mean seasonal concentration of TGM depends largely on meteorological variables. The study aims to investigate that during the day concentration of TGM is strongly change by solar radiation, evaporation and weather patterns. The main objectives of this study are to assess the Seasonal variability of atmospheric Total Gaseous Mercury (TGM) in high-altitude background station (Kodaikanal) and coastal urban city (Chennai) in India, to identify the potential sources and sinks of atmospheric mercury in the study areas and the influence of changing weather conditions on the atmospheric mercury distribution. Further, to compare the concentration of mercury in the past and recent findings of mercury at different locations around the world.

2. Materials and methods

In this study, monitoring sites are centrally located in high-altitude background station (Kodaikanal) and coastal urban city (Chennai) in South India (**Figure 1**). Kodaikanal is situated on a plateau on the southern ridge of upper Palani hill, at 2133 m (6998 ft) between the valleys of Parappar and Gundar. Such hills surround the Western Ghats mountains on the western side of South India. Kodaikanal region covering the whole of Kodaikanal taluk is located between 10° 7'56" N latitude and 10°26' and 77°15' East and 77°42' East longitude. Such hills shape the western Ghats on the west portion of South India's eastward slope. Kodaikanal is located on the east coast of the Western Ghats, at the southern end of the elevated hills of Palani of Dindigul district, in the state of Tamil Nadu. For a long time in Kodaikanal, despite reports of extensive mercury contamination, the closure of a mercury factory owned by the Indian Unilever company Hindustan Unilever became a big concern. There are 35,021 residents in Kodaikanal.

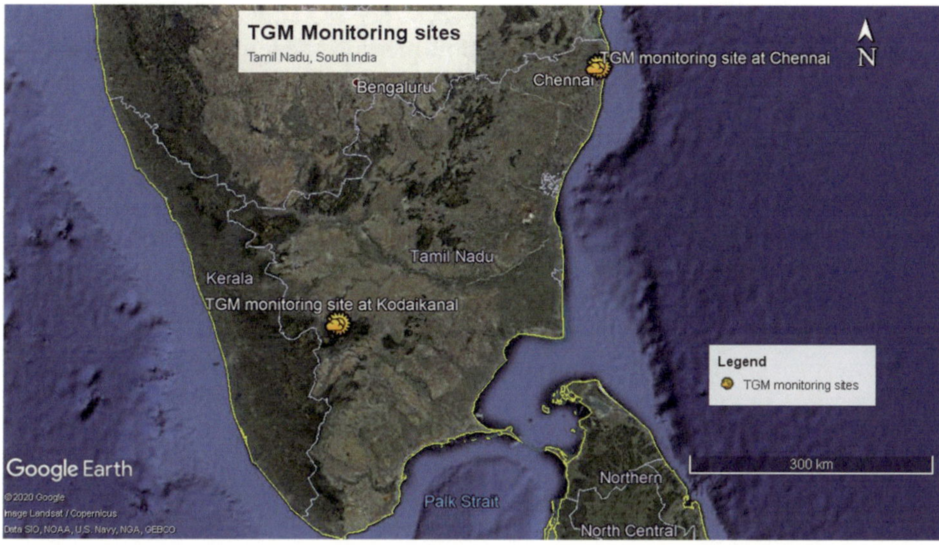

Figure 1.
Atmospheric Total Gaseous Mercury (TGM) monitoring sites in high-altitude background station (Kodaikanal) and coastal urban city (Chennai) in South India.

Chennai, situated on the South East Coast of India is the capital of Tamil Nadu. The Chennai city houses large scale enterprises like Petrochemical businesses, Thermal power plants, Rubber Factories and also many small-scale industries are prospering in and around the city. Chennai Metropolitan falls in the tropical wet and dry climatic condition, with the average barometrical temperature of around 25 to 40°C. The normal yearly precipitation of the city is approximately 140 cm. Because of its varied industrial and domestic setting, Chennai Metropolitan is a suitable site for studying the variations in the concentration of TGM in the air. The computerized mercury vapour analyzer (Model No - Tekran 2537B) placed at Anna University, Guindy Campus, Chennai (13° 0'45.05" N - 80° 14'2.66" E; MSL – 49 ft) was used for TGM measurements. TGM measurement and dataset were collected from the top of the Institute for ocean management, building in the Guindy campus of Anna University. The sampling height is about 50 m above the ground level, and the sample inlet was fixed in 1m above the floor of the sampling site. Many significant roads crossed in the nearby observation sites it creating vehicular pollution, with no significant sources of massive industrial pollution within 10 km radius.

3. Sampling methods and materials

The Total Gaseous Mercury (TGM) estimation was carried out using a Tekran™ 2537B utilizing an in-situ automated ambient mercury vapour analyzer. Tekran mercury vapour analyzer (2537B) continuously measured the TGM every 5-minutes from January 2015 to December 2016 at high-altitude background station (Kodaikanal) and coastal urban city (Chennai) in South India. The meteorological information was acquired from computerized weather stations such as the Central Pollution Control Board (CPCB) Chennai station, Indian Meteorological Department and World weather online. Consistent informational collections of the above said parameters were recorded for each 15-minute interim and per day averages. Using a Mercury Vapor Automated Analyzer (Model No–Tekran 2537B), the addition of total gaseous mercury (TGM) was studied. To this achieve, using cold vapour atomic fluorescence spectroscopy (CVAFS) techniques and the minimum detection limit 0.1 ng/m^3, which are described. When the ambient air was analyzed, a 47 mm Teflon filter was inserted in the whole measurement method of usage of the experiment. The flow rate is constant 1 L min^{-1} during all sampling periods. The implicit two gold cartridges, on the other hand, gather and thermally desorb mercury. The analyzer measured the Hg concentrations with intervals of 5 minutes automatically every 24 hours, and it was calibrated with its internal permeation sources. Present measurements at atmospheric TGM concentrations include other parameters, temperature, relative humidity, density, rainfall intensity, the direction of wind and velocity of the wind. The analyzer is automatically adjusted every 24 hours for each cartridge utilizing the internal ZERO and SPAN Permeation Processes. The peak areas for both cartridges during the calibration cycle are ensured during the ZERO process and under an error of less than ± 10% during the SPAN process. Computerized day-to-day alignments were performed every 24 hours (3.10 p.m. and 3.40 p.m.) using the instrument's internal adjustment source [35, 36]. The periodical inner alignment expels both in traverse and zero that are caused for the most part by temperature and maturing of the fluorimeter light. The tested air was estimated in each five-minute time interim at a stream rate of 5 L min^{-1}. The detail of the inspecting air and the precision status of the instrument is clarified by Mao et al. The recognition furthest reaches of the TGM are< 0.1 ng m^{-3}. The precision of the estimation and the task is ± 5 %. Zero air was utilized as straightforward for the instrument. Airstream was gathered through PFA Teflon tube, which was tried with an aftereffect of around 100 % RGM passing proficiency (vacillation of RGM is once in a while < 2 %). However, this method is still the most accurate to date and is widely used for the observation of speciated Hg in ambient air.

4. Results and discussion

4.1 Characteristics of TGM in the high-altitude background station in South India

In the meteorological variables at the high-altitude ground station at Kodaikanal, India continued measurement of total gaseous mercury (TGM= Gaseous Elemental Mercury (GEM) + Reactive Gaseous Mercury (RGM) was performed from Jan 2015 to December 2015). The mean concentration for TGM was 1.49 ng m^{-3} with a range of 1.1–2.10 ng m^{-3} is shown in **Figure 2**. The Global Mercury Observation System (GMOS) ground-based monitoring sites in India are also the highest altitude monitoring location in the GMOS network at Kodaikanal (South India). Such measurement positions constitute a major addition to the GMOS network and improve the understanding of atmospheric Hg species in this world region. The statistical summary of TGM concentration along with the meteorological parameters in the ambient air of Kodaikanal during the study period provided in **Table 1**. **Figure 2** shows the hourly average, daily average, monthly variation of TGM concentration in high-altitude background station (Kodaikanal) in South India. The maximum hourly and daily average concentrations were 2.55 ng/m^3 and 1.95 ng/m^3, respectively. The TGM concentration was occurring at every day for a month, evening time (3.00 am to 6.00 pm; the maximum concentration within the whole-time frame) it is shown in **Figure 2**. This finding was identical to previous observations of [37], at high altitude, remote area of the region of Mt. Changbai, northeast China. Mean annual TGM concentrations at the site of Kodaikanal were recorded at 1.52 ± 0.24 ng/m^3; between 0.77 ng/m^3 and 3.35 ng/m^3. These observable values of mean TGM concentrations were strongly linked to previous observations [35]. The average TGM values in the study area have also been compared with those reported from the high-altitude rural areas, but lower than in the Asian coastal regions [30, 38]. The highest monthly average TGM was reported in April 2015 (2.07 ng/m^3), while in July 2015 the lowest monthly average was 1.08 ng/m^3. The TGM concentrations range from 0.7 to 2.0 ng/m^3, accounted for approximately 96% of the overall TGM. The annual mean TGM values were usually higher during the day time (1.57 ng/m^3) compared tonight (1.08 ng/m^3) it shows in **Figure 2**. The day-night fluctuations in the TGM level may be induced by temperature variations and thus condensation levels and soil volatilization. A rural site with a similar altitude (~2800 m) in the south of France, where the estimation of TGM in the Pic du Midi Observatory [39], with equivalent techniques, recorded an average of 1.86 ± 0.27 ng/m^3. The geogenic mercury emissions are almost ~0.5 kilotonnes per year (kt y^{-1}) and re-emission of Hg ~1.6 kt y^{-1} from the sources of plants and biomass burning [40]. The mercury deposition can be influenced by organic substances complexation, binding to Fe-Mn oxides, hydrothermal pollutants, sulfide interaction and methylation, as well as world proximities such as river drainage, waste sources, etc. [9, 10, 19, 41]. Meteorological conditions of high-altitude background station (Kodaikanal) in South India studied during the period under report are presented in **Figure 3**. The rose diagram graphically displays wind speed, and wind direction graph indicates that West, ENE direction has the maximum value of frequency fall in 20% with a wind speed range of 4–5 m/s at December to March. The minimum wind speed ranges 2–3 m/s falls in during May to November in Kodaikanal site (**Figure 3**). The relative humidity values increased from June to November; also, the TGM

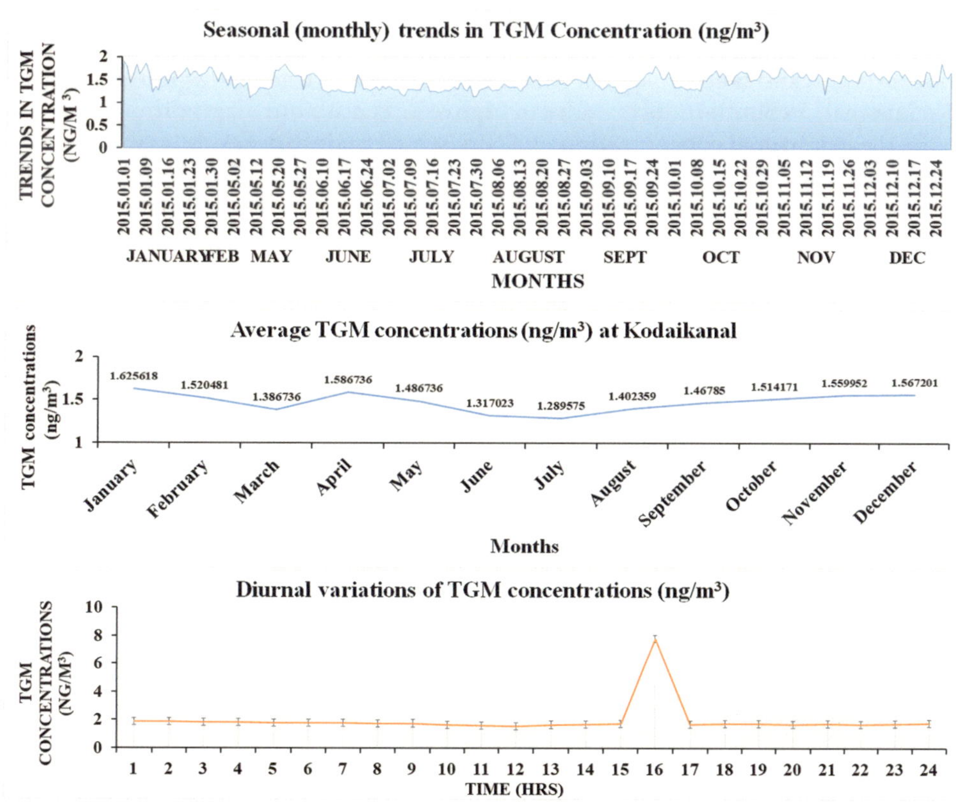

Figure 2.
Diurnal and monthly variation of TGM concentration in high-altitude background station (Kodaikanal) in South India.

Sites	Air Quality Parameters	Seasons			
		Autumn	Spring	Summer	Winter
Chennai	Wind Speed (m/s)	1.13	1.38	1.41	1.11
	Wind Dir (Deg)	198	131	157	187
	Temp (Â°C)	29	30	31	27
	RH (%)	71	73	67	73
	SR (W/sq.m)	199	254	213	198
	TGM (ng/m^3)	4.69	5.40	3.62	5.39
Kodaikanal	Wind Speed (m/s)	8.60	6.70	8.40	9.20
	Wind Dir (Deg)	177	258	129	150
	Temp (Â°C)	17.6	17.4	20.5	18.1
	RH (%)	92.2	91.1	82.6	80.3
	SR (W/sq.m)	290	304	346	322
	TGM (ng/m^3)	1.54	1.38	1.62	1.59

Table 1.
Statistical summary and Seasonal variations of ambient air quality parameters sites in high-altitude background station (Kodaikanal) and coastal urban city (Chennai) in South India.

concentrations were decreased. But relative humidity values decreased from December to May; similarly, the TGM concentrations were increased. Between November and May (dry season), the TGM concentration difference was relatively higher than between June and August (wet season). The Correlation trends of TGM concentration and meteorological parameters in high-altitude background station (Kodaikanal) in South India it shows in **Figure 3**. There were major differences in the mean seasonal concentration of TGM, which mainly depends on weather conditions, and found to be the following: Summer > Winter > Northeast monsoon or Autumn > South-West monsoon or Spring it is given in **Table 1**. This research also showed that solar radiation, evaporation and rainfall strongly changed the daytime TGM concentration. The seasonal variation is influenced by meteorological conditions and other external sources [4, 14]. The gaseous elemental mercury is an important pathway from soil to atmosphere at the forest and to the environment [19]. Also, in the Kodaikanal region, the mean annual TGM value in the Northern Hemisphere in Kodaikanal is well within the ranges of the recorded TGM background for the area (1.5–1.7 ng/m^3). These ground stations mainly track the remote background at high altitude sites and sea levels. The meteorological conditions are significantly influenced by the topsoils and vegetation to release mercury in nature environments [8, 26]. The findings were also significantly affected by long- transport of improved Hg air masses from the eastern part of Gansu, the west of Shanxi, the west of Ningxia as well as northern India [37]. Furthermore, these studies have shown that natural source emissions in summer are higher than in winter.

4.2 Temporal variability of atmospheric mercury in Chennai coastal and urban region

Diurnal and monthly variation of TGM concentration and meteorological parameters in the coastal urban city (Chennai) in South India were estimated in-situ. In the overall monitoring period, day by day, TGM esteem ranges from 0.07 to 638.74 with a mean estimation of 4.68 ng/m^3. The highest concentration of total

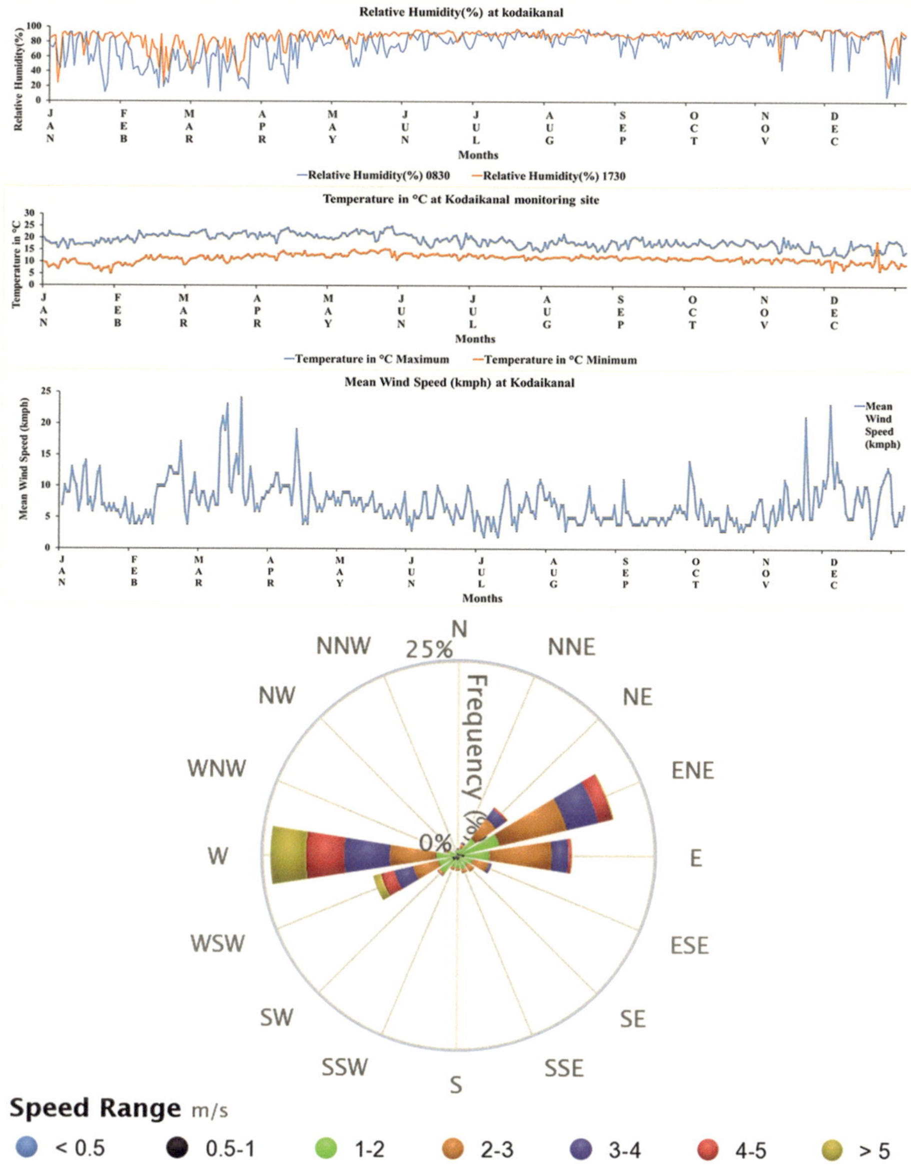

Figure 3.
Trends of TGM concentration and meteorological parameters in high-altitude background station (Kodaikanal) in South India.

gaseous mercury was recorded in June 2016 (638.74 ng/m^3), and lowest concentration was recorded in August 2016 (0.07 ng/m^3) at the coastal urban city (Chennai) in South India (**Figure 4**). The measured values of TGM are having a higher range than the Northern Hemisphere foundation concentration (1.50–1.75 ng/m^3) [9, 42]. TGM concentration occurring at every day, night or early morning (2.00 am to 7.00 am; the maximum concentration within the whole-time frame) is shown in **Figure 4**. Similarly, TGM concentrations were higher in the early in the morning and mid-night times reported by Schmolke et al. [36]. Such night-time maximums of TGM concentration [33, 36, 43, 44] have been due to mercury releases in the night-time inversion layer from surface accumulations. The potential sources of TGM in the investigation ground are from coal-based power plants, vehicular discharge, and

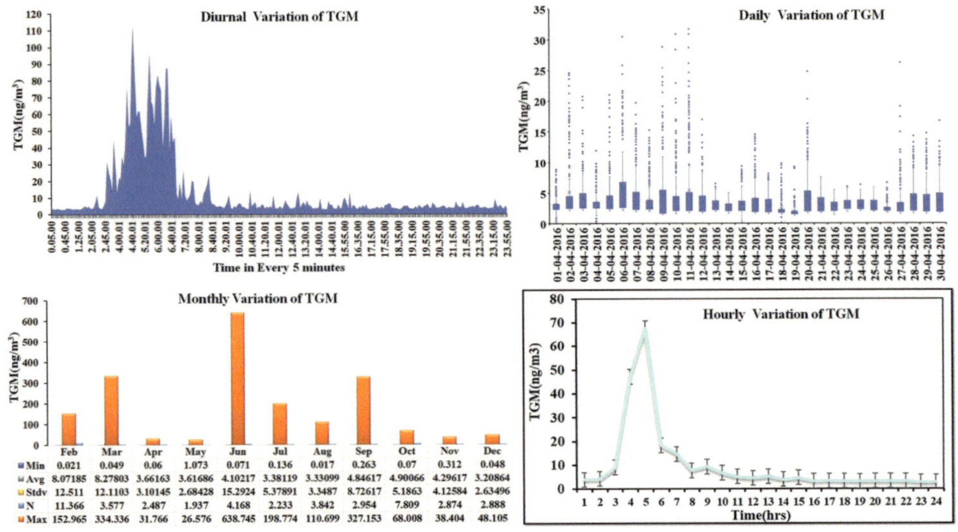

	Feb	Mar	Apr	May	Jun	Jul	Aug	Sep	Oct	Nov	Dec
Min	0.021	0.049	0.06	1.073	0.071	0.136	0.017	0.263	0.07	0.312	0.048
Avg	8.07185	8.27803	3.66163	3.61686	4.10217	3.38119	3.33099	4.84617	4.90066	4.29617	3.20864
Stdv	12.511	12.1103	3.10145	2.68428	15.2924	5.37891	3.3487	8.72617	5.1863	4.12584	2.63496
N	11.366	3.577	2.487	1.937	4.168	2.233	3.842	2.954	7.809	2.874	2.888
Max	152.965	334.336	31.766	26.576	638.745	198.774	110.699	327.153	68.008	38.404	48.105

Figure 4.
Diurnal and monthly variation of TGM concentration in the coastal urban city (Chennai) in South India.

squander burning [11, 12, 43]. The short-term measurements of TGM in china report recommend that the TGM ranges from 2.5 to 3.5 ng/m^3 for east beach front territories of China, 1.94 to 3.22 ng m^{-3} for Indochina peninsular regions [29]. Ci et al. [45] revealed that sea occasions are effectively engaged with the conveyance of the GEM along with the beachfront territories. Globally an average of 1.5 ng/m^3 of gaseous mercury is found in the atmosphere and Chennai; the average is 4.68 ng/m^3. The present-day a large source of atmospheric mercury obtains from the ocean the mostly in Hg0 (approximately ranges 1900–4200 Mg/year). The datasets of meteorological parameters versus TGM were plotted in **Figure 5**. Amidst the whole investigation time frame, the most extreme aggregate recurrence of wind rose was seen between 35 to 65° (NE) and 195 to 275° (SSE to WSW), and this focus is around 39% of the aggregate TGM outflow from the coastal urban city (Chennai) in South India **Figure 5**. TGM fluctuations were observed seasonally and diurnally, which suggested differences in source intensity, deposition processes and meteorological influences. The meteorological data set observed used to compare total gaseous mercury variation in the coastal urban city (Chennai) in South India, and it shows in **Figure 5**. The annual rose diagram graphically displays wind speed, and wind direction graph indicates that NE direction has the maximum value of frequency fall in 14% with a wind speed range of more than 5 m/s at Chennai urban environments. It is observed that when the temperature (27 centigrade) is low, the total gaseous mercury is found to be maximum (8.07 ng/m^3) for February (**Figure 5**). olar radiation, temperature, relative humidity and the wind speed increased a month of April, but the TGM concentration was in declined it shows in **Figure 5**. The TGM concentration was in positively correlated in barometric pressure and wind direction. The TGM concentrations continuously decreased in the following months, April to August; similarly, the barometric pressure and relative humidity also decreased (**Figure 5**). The meteorological parameters play a vital role in regulating atmospheric total gaseous mercury concentrations [15, 46]. The leading cause of pollution in megacities India is affected by the coal-fired power plants, transportation, industrial activity and also urban solid waste [47]. The peak concentration of total gaseous mercury was observed in Chennai urban environments during Winter as 5.64 ng/m^3, and the lowest concentration occurred during South-West

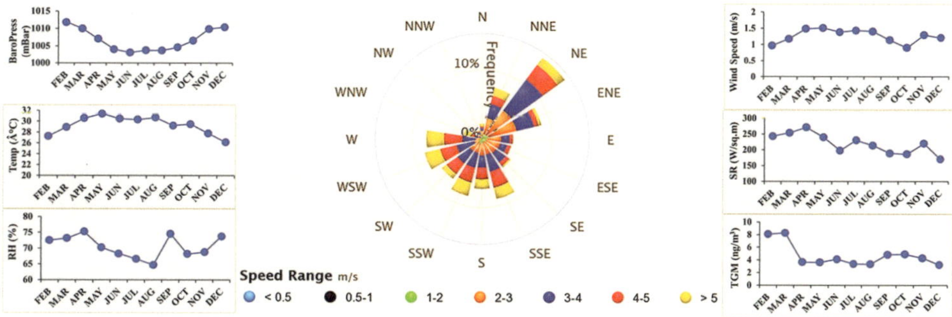

Figure 5.
Trends of TGM concentration and meteorological parameters in the coastal urban city (Chennai) in South India.

Monsoon, which is 3.91 ng/m^3. The highest concentration is observed during Winter and Summer due to long-range transportation of total gaseous mercury compared to autumn and spring seasons. The concentration of total gaseous mercury for the four seasons are arranged in the following order: Winter > Summer > Autumn > Spring for the coastal urban city (Chennai) in South India it is given in **Table 1**.

4.3 Mercury assessment of South Asia and global perspective

The influence of seasonal shift is very predominant in high-altitude background station (Kodaikanal) and coastal urban city (Chennai) in South India. The meteorological parameters (Wind speed, wind direction, solar radiation, Atmospheric temperature, Relative humidity) and TGM focus were connected to decide the relationship connection between the informational indices of the monitoring sites. Total gaseous mercury concentration varies significantly with wind speed and wind direction and other meteorological parameters, a concentration which changes with the seasons as given in **Table 1**. China and India where less attention of recycling the waste and increased production of coal combustion, metals, chlorine, and cement production. In India imported mercury users of Chlor-alkali plants, thermometers, batteries, Hg-Zinc, Zn-Carbon, fluorescent lamps, thermostat switches, alarm clocks, and hearing aids a total mercury user of 129.32 (Mg) reported by Mukherjee et al. [47]. A total of 6500 tones year^{-1}, adapted from, was measured for mercury emissions from biomass combustion, geogenic activities, and soil/vegetation/ocean emissions. The atmospheric mercury emissions approximately one-third from the sources of anthropogenic emissions similarly, natural emissions 70% and Oceanic emissions from 36% [31]. The primary anthropogenic sources such as combustion of fossil fuels for 24% and coal-burning (21%) at worldwide estimated emissions [5]. The approximately 2320 Mg of mercury is released yearly to the worldwide atmosphere (31%) for the primary sources of anthropogenic emission [8]. The world's leading mercury reservoirs, a unit of the Earth's measurement system and still an ecosystem suffering from anthropogenic activity, encompass the atmosphere (4.4 to 5.3 Gt), the terrestrial environment (in particular soils: 250 to 1000 Gg) and aquatic ecosystems (e.g. oceans: 270 to 450 Gg) [48]. The sustainability of mercury monitoring networks is an essential factor affecting the effectiveness of monitoring efforts.

In a global mercury assessment in 2013, mercury reported to dental usage measured at roughly 270-341 tons in 2010 [49], which represents 10% of global consumption of mercury **Figure 6**. Recently, the United Nations Environment Programme (UNEP) report 2018 to estimate the anthropogenic sources of anthropogenic sources in 2015 were about 2220 tons. Such sources constitute respectively 25 to 37 percentages of overall worldwide mercury emissions, measured at approximately 2000 tons. The TGM concentration in South Asia (India) are compared with other nations, the TGM levels are similar to the east, and southeast Asian countries and also Europe, Sub-Saharan Africa and North America are the averages and maximum concentration generally smaller. Mercury emission estimated (kg) in global in south Asia was in the second-largest nation in the worldwide it shows **Figure 6** [49]. Recent assessments of emissions of mercury into the environment (the 2010 targets) indicate that the primary anthropogenic sources of mercury pollution into the environment are artisanal and small-scale gold mining and fossil fuels (primarily coal) for power plants and industrial boilers for the generation of heat and electricity. In India majority of mercury releases from coal-burning (89,444 kg) followed by non-ferrous metal production (22,536 kg), waste from products (13,692 kg), cement production (13,421 kg) non-ferrous metal production, combustion of fossil fuels and artisanal small scale gold mining was in less than 1000 kg etc. it shows in **Figure 6** [49]. The most important natural sources and sources of re-emissions assessed within the GMOS project are oceans, which contribute 36% of the emission of mercury, followed by biomass (9%), deserts, metal and non-vegetation areas (7%), tundra and grassland (6%), forest (5%) and evasion after the events of mercury depletion (3%) [25, 40]. The majority of mercury releases worldwide estimated by fossil fuel combustion (11%), small-scale gold mining (5%), non-ferrous metal production (4%), cement production (3%),

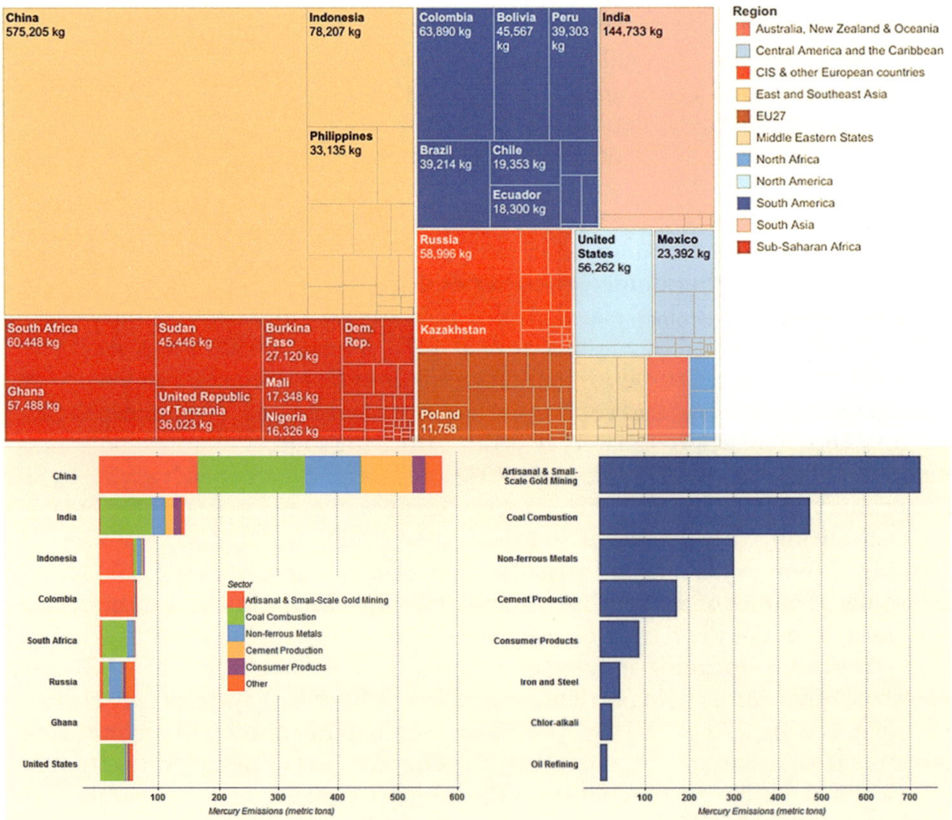

Figure 6.
Country and sector-wise mercury emission and sources of emission sectors in India and global (data source: UNEP [49]).

caustic soda production (2%), waste incineration (2%) and pig iron production (1%) [8]. Total mercury emissions are dominant in Asian countries, particularly China and India, and this information on the above factors and detailed estimates for mercury can be found in AMAP/UNEP [49].

Mukherjee et al. [47] reported the mercury contamination in India its mainly from industrial mercury emissions from coal combustion, the iron and steel industry, non-ferrous metallurgical plants, chloralkali plants, cement industry, waste disposal and other minor sources (i.e. brick production). The largest contributors to the source categories are coal combustion (52%) and waste incineration (32%) as shown in **Figure 6**. In general, TGM concentrations in urban and suburban areas are higher than in rural areas [49]. Mercury emission estimated (kg) in global in south Asia was in the second-largest nation in the worldwide [49]. However, measurements from global urban sites, which are also situated in the same region Asia, showed less than half of the mean concentration from our site. One of the main reasons for our study area is located in the coastal region was that episodically diluting with cleaner marine air and TGM with oceanic bromine will reduce pollution [50]. The possible sources of TGM in India are coal-fired power plants, vehicular emission, manufacture of ferrous and non-ferrous metals, waste incinerating sites, domestic fuel use from residents within the Informal villages around the Landward side, and ocean origin sources. Also, The Asian countries emissions are dominated in the global anthropogenic mercury emissions [21]. Current estimations on mercury emissions and re-emissions of primary natural mercury, including, mercury leakage cases, were measured at 5207 tonnes year^{-1}, which accounts for approximately 70% of the GMOS programme [5]. This pollution estimate is accurately compared to the information given by Cohen et al. Various additional lines of study and measurement are necessary to improve inventories of mercury and improve the ability to assess control options.

5. Conclusions

India is known to be the second-highest mercury (Hg) contributor to the global Hg budget for the environment. The present study is focused on the hourly, daily, and seasonal variations of the TGM concentration and meteorological parameters investigated at high-altitude background station (Kodaikanal) and coastal urban city (Chennai) in India. The mean total gaseous mercury concentration in Chennai is 4.68 ng/m^3, which is higher when compared to Kodaikanal, where it's approximately 1.53 ng/m^3. TGM concentrations exhibit an obvious diurnal pattern at Chennai urban region. All peak values appear between 3:00 am, and 8:00 am in all the seasons. This is probably the result of the change in the height of the atmospheric boundary layer that occurs between day and night. This is in large relation to global averages, but slightly less than in semi-industrial/urban areas in India.

The reason behind the higher concentration of total gaseous mercury in Chennai region is the high pollution due to anthropogenic sources, for example, industrial and vehicular emissions, which essentially improves vaporous mercury and also significantly enhances the atmospheric mercury level. Among the seasons, concentrations of TGM were higher during winter season both in Chennai and Kodaikanal indicating dry air with lower humidity aggregates higher pollutants in an urban environment. Total gaseous mercury concentration during the winter season is observed to be maximum in both regions. The average TGM concentrations during four monitoring seasons were ordered as Winter > Summer > Autumn > Spring. The average TGM concentrations in Chennai during the four monitoring seasons were ordered as Winter (5.64 ng/m^3) > Summer (5.16 ng/m^3) > Autumn (4.59 ng/m^3)

> Spring (3.92 ng/m^3). The average concentration of total gaseous mercury in the high-altitude background station (Kodaikanal) for the four seasons are arranged in the following order: Winter (1.61 ng/m^3) > Autumn (1.53 ng/m^3) > Summer (1.51 ng/m^3) > Spring (1.36 ng/m^3). From the results, it is clear that meteorological parameters play a vital role in the variation of total gaseous mercury. Factors such as the re-emission of concentrated mercury through Earth soils, vertical mixing and long-range transport influenced the seasonal variability of TGM at the monitoring sites. Moreover, it is clear that in the future if these meteorological parameters changes, it will change the concentration of total gaseous mercury in the observation regions. The present study can be extended by quantifying the total mercury emission from the earth systems and its impact on environments and human health in the Chennai urban region. There is a shortage of essential information and pollution factors for Asian countries to complete this analysis to address this situation. Recent work has used TGM and meteorological parameters, although the impact of wind speed, wind direction, and solar radiation on pollutant behaviour are well known, and these factors can be more easily approached in future research.

Acknowledgements

The authors express their gratitude towards The Global Mercury Observation System (GMOS), European Commission, for providing instrumental (Grant Agreement no. 265113) and technical support and Central Pollution Control Board (CPCB), Chennai, for providing meteorological data.

Author details

Manikanda Bharath Karuppasamy[1*], Srinivasalu Seshachalam[1], Usha Natesan[2] and Karthik Ramasamy[3]

1 Institute for Ocean Management, Anna University, Chennai 600025, Tamil Nadu, India

2 Centre for Water Resource, Anna University, Chennai 600025, Tamil Nadu, India

3 National Centre for Sustainable Coastal Management, Ministry of Environment, Forest and Climate Change, Chennai 600025, Tamil Nadu, India

*Address all correspondence to: krmanibharath93@gmail.com

References

[1] Johannes, Bieser., Hélène, Angot., Franz, Slemr., Lynwill, Martin.: Atmospheric mercury in the Southern Hemisphere – Part 2: Source apportionment analysis at Cape Point station, South Africa. Atmos. Chem. Phys. 63 (2020). https://doi.org/10.5194/acp-2020-63

[2] Lyman, S.N., Cheng, I., Gratz, L.E., Peter, Weiss-Penzias., Leiming, Zhang.: An updated review of atmospheric mercury, Sci. Total Environ. (2019). https://doi.org/10.1016/j.scitotenv.2019.135575.

[3] Fitzgerald WF, Engstrom DR, Mason RP, Nater EA. The case for atmospheric mercury contamination in remote areas. Environ. Sci. Technol. 1998;**32**(1):1-7. DOI: 10.1021/es970284w

[4] Tripathee Lekhendra., Guo Junming., Kang Shichang., Paudyal Rukumesh., Sharma Chhatra Mani., Huang Jie., Chen Pengfei., Sharma Ghimire Prakriti., Sigdel Madan., Sillanpää Mika.: "Measurement of mercury, other trace elements and major ions in wet deposition at Jomsom: The semi-arid mountain valley of the Central Himalaya". Atmos. Res. Volume 234, article id. 104691 (2020). DOI:10.1016/j.atmosres.2019.104691

[5] UNEP.: Global Mercury Assessment. United Nations Environment Programme, Geneva (2018).

[6] Liuwei, Wang., Deyi, Hou., Yining, Cao., Yong, Sik OK., Filip, M.G., Tack Jörg, Rinklebe., David O'Connor.: "Remediation of mercury contaminated soil, water, and air: A review of emerging materials and innovative technologies". Environ. Int. 134, 105281 (2020). https://doi.org/10.1016/j.envint.2019.105281

[7] Slemr, F., Weigelt, A., Ebinghaus, R., Bieser, J., Brenninkmeijer, C.A., Rauthe-Schöch, A., Hermann, M., Martinsson, B.G., van Velthoven, P., Bönisch, H., Neumeier, M., Zahn, A., Ziereis, H.: Mercury distribution in the upper troposphere and lowermost stratosphere according to measurements by the IAGOS-CARIBIC observatory: 2014-2016. Atmos. Chem. Phys. 18, 12329-12343 (2018)

[8] Pirrone N, Stracher GB, Cinnirella S, Feng X, Finkelman RB, Friedli HR, et al. Global mercury emissions to the atmosphere from anthropogenic and natural sources. Atmos. Chem. Phys. 2010;**10**:5951-5964

[9] Yi, Hui., Tong, Lei., Lin, Jia-mei., Cai, Qiu-liang., Wang, Ke-qiang., Dai, Xiao-rong., Li, Jian- rong., Chen, Jin-sheng., Xiao, Hang.: "Temporal variation and long–range transport of gaseous elemental mercury (GEM) over a coastal site of East China". Atmos. Res. Volume 233, article id. 104699 (2020). DOI:10.1016/j.atmosres.2019.104699

[10] Cooke, C.A., Martínez-Cortizas, A., Bindler, R., Sexauer Gustin, M.: Environmental archives of atmospheric Hg deposition – A review. Sci. Total Environ. (2019). DOI: https://doi.org/10.1016/j.scitotenv.2019.134800

[11] Huan, Zhang., Zhangwei, Wang., Chunjie, Wang., Xiaoshan, Zhang.: Concentrations and gas-particle partitioning of atmospheric reactive mercury at an urban site in Beijing, China. Environ. Pollut.Vol: 249, Page: 13-23 (2019). https://doi.org/10.1016/j.envpol.2019.02.064

[12] Jiaoyan, Huang., Matthieu, Miller B., Eric, Edgerton., and Mae Sexauer, Gustin.: "Deciphering potential chemical compounds of gaseous oxidized mercury in Florida, USA". Atmos. Chem. Phys. 17, 1689-1698 (2017). DOI:10.5194/acp-17-1689-2017.

[13] Poissant L, Pilote M, Beauvais C, Constant P, Zhang HH. A year of continuous measurements of three atmospheric mercury species (GEM, RGM and Hgp) in southern Québec, Canada. Atmos. Environ. 2005;**39**:1275-1287

[14] Cheng, Z., Tang, Y., Li, E., Wu, Q., Wang, L., Liu, K., Wang, S., Yongmei Huang., Lei Duan.: Mercury accumulation in soil from atmospheric deposition in temperate steppe of Inner Mongolia, China. Environ. Pollut. Vol: 258, Page: 113692 (2019). https://doi.org/10.1016/j.envpol.2019.113692

[15] Guor-Cheng Fang, Kai-Hsiang Tsai, Chao-Yang Huang, Kuang-Pu OuYang, You-Fu Xiao, Wen-Chuan Huang, Yuan-Jie Zhuang.: "Seasonal variations of ambient air mercury species nearby an airport". Atmos. Res. 202, Pages 96-104 (2018). https://doi.org/10.1016/j.atmosres.2017.11.008

[16] Fang F, Wang Q, Li J. Urban environmental mercury in Changchun, a metropolitan city in Northeastern China: source, cycle, and fate. Sci. Total Environ. vol. 2004;**330**:159-170

[17] Syed, Abdul Rehman Khan., Yu, Zhang., Anil, Kumar., Edmundas, Zavadskas., Dalia, Streimikiene., "Measuring the impact of renewable energy, public health expenditure, logistics, and environmental performance on sustainable economic growth," Sustainable Development, John Wiley & Sons, Ltd., vol. 28(4), pages 833-843, (2020), DOI: 10.1002/sd.2034.

[18] Weiss-Penzias P, Jaffe D, Swartzendruber P, Hafner W, Chand D, Prestbo E. Quantifying Asian and biomass burning sources of mercury using the Hg/ CO ratio in pollution plumes observed at the Mount Bachelor Observatory. Atmos. Environ. 2007;**41**:4366-4379

[19] Jun, Zhou., Buyun, Du., Lihai, Shang., Zhangwei, Wang., Hongbiao, Cui., Xingjun, Fan., & Jing, Zhou.: Mercury fluxes, budgets, and pools in forest ecosystems of China, A review. Crit. Rev. Eng., Sci. Technol. (2019). DOI: 10.1080/10643389.2019.1661176

[20] Liu L, Zhang W, Lu Q, et al. Variations in the Sensible Heating of Tibetan Plateau and Related Effects on Atmospheric Circulation Over South Asia. Asia-Pacific. J. Atmos. Sci. 2020. DOI: https://doi.org/10.1007/s13143-020-00207-0

[21] Pacyna EG, Pacyna JM, Sundseth K, Munthe J, Kindbom K, Wilson S, et al. Global emission of mercury to the atmosphere from anthropogenic sources in 2005 and projections to 2020. Atmos. Environ. 2010;**44**:2487-2499

[22] Horowitz HM, Jacob DJ, Zhang Y, Dibble TS, Slemr F, Amos HM, et al. A new mechanism for atmospheric mercury redox chemistry: Implications for the global mercury budget. Atmos. Chem. Phys. 2017;**17**:6353-6371

[23] Streets DG, Horowitz HM, Jacob DJ, Lu Z, Levin L, TerSchure AFH, et al. Total mercury released to the environment by human activities. Environ. Sci. Technol. 2017;**51**:5969-5977

[24] Amos HM, Jacob DJ, Streets DG, Sunderland EM. Legacy impacts of all-time anthropogenic emissions on the global mercury cycle. Global Biogeochem. Cycles. 2013;**27**(2):410-421. DOI: 10.1002/gbc.20040

[25] Ghazvini MV, Ashrafi K, Shafiepour Motlagh M, et al. Simulation of atmospheric mercury dispersion and deposition in Tehran city. Air Qual. Atmos. Health. 2020;**13**:529-541 https://doi.org/10.1007/s11869-020-00813-x

[26] Lynam M, Dvonch JT, Barres J, et al. Atmospheric wet deposition of

mercury to the Athabasca Oil Sands Region, Alberta, Canada. Air. Qual. Atmos. Health. 2018;**11**:83-93 https://doi.org/10.1007/s11869-017-0524-6

[27] Slemr F, Brunke EG, Ebinghaus R, Kuss J. Worldwide trend of atmospheric mercury since 1995, Atmos. Chem. Phys. 2011;**11**:4779-4787

[28] Sprovieri F, Pirrone N, Bencardino M, D'Amore F, Carbone S, Cinnirella F, et al. Atmospheric mercury concentrations observed at ground-based monitoring sites globally distributed in the framework of the GMOS network. Atmos. Chem. Phys. 2016;**16**:11915-11935

[29] Fu XW, Feng XB, Dong ZQ, Yin RS, Wang JX, Yang ZR, et al. Atmospheric gaseous elemental mercury (GEM) concentrations and mercury depositions at a high-altitude mountain peak in south China. Atmos. Chem. Phys. 2010;**10**:2425-2437

[30] Nguyen HT, Kim KH, Kim MY, Kang CH, Shim SG. Mercury in air in an area impacted by strong industrial activities. Chemosphere. 2008;**71**:2017-2029

[31] Pacyna, J.M., Travnikov, O., Simone, F.D., Hedgecock, I.M., Sundseth, K., Pacyna, E.G., Steenhuisen, F., Pirrone, N., Munthe, J., Kindbom, K.: Current and future levels of mercury atmospheric pollution on a global scale. Atmos. Chem. Phys. 16, 12495-12511 (2016)

[32] Pandey SK, Kim KH, Yim UH, Jung MC, Kang CH. Airborne mercury pollution from a large oil spill accident on the west coast of Korea. J. Hazard. Mater. 2009;**164**:380-384

[33] Schiavo B, Morton-Bermea O, Salgado-Martinez E, et al. Evaluation of possible impact on human health of atmospheric mercury emanations from the Popocatépetl volcano. Environ. Geochem Health. 2020. DOI: https://doi.org/10.1007/s10653-020-00610-6

[34] Wang Z, Chen Z, Duan N, Zhang X. Gaseous elemental mercury concentration in atmosphere at urban and remote sites in China. J. Environ. Sci. 2007;**19**:176-180

[35] Karthik R, Paneerselvam A, Ganguly D, Hariharan G, Srinivasalu S, Purvaja R, et al. Temporal variability of atmospheric Total Gaseous Mercury and its correlation with meteorological parameters at a high-altitude station of the South India. Atmos. Pollut. Res. 2016;**8**(1):164-173

[36] Schmolke SR, Schroeder WH, Kock HH, Schneeberger D, Munthe J, EbinghausR.Simultaneousmeasurements of total gaseous mercury at four sites on an 800 km transect: spatial distribution and short-time variability of total gaseous mercury over central Europe. Atmos. Environ. 1999;**33**:1725-1733

[37] Fu, X.W., Feng, X., Shang, L.H., Wang, S.F., and Zhang, H.: Two years of measurements of atmospheric total gaseous mercury (TGM) at a remote site in Mt. Changbai area, Northeastern China. Atmos. Chem. Phys. 12, 4215-4226 (2012)

[38] Nguyen HL, Leemakers M, Kurunczi S, Bozo L, Baeyens W. Mercury distribution and speciation in Lake Balaton, Hungary. Sci. Total Environ. vol. 2005;**340**:231-246

[39] Fu, X., Marusczak, N., Heimburger, L.R., Sauvage, B., Gheusi, F., Prestbo, E.M., and Sonke J.E.: Atmospheric mercury speciation dynamics at the high-altitude Pic du Midi Observatory, Southern France. Atmos. Chem. Phys. 16, 5623-5639 (2016)

[40] Outridge, P.M., Mason, R.P., Wang, F., Guerrero, S., Heimbürger-Boavida,

L.E.: Updated Global and Oceanic Mercury Budgets for the United Nations Global Mercury Assessment. Environ. Sci. Technol. (2018). https://doi.org/10.1021/acs.est.8b01246.

[41] Lee DS, Dollard GJ, Pepler S. Gas-phase mercury in the atmosphere of the United Kingdom. Atmos. Environ. 1998;**32**:855-864

[42] Zhang, Y., Khan, S.A.R., Kumar, A., Golpîra, H., Sharif, A., Is tourism really affected by logistical operations and environmental degradation? An empirical study from the perspective of Thailand. J. Clean. Prod. 227, 158-166, 2019.

[43] Lee T, Shin U, Park S. Atmospheric Structure for Convective Development in the Events of Cloud Clusters over the Korean Peninsula. Asia-Pacific. J. Atmos. Sci. 2020. DOI: https://doi.org/10.1007/s13143-020-00211-4

[44] Syed Abdul Rehman Khan., Arshian, Sharif., Hêriş, Golpîra., Anil, Kumar., "A green ideology in Asian emerging economies: From environmental policy and sustainable development," Sustainable Development, John Wiley & Sons, Ltd., vol. 27(6), pages 1063-1075, (2019). DOI: 10.1002/sd.1958.

[45] Ci ZJ, Zhang XS, Wang ZW, Niu ZC, Diao XY, Wang SW. Distribution and air-sea exchange of mercury (Hg) in the Yellow Sea. Atmos. Chem. Phys. 2011;**11**:2881-2892

[46] Penuelas J, Sardans J. Developing holistic models of the structure and function of the soil/plant/atmosphere continuum. Plant Soil. 2020. DOI: https://doi.org/10.1007/s11104-020-04641-x

[47] Mukherjee, A.B., Bhattacharya, P., Sarkar, A., and Zevenhoven, R.: Mercury emissions from industrial sources in India and its effects in the environment, Springer, New York, USA, chap. 4, 81-112 (2009). DOI: 10.1007/978-0-387-93958-2_4

[48] Obrist D, Johnson DW, Edmonds RL. Effects of vegetation type on mercury concentrations and pools in two adjacent coniferous and deciduous forests. J. Plant Nutr. Soil Sci. 2012;**175**(1):68-77. DOI: 10.1002/jpln.201000415

[49] AMAP/UNEP.: Technical Background Report for the Global Mercury Assessment2013, Arctic Monitoring and Assessment Programme, Oslo, Norway/UNEP Chemicals Branch, Geneva, Switzerland. vi + 263 pp, (2013).

[50] Zhu J, Wang T, Talbot R, Mao H, Hall CB, Yang X, et al. Characteristics of atmospheric Total Gaseous Mercury (TGM) observed in urban Nanjing. China. Atmos. Chem. Phys. 2012;**12**:12103-12118 https://doi.org/10.5194/acp-12-12103-2012

Chapter 5

Health Impacts of Air Pollution

Muhammad Ikram Bin A Wahab

Abstract

Urban air pollution has become a salient environmental issue in many Asian countries due to their rapid industrial development, urbanization, and motorization. Human-induced air pollution has been and continues to be considered a major environmental and public health issue. Its severity lies in the fact that high levels of pollutants are produced in environments where damage to human to concentration, duration of exposure health and welfare is more likely. This potential is what makes anthropogenic air pollution an important concern. Extreme air pollution episodes were reported for the Meuse Valley, Belgium, in 1930; Donora, PA, and the Monongehela River Valley in 1948; and London in 1952. These episodes are significant in that they provided solid scientific documentation that exposure to elevated ambient pollutant levels can cause acute illness and even death. The most devastating events contributed to important efforts to control ambient air pollution. The International Agency for Research on Cancer (IARC) assessment concluded that outdoor air pollution is carcinogenic to humans, with the particulate matter component of air pollution mostly associated with increasing cancer incidence especially lung cancer. Pollutant effects typically occur in some target organs. These can be straightforward; i.e. pollutants come into close contact with the affected organ.. Such is the case for eye and respiratory irritation. Effects may be indirect. For example, Pollutants can enter the bloodstream from the lungs or gastrointestinal system through the respiratory route. Effects may then be distant from the immediate organ of contact. A target organ can have no immediate and intimate contact with atmospheric contaminants.. The primary organs or target organs are the eyes and the respiratory and cardiovascular systems.

Keywords: Urban Air Pollution, Human Health Effects, Respiratory and Cardiovascular Disease

1. Introduction

Air pollution is the presence of unwanted substances in the air in sufficient quantities to produce adverse effects. Undesirable substances can affect human health, vegetation, human property or the global environment including creating esthetic slurs in the form of brown or foggy air or offensive odors. Outdoor or ambient air pollution has been recognized as one of the major concerns that have high potential for its deleterious effects on health. Increased urbanization, human activities and changing urban setting in the country have resulted in elevated air pollution and the occurrence of urban heat islands.

The classification of air pollutants is based mainly on the sources producing pollution. There are four main air pollution sources namely major, area, mobile and natural sources. Major sources include the emissions from power stations, refineries, petrochemicals, manure industries, metalworking and other industrial facilities, and municipal incineration. Indoor sources include household cleaning operations, printing facilities and gas stations. Mobile sources include motor vehicles, cars, rail lines, airways and others. Last but not least, natural sources include physical disasters such as forest fires, volcanic eruptions, dust storms, and agricultural burning [1].

Particulate matter air pollution less than 2.5 μm has received great international attention due to its diverse contribution to the global burden of disease. Study done by Brauer et al [2] mentioned that majority of the planet still resides in areas where the World Health Organization Air Quality Guidelines of 10 $\mu g/m^3$ (annual) and 25 $\mu g/m^3$ (24hrs) is exceeded. Ground level measurement of $PM_{2.5}$ or lower are still limited in most of the places in the world. Therefore, studies are required to evaluate and provide insight in high risk as to the exposure and risk as many of the form of air pollution are beyond the control of individual and require policy at national and international levels.

Studies over the past two decades have assessed the relationship between size distribution of particulate matter and trace metal concentrations in urban areas. In the past, numerous researchers in Europe have carried out experiments to investigate particle-based metal particle size distribution [3–5].

The inorganic components constitute a small portion by mass of the particulates; however, it contains some trace elements such as As, Cd, Co, Cr, Ni, Pb and Se which are human or animal carcinogens even in trace amounts [6, 7]. The high level of Pb can induce severe neurological and hematologic effects on the unprotected population particularly in children, whereas both Cd and Ni are known to cause cancer-causing effects on humans by inhalation. Workplace exposure to Cd is an important risk factor for chronic pulmonary diseases [8]. Cr (VI) have been recognized to cause toxicity and carcinogenicity in the bronchial tree [9, 10]. Exposure to Mn is associated with an increase in neurotoxic deficiencies [11]. Elevated levels of Cu may cause respiratory irritation [10, 12].

PAHs have attracted a substantial amount of attention due to their persistent, bio-accumulative, carcinogenic and mutagenic properties related to health problem such as cataracts, kidney and liver damage, and jaundice [13]. The maximum concentrations of airborne PAHs are typically occur in the urban environment due to increased vehicle traffic and the spread of air pollutants. Given the high urban population density, the risk from human exposure to airborne PAHs in highest [14].

Global Burden of Disease Study (GBD) 2015 indicated that ambient particulate matter pollution accounted for 4.2 million deaths and 103 million healthy life-years lost in 2015, representing 7.6% of total global mortality and making it the fifth-ranked global risk factor.

2. Particulate matter

The term particulate matter (PM) or atmospheric aerosols is a mixture of solid particles and/or liquid droplet that may vary in concentration, composition and also size distribution. Aerosols can be defined as suspensions of solid or liquid in a gas. Therefore, aerosols include both the particle and the gas in which they are suspended. Although aerosol and particle are different, they are often used interchangeably throughout the literature to refer to the particle only.

Aerosol particulates in the atmosphere come from a wide mixture of natural and anthropogenic sources. Primary particulates are released directly as liquids or solids from sources such as biomass combustion, incomplete fossil fuel combustion, volcanic eruptions and suspension of road, ground and mineral dust, sea salt and biological materials caused by wind or traffic. Secondary particulates, on the other hand, are formed by the conversion of gas into particulates in the atmosphere. Primary gaseous species might undergo chemical reaction to produce low-volatility products. Then, these products partitions to the particulate phase, i.e. new particles are formed by nucleation and condensation of gaseous precursors [15].

Small aerosol particles predominantly contribute to number concentrations; however they only play a smaller role for the volume distribution. On the other hand, larger particles play a role for the volume distribution but do not contribute substantially to the number concentration [16]. Particles are conventionally divided into different size of fractions; based on the physical and chemical processes involved in the particle formation and growth. The different size of fractions is generally called "modes" [17].

The nucleation (or ultrafine) mode resides in the range below 0.02 µm of particle diameter and usually presents its maximum number-density around 5-15 nm of particle diameter. H_2SO_4, NH_3 and H_2O are examples of precursor gases to form new particles by homogenous nucleation in the ambient atmosphere. However, due to the condensation by other condensing gases, organic and inorganic components, the newly formed particles rapidly grow bigger. These particles have hours lifetime in the atmosphere as they rapidly coagulate with larger particles or grow into larger sizes due to condensation. Classical nucleation theory shows that the nucleation highly depends on the concentrations of the gaseous precursors, relative humidity and temperature. In particular, the nucleation is favored by decreases in the temperature and/or increases in the relative humidity [18].

Aitken mode particles range from 0.02 to 0.1 µm and originate either from primary particles, natural and anthropogenic, or by growth of nucleation mode particles. Secondary Aitken mode particles are likely to be formed by coagulation of ultrafine particles, by condensation and by liquid phase reactions. Combustion process is a primary source that has very large emissions of Aitken mode particles. Aitken mode particles are present at relatively stable concentration in the atmosphere; indicating a long residence time of Aitken particles at ambient atmosphere. The accumulation mode covers the range between 0.1 and up to 1 µm. Aitken mode particles have a tendency to grow to accumulation mode particles due to coagulation and liquid phase reactions occurring in cloud droplets. Hoppel et al. [19] stated that the mass transfer by condensation and/or nucleation/coagulation is not enough to cause any significant change in particles size compared with the observed growth.

3. Health effects of particulate matter

Particulate pollution includes particulate matter with a diameter of 10 micrometers (µm) or less, referred to as PM10, and extremely fine particulate matter with a diameter of 2.5 micrometers (µm) or less. Particles contain tiny liquid or solid droplets which may be inhaled and cause adverse health effects. PM10 when inhale can enter the lungs and the bloodstream. Fine particulate matter, PM2.5, represents a greater health risk (**Table 1**) [1].

There is consistent evidence for the relationship between atmospheric particulate matter and public health outcomes for adverse health effects [20]. The range of effects is extensive, including effects on the respiratory and cardiovascular systems that extend to children and adults in the general population [20–22], but also including lung cancer [21, 22].

The risk for various outcomes has been shown to increase with exposure and there is little evidence for a threshold below which no adverse health effects would be anticipated [20]. In one WHO report [23, 24], the importance to public health

Particle size	Penetration degree in human respiratory system
11 μm	Passage into nostrils and upper respiratory tract
7–11 μm	Passage into nasal cavity
4.7–7 μm	Passage into larynx
3.3–4.7 μm	Passage into trachea bronchial area
2.1–3.3 μm	Secondary bronchial area passage
1.1–2.1 μm	Terminal bronchial area passage
0.65–1.1 μm	Bronchioles penetrability
0.43–0.65 μm	Alveolar penetrability

Table 1.
Penetrability according to particle size.

of the long-term effects of particulate matter exposure outweighs the importance of short-term effects. The short-term effects of exposure have been recognized in many time series studies, and short-term (24-hour) and long-term (yearly average) guidelines are suggested [24]. PM2.5 and PM10 are recommended for assessment and control as fine and coarse particles have diverse sources and can have multiple effects. With respect to ultrafine particulate matter, there is insufficient information to support a quantitative assessment of the potential health effects of exposure [25].

Particle size and surface area are important characteristics from a toxicological point of view. Pope & Dockery [21]; Schlesinger et al. [22] recently review the current understanding of particulate matter, its size and its health effects.

Study done by Pope & Dockery [21]; Schlesinger et al. [22] proved that the relationship between fine particulate matter, PM2.5, and most health effects is greater than between PM10 and health effects. Ultrafine particulate matter contributes little to the mass concentration of particulate matter; however, it affects the surface in large numbers and is of interest to toxicological studies [22]. Study done by Nel et al. [26] concluded that as the size of a particle decrease, its surface area increases, thus allowing a greater proportion of its atoms or molecules to be displayed on the surface. Furthermore, particle size is crucial for penetration and deposition efficiency into human lungs, since ultrafine particles ability to penetrate the membranes of the respiratory tract and enter the blood circulation.

Particulate organic matter in ambient air is a complex mixture of chemicals, i.e. polycyclic aromatic hydrocarbons (PAHs). The International Agency for Research on Cancer (IARC) classified some PAHs, i.e. benzo[a]pyrene, as human carcinogens [27]. Another group that has gained interest in recent years is metals. Schlesinger et al. [22] have summarized current knowledge regarding trace metals and their health impact. The assessment of the risk associated with exposure to PAHs is still a challenge. PAHs present in ambient air mainly as a complex mixture and the interactions among components may lead to additively, synergistic or antagonistic effects. In addition, a study conducted by Topinka et al. [28] shows that organic matter extracted from $PM_{2.5}$ are likely to induce DNA adducts and oxidative DNA damage in in-vitro cell-free assay experiments.

The toxicological findings strongly suggest that transition metals such as V, Cr, Mn, Fe, Ni, Cu and Zn are components in PM with toxic capability based on their potential for oxidative activity and the production of reactive oxygen species [22]. Soil dust is composed of crust elements (Si, Ca, Al and Mg) and is existent in ambient PM2.5, but bigger segments are present in coarse mode.

Windblown soils, dust, at least in rural areas, are not considered a significant health risk however natural crusty materials may be contaminated with road dust generated by moving vehicles. Contaminants may contain a range of constituents, including PAHs and various metals that can alter the toxicology of crust particles.

Several studies indicated that particles produced from burning of biomass for domestic heating have increased and become one of the major sources of $PM_{2.5}$ in Europe and other parts of the world [29, 30]. Incomplete combustion of biomass produces a multitude of different organic species as well as trace metals. In a assessment of some epidemiological studies carried out in areas where wood smoke is prevalent. Boman et al. [31] concluded that particulate matter from wood smoke appears to be at least as deleterious as particulate matter from other sources.

In 2016, air pollution was ranked sixth in terms of global burden of disease and that 7.5% of all deaths were attributable to ambient air pollution (4.1 million deaths) [32]. Globally, ambient air pollution is responsible for about 27.5% of all lower respiratory infection deaths and 27% of all deaths due to chronic obstructive pulmonary disease [32]. The World Health Organization (WHO) estimates that about 26% of deaths due to ambient air pollution in 2012 occurred in South East Asia [33].

South East Asia population is not only exposed to local sources of air pollutants (for example, industrial and traffic related air pollutants) but it also exposed to regional smoke from the seasonal forest fires (or 'haze') that occur in Sumatra and Kalimantan. The fine particulate pollution from these forest and peat fires can reach hazardous levels for extended periods of time, posing severe health risks to millions of people across Malaysia. These forest fires are also a major cause of transboundary air pollution throughout South East Asia. The forest fires occur in the dry season, (known as the South-west monsoon season, April to September) and tend to be more severe in the El Nino years. During this monsoon season, the prevailing wind direction is south westerly, and the smoke from forest fires in Sumatra are commonly transported towards Malaysia. The most severe haze episode from these forest fires occurred in 1997. Since then, there have been severe haze episodes from forest fires every two to three years, for example, in 2005, 2006, 2009, 2010, 2011, 2012, 2013, 2014 and 2015 [34]. The last major transboundary haze episode occurring in 2015 has been estimated to have resulted in 91,600 excess deaths in Indonesia, 6,500 deaths in Malaysia and 2,200 deaths in Singapore [35].

During the haze episodes, the main air pollutant of concern is PM2.5. The health impacts of PM2.5 are related to concentration, duration of exposure and individual susceptibility. PM2.5 pollution is causally related to cardiovascular (myocardial infarction, hypertension, heart failure, arrhythmias) and respiratory disease (chronic obstructive pulmonary disease and lung cancer [32]. Long-term exposure to PM2.5 has also been linked to adverse birth outcomes, childhood respiratory disease and possibly neurodevelopment and cognitive function, and diabetes [36, 37]. PM2.5 is also associated with systemic inflammation with associated increases in acute phase proteins such as C-reactive protein and fibrinogen [38]. These biomarkers have been consistently linked to subsequent cardiovascular disease and death [37].

4. Trace metals

Toxic heavy metals are one of the major hazardous that affects us today. Atmospheric deposition of toxic heavy metals has stimulated profound research all over the world due to effects on living organisms. Toxic metals like lead (Pb), cadmium (Cd), arsenic (As) and chromium (Cr) can essentially attack specific areas in the human body upon exposure [39]. Their dispersal and transport through

Natural sources	% released into the atmosphere
Biogenic	• 50% Hg, Mo • 30–50% As, Cd, Cu, Mn, Pb and Zn
Volcanic Gas	• 40–50% of Cd and Hg • 20–40% of the As, Cr, Cu, Ni, Pb, and Sb
Sea aerosols	• 10% various heavy metals

Table 2.
Share of releases of trace metals from natural sources.

the atmosphere can be worsened by anthropogenic and natural phenomena. High levels of these trace metals in air can cause ecotoxic effects on plants, animal as well as humans.

In the air, trace metals are combined with sediment dust and respirable airborne particulate matter [40]. They can be released from air through precipitation or direct dry deposits in various environmental compartments near or away from their source. An in-depth investigation of the levels of trace metals in the atmosphere is therefore both essential and fundamental to ensure a more secure environment.

Naturally trace metals can be in the form of vapor ions dissolved in water and in the form of minerals or salts in rock, soil and sand. Pacyna [41], stated biogenic sources represent over half of the Hg and Mo emitted to the atmosphere, and approximately 30–50% of the As, Cd, Cu, Mn, Pb and Zn freed (**Table 2**). Soils can be responsible for substantial releases of trace metals to the atmosphere.

Trace metals are not only brought into the environment from natural sources, they are also introduced through human activities [42]. Waste incineration sites, agricultural runoff, vehicle emissions and urban effluents release trace metals to the environment [43]. Industrial sources can pose a significant threat, particularly in densely populated areas. Combustion of fossil fuels from stationary sources can be important in dispersing trace metals into the atmosphere [44].

5. Metal toxicity

A toxic material is a substance that has an adverse effect on health. Many chemicals could be classed as toxic, but some are more toxic than others. When metals bind to the sulfhydryl groups in proteins, they may move essential elements, interrupt function or obstruct the activity of the sulfhydryl group leading to toxicity [45]. Trace metals like mercury, lead, arsenic, zinc, copper, gold, cobalt and cadmium can be highly toxic and easily accessed by living organisms. Plants may exhibit different signs of toxicity according to type of metal and plant. Generally, in humans, heavy metal toxicity may range from reproductive defects to lung damage [46, 47] while in plants, they may cause damage to root systems, disrupt the proper functioning of the stomata and inhibit growth [48, 49],

Over the last 50 years, there has been a rapid rise in the number of significant trace elements associated with water pollution incidents. The release of cadmium into the Jinstu River in Japan resulted in serious bone damage [50].

Even chromium is an essential micronutrient, chromatic salts can have serious effects on the skin, nasal septum and lungs [46]. Homeostatic mechanisms generally work to control gastrointestinal absorption, so entering the body through other routes bypasses this control. Another example is Zn, whereby inhalation of zinc oxide vapors can cause an allergic reaction leading to metal smoke fever [51].

Large number of metals such as Cd, Pb and Hg have no recognized biological significance in human and exposure of small amounts may be toxic [52]. "Trace metals" is a broad collective term that includes any metal that is toxic and has a relatively high specific density as well [52]. Common indications of intoxication (Cd, Pb, As, Hg, Zn, Cu and Al) encompass gastrointestinal disturbances, diarrhea, stomach ulser, hemoglobinuria, ataxia, paralysis, vomiting and seizures, depression and pneumonia [52].

Metals intrude with the biochemistry of the organism during common metabolism processes. In the acidic middle of the stomach, they transform into stable oxidation states and merging with the body's biomolecules like proteins and enzymes to establish solid and stable chemical bonds, replacement of hydrogen or essential metals in an enzyme and, therefore, inhibit its functioning [52].

Metallic ions in the body's metallo-enzymes can be easily substitute by another similar sized metallic ion. Occasionally, the enzymes of a whole sequence coexist in the form of a single multi-enzymatic complex hence replacing an essential metal by an interfering metal blocks the biological reaction of the enzyme function. One example is the substitution of Zn by Cd and caused detrimental effects on body chemistry [53].

Therefore, the metal remains incorporated in the tissue and will effect in a variety of bio-malfunctions, not all evenly severe [52]. The most toxic forms of interference metals are their most stable oxidation states, since these very stable forms of biotoxic compounds are difficult to isolate with detoxification therapies [52].

6. Trace elements in whole blood

Heavy metals and trace elements can be measured in whole, serum, urine and other tissues. In blood, metals are distributed between the non-cellular (plasma/serum) and intra-cellular compartment (predominantly erythrocytes). For examples, lead (Pb) is known to have a strong affinity for erythrocyte [54]. Thus Pb is measured primarily in whole blood. Levels of metals in blood and serum have been extensively studied in different samples to identify the exposure of human population. Pregnant mother and their growing fetuses are especially vulnerable to exposure to pollutants. Air and water pollution, exposure to toxic elements, and exposure to persistent organic compounds have been associated with adverse pregnancy and developmental outcome [55, 56].

In addition to that, trace elements are also required in the body for its normal function. Macrominerals refer to minerals that adults need in quantities greater than 100 mg/day. The principal (macro-nutrients) consist of sodium, potassium, chloride, calcium, magnesium and phosphorus. Trace elements (trace mineral) are commonly stipulated as minerals required by adults between 1 and 100 mg/day. The group of trace minerals is composed of iron, copper and zinc. Ultra-trace minerals are defined as minerals that are needed in quantities under 1 mg/day. These consist of chromium, manganese, fluoride, iodide, cobalt, selenium, silicon, arsenic, boron, vanadium, nickel, cadmium, lithium, lead and molybdenum [57].

7. Polycyclic aromatic hydrocarbon (PAHs)

PAHs comprise diverse groups of compounds whose structure comprises at least two benzene groups and assorted functional groups that may include more than one element. They can be eliminated or converted to even more toxic compounds via chemical reactions such as sulfonation, nitration or photooxidation. For example, under certain conditions, traces of nitric acid may convert some PAHs into nitro-PAHs [58].

Organic compounds can be released from their sources in gas phase or can be associated with particles by nucleation and condensation, forming particulate matter. PAHs can be found in the particulate and gaseous phases, depending on their volatility. Low molecular weight PAHs (LMW PAHs) with two or three aromatic rings are released in the gas phase. High molecular weight PAHs (HMW PAHs) of five or more rings are generated in the particulate phase.

The particulate form of PAHs is first found in the high-temperature gas phase. Nevertheless, when the temperature drops, the gas-phase PAHs adsorb or settle on the fly ash particles. Smaller particles offer more surface area for PAH adsorption. Environmental temperature is very important for the distribution of PAH gas particles. PAH can be formed in any incomplete combustion or high temperature pyrolytic process involving fossil fuels, or more generally, materials containing C and H [59].

The mechanisms by which PAHs are formed and emitted can be divided into two processes: pyrolysis and pyrosynthesis in any fuel combustion system. Pyrolysis is the development of smaller, unstable fragments from a heated an organic compound. Fragments are extremely reactive free radicals whose average life expectancy is very short. Through recombinant reaction, these free radicals result in more stable PAHs and this process is known as pyrosynthesis. For instance, B (a) P and other PAHs are formed by pyrolysis of methane, acetylene, butadiene and other compounds [60].

The formation of PAHs in pyrolysis oils was attributed by Diels-Alder responses of alkenes to form cyclic alkenes. In cyclic alkene dehydrogenation responses, stable rings of aromatic compounds from which PAH compounds are formed. Nonetheless, complex hydrocarbons need not necessarily decompose into small fragments prior to the recombinant process. Compounds with multiple rings are susceptible to partial cracking. Moreover, phenyl radicals also play a significant role in addition to intermolecular and intramolecular hydrogen transfers at intermediate constituent in high-temperature process which result in PAH formation [60].

8. Toxicity and carcinogenicity of PAHs

Elevated levels of PAHs in air cause numerous adverse effects on different types of organisms, inclusive plants, birds and mammals. A few studies have shown a significant positive association between lung cancer mortality in humans and PAH exposure from coke oven exhaust, cover tar and tobacco smoke. Concurrently, certain PAHs were shown to react with near ambient levels of NO_2 + HNO_3 and with O_3 in synthetic atmospheres, to form directly mutagenic nitro-PAH and oxy-PAH [61].

Some of the PAHs and their metabolites can induce stable genetic alterations that have the potential to irreversibly alter the control of cell division. This may result in tumor growth and cancer in fish and mammals. Due to solubility of PAHs in fatty tissue, they may bioaccumulate and transferred in the food chain. Certain PAHs have been specified as possible or probable cancer causing in humans, notably benzo (a) anthracene, chrysene, benzo (b and k) fluoranthene, benzo (a) pyrene and others [62]. Epidemiological studies have demonstrated that individuals exposed to mixtures containing PAHs have increased lung cancer rates [63].

It is known that the lower molecular weight PAHs is less harmful. They are predominantly discovered in the vapor phase in an urban air where they can react with other pollutants (O_3 and NO_x) to form more toxic derivatives. For example, PAHs react with NO_3 will form carcinogenic nitro-derivatives [64].

Particles smaller than 10 µm are more potent to incorporate larger quantities (unit mass) of PAHs because of their large area-to-volume ratio. That is a major concern because smaller particles are retained by the lung. In the human respiratory system, particles greater than 10 µm in diameter could not reach the thorax. Particles between 2.1 and 10 µm are preferably trapped by the pharynx, trachea and bronchi while particles less than 2.1 µm can reach bronchial and terminal alveoli. Consequently, harmful physical action of inhalable particles (i.e., development of lung emphysema) is ascertained with chemical impact as a result of their toxicity [65].

Humans may be exposed to PAHs by inhaling contaminated air or cigarette smoke, ingesting food with PAHs, and skin absorption of soil or materials that contain PAHs. Ingestion and inhalation are the two principal routes of exposure to the general population [27, 66]. For some professions, such as coal tar roofers and coking plant workers, skin absorption may be the primary route of exposure to PAHs [67].

Once PAHs have penetrated the human body, they undergo a series of biotransformation processes. During Stage I metabolism, PAHs are oxidized by cytochrome P450 enzymes to develop reactive epoxy intermediates, followed by hydrolysis to form hydroxylated derivatives (OH-PAHs). In Phase II metabolism, PAHs-OH are combined with glucuronic acid and/or sulphate to increase metabolite solubility in water and for ease of removal by urine, bile or stool [68]. Urinary OH-PAHs have been used as biomarkers to assess human exposure to PAHs, with 1-hydroxypyrene (1-PYR) as the most commonly used indicator in biomonitoring studies [69].

9. PAHs in whole blood

Blood measurement for DNA adducts of PAHs have been employed as a markers for PAHs exposure in human populations. Environmental monitoring documents the presence of a environmental pollutants, but the biological consequences of exposure to the organism are found only intracellularly [70, 71]. Protein-adduct formation is considered a surrogate of DNA adducts formation, but only the latter results in critical mutagenic changes [72].

Benzo(a)pyrene B(a)P is a potent carcinogen that have been used a proxy marker for PAHs in environmental sample. Individual metabolism transforms B(a)P to benzo epoxide (a) pyrene diol (BPDE), that establish adducts with DNA and proteins and hydrolysed to BPDE tetrols that are eliminating from the individual body [73]. The concentrations of human serum albumin (HSA) is more than a thousand fold greater than that in DNA in human blood and unlike DNA adducts, HSA adducts are not repair. Therefore, BDPE-HSA should be relatively stable biomarker of B(a)P exposure, with human half life of 20 days that ensures integration B(a)P exposure over about a month [74].

Over the years many studies have contributed to validation of PAH-DNA adduct measurements in human DNA using immunoassays, the most-sensitive of which is the chemiluminescence immunoassay (CIA) [75]. A cohort study of US military soldiers displaced from an area with a clean environmental zone to a much more polluted area showed a substantial increment in the levels of HAP-DNA adducts in blood cells for each individual [76, 77]. In addition, a tremendously higher rate of PAH-DNA adducts for blood cells was observed in young adults sampled in Mexico City during the dry season when atmospheric PAH concentrations were higher, in comparison to the rainy season, when PAH levels were decrease [78].

10. Volatile organic compounds

Volatile organic compounds (VOCs) are defined as photochemically reactive organic species with a high vapor pressure in the Earth's atmosphere [79]. VOCs include a wide range of compounds such as carbonyls, organic acids, alcohols, alkanes, alkenes, aldehydes, esters, paraffins, ketones and aromatic hydrocarbons [80]. The physical and chemical properties of VOCs and their residence times in the atmosphere (ranging from a few minutes to several months) often lead to threats towards the environment and human health [81].

VOCs originating from biogenic sources play crucial roles in atmospheric chemistry because they are strong ozone precursors that supply essential OH radicals in ambient air for the formation of tropospheric ozone [82, 83]. VOCs has been identified for formation of photochemical smog, stratospheric ozone depletion and the formation of organic acids which contribute to environmental acidification by lowering the pH of rainwater [84–86]. From a human health perspective, their toxic nature and ability to form fine aerosols pose that contribute to health risks, such as asthma, headaches, dizziness, visual disorders and memory impairment [87–90].

11. Health effects of VOCs

The volatilization characteristics of VOCs allow them to enter living organisms through three main routes; by inhalation, dermally and orally through contaminated water or food in order of importance [81]. Toxicokinetic studies showed distribution to lipid-rich tissues such as the brain, bone marrow and body fat [91]. Children and the elderly are the most vulnerable groups due to their higher metabolic rate and weak immune systems [92]. BTEX and carbonyl compounds are classified as toxic air pollutants able to cause adverse health effects even at low concentrations by affecting different target organs e.g. central nervous system, respiratory system, liver, kidneys and reproductive system [93]. Benzene can be absorbed through various tissues such as tissues in the brain, bone marrow cells and also tissues containing high amounts of lipid, with succeeding genotoxic action. Prolonged exposure to benzene may have effects on neurological, immunological, endocrine and blood disease disorders such as aplastic anemia and myeloid leukemia [94]. As a result benzene is the most regulated substance in the world being classified as 'Group 1, carcinogenic to humans', by the International Agency for Research on Cancer (IARC).

When human exposed to toluene, the vaporized toluene will be absorbed to the respiratory tract. According to Pierrehumbert [95], inhalation is the most important route for toluene exposure. About 80% of toluene vapor will be absorb in the first exposure and the absorption is decreasing afterwards. It is known that pulmonary absorption is faster than oral absorption. Oral absorption takes around 2–3 hour to get contact with blood, compare to inhalation absorption which takes only 20–3-minutes at the concentration of 100 ppm. The half life of toluene in human blood is around 3.4 hour but likely to increase to 0.5–2.7 days with the greater amounts of body fat. Due to its solubility, toluene can passess through the placenta, amniotic fluid, human neonates then end up in edipose tissues. This was proved by a study from Fabietti et al. (2—24) which found 0.76 ug/kg concentration of toluene in human breast milk. Apart from the neurotoxicity of toluene, recent studies found that toluene has been linked with color vision loss [96]. The most evidences are the effect on blue yellow discrimination but in few cases, the red green discrimination might be experiences [97]. On top of that, the prolonged exposure of small dose to toluene may effect the lens eyes and outer retinal layer [98, 99].

IARC [100] assessed ethylbenzene as demonstrating sufficient evidence of carcinogenicity in animals and therefore is classified as a possible human carcinogen. Rodent bioassays by Scott et al. [101], clearly showed enough evidence of carcinogenic activity following inhalation exposure in male rate. The carcinogenicity is depends on the incidences of renal neoplasms after exposure to dose up to a 750 ppm. In human body, ethylbenzene is effectively absorbed through the inhalation and expeditiously circulated. Both in human and rodents, ethylbenzene is metabolized through hydroxylation to produce phenylethanol and will excrete via urination [100]. Urinary excretion is believed to be the main pathway of ethylbenzene elimination both in human and animal studies following inhalation exposure. Human appears to have acute symptom of ethylbenzene on ocular irritation and respiratory tract. Furthermore, affects on hematological changes might be experiences. Acute toxicity studies on animal are almost the same as have been detected in human, however the differentiated between human is that acute toxicity in animal demonstrated by neurobehavioral and neurological effects.

Respiration appears to be the major pathway of exposure to xylene. Approximately 60% of inhaled xylene goes to lungs. As xylene is well metabolized in human bodies; more than 90% is bio transformed to methylhippuric acid, where it is excreted in urine. Vapors xylene in the lung alveoli dispersed into the blood and are distributed across human body by the circulatory system [102]. Nevertheless, xylene do not have tendency to accumulate in human body. Acute exposure of xylene results in irritation of eyes, nose, throat as well as gastrointestinal effects, and neurological effects. Meanwhile, CNS, kidney, cardiovascular and respiratory effects have been discovered due to xylene exposure. In addition, a study conducted by Adams et al. [102] suggest that chronic exposure to xylene as also associated to a variety of disease, leukopenia, electrocardiogram abnormalities, dyspnoea and cyanosis.

Table 3 shows the carcinogenic and non-carcinogenic effects of selected VOC compounds, target organs and their critical health effects.

Compounds	USEPA cancer classification	Target organ	Precursor effect/ tumor type	Critical effects
Benzene	A	Blood	Leukemia	Decreased lymphocyte count
Toluene	D		—	Neurological effects
Ethylbenzene	B2	Kidney	Tumors	Developmental toxicity
Xylene	D		—	Impaired motor coordination
Formaldehyde	B1	Nasal Cavity	Squamous cell carcinoma	
Acetaldehyde	B2	Nasal	Nasal squamous cell carcinoma or adenocarcenoma	Olfactory degeneration

Adapted from Kitwattanavong et al. [103].
Sources: Integrated Risk Information System. The Risk Assessment Information System [104].
USEPA cancer classification: A = human carcinogen; B1 = probable human carcinogen; B2 = probable human carcinogen; C = possible human carcinogen; D = not classifiable as to human carcinogenicity; E = evidence of non-carcinogenicity for humans.

Table 3.
The health effects of the main BTEX and carbonyl compounds.

12. Ozone

Ozone is developed upon the reaction of the dioxygen and a single oxygen in the existence of a molecule of the third body that can absorb the heat of the reaction. The unique highly responsive and short-lived oxygen (O) can be produced by photolysis of nitrogen dioxide (NO2) or by ionization of O2.

The stratosphere and troposphere are composed of background ozone. Stratospheric ozone is confined to the tropopause (between 8 and 15 km high) a region it is known as the ozone layer. Stratospheric ozone is referred to as the "good" ozone, considering the ozone layer is crucial for the absorption of life-threatening ultraviolet (UV-B) rays to human health. Given that immediate contact with ground-level ozone can induce detriment to living cells, organs and species, including individual, animals and plant life, ground-level ozone is considered to be a "bad" ozone.

According to Nuvolene et al. [105] and Koman & Mancuso [106], the documented health effects of ozone are

- Ozone can create undesirable respiratory effects like difficulty breathing and inflammation of the respiratory tract in the general individual (breathlessness and pain during deep breathing). These effects can worsen pulmonary illnesses such as asthma, emphysema and chronic bronchitis.

- Prolonged exposure to ozone is likely one of many contributing factors to the development of asthma.

- Exposure to ozone is likely to result in premature death, and there is stronger evidence for mortality from respiratory diseases.

- Children are at increased risk from ozone exposure, as children have a relatively higher dose per body mass and children's lung is still developing.

Even at really low levels, tropospheric ozone result of a vary of health problem including worsened asthma, decreased pulmonary ability, and increased sensitiveness to respiratory diseases including pneumonia and bronchitis. Persistent exposure to ozone over several months can permanently damage the lungs [107]. Ozone can irritate lung airways and cause inflammation much like a sunburn [108, 109]. Other symptoms include wheezing, coughing, pain when taking a deep breath, and breathing difficulties during exercise or outdoor activities [109].

13. Carbon monoxide

Carbon monoxide is produced from the incomplete combustion of fossil fuel such as petrol, coal, wood, and natural gases. The health effects of CO breathing include headache, vertigo, nausea, vomiting and eventually loss of consciousness.

The affinity of carbon monoxide to hemoglobin is far superior to that of oxygen. Along this vein, severe intoxication may occur in individuals susceptible to elevated levels of carbon monoxide for an extended period of time. As a result of oxygen loss due to competitive binding of carbon monoxide, hypoxia, ischemia and cardiovascular diseases are discovered [1].

14. Environmental burden of disease of air pollution and tempereture

Air pollution is currently considered as the most significant environmental cause of disease, whereby kills about 3 million people annually and majorly affected the Western Pacific and South East Asia regions [33]. Short-term exposure to air pollutants are associated with Chronic Obstructive Pulmonary Disease (COPD), cough, shortness of breath, wheezing, asthma, respiratory disease, and high morbidity (hospitalization). Meanwhile, the long-term effects associated with air pollution are chronic asthma, pulmonary insufficiency, cardiovascular diseases, and cardiovascular mortality [1].

The differential effect of air pollution on health, which is particularly deleterious for older people, children and people with limited resources, is of major concern to global health populations. It is estimated that airborne fine particulate pollution is responsible for approximately 3% of adult cardiopulmonary mortality worldwide [110]. Air pollution is expected to have similar negative impacts in developing countries, with Asian countries accounting for about two-thirds of the world's burden [110]. Beelen et al. [111] had found a relationship between mortality and long-term exposure to particulate matter, fine particles, and nitrogen compounds.

Elevated air pollution is usually related with extreme events such as increasing in ambient temperature. Engardt et al. [112] indicated that ozone levels are directly driven by weather since ozone-generating photochemical reactions of air pollutants (nitrogen oxides; methane; volatile organic compounds, VOCs) need high temperatures and bright sunshine.

The associations between ambient temperatures and human health have been widely studied, and growing evidences have revealed that exposure to ambient temperatures may increase the risks of a range of respiratory diseases, cardiovascular diseases, and other disease [113–115]. Although most previous studies found significant relationships between ambient temperature and morbidity or mortality [116, 117], only few had assessed the disease burden attributable to ambient temperatures. It was demonstrated that heat-related mortality and morbidity rapidly increase when temperatures were above optimal. Recently study done by Yiju Zhao et al. [118] highlighted high and low temperature increase the morbidity risk of respiratory disease. Meanwhile, study done by Chung et al. [119] discovered excess mortality due to high ambient temperature was expected to be profound in Korea.

The disease burden of a population and how the burden is distributed across different subpopulations (e.g infants, women) are important pieces of information for defining strategies to improve population health. Burden of disease (BOD) is a comprehensive measurement of mortality and morbidity in a single index, which is manifested as disability-adjusted life years (DALY). DALY is estimated by summing the years of life lost (YLL) and years lost due to disability (YLD). Estimation of burden of disease had advantaged over other epidemiological indexes because of its simplicity, comprehensiveness, and applicability for policy-making process.

The Environmental burden of Disease (EBD) series continues the effort of BOD to generate reliable information, by presenting methods for assessing the environmental burden of outdoor air pollution at national and local levels, as what had been described in World Health Report [120]. Worldwide global burden of disease between 1990 and 2015 shows COPD mortality rates and DALYs were observed in South-East Asia region with 39 deaths per 100,000 people and 791 DALYs per 100,000 people in 2015, followed by Western Pacific region with 25 deaths per 100,000 people and 421 DALYs per 100,000. Ischemic stroke, Trachea, Bronchus and Lung cancer, Ischemic heart disease's DALYs and death rate were generally higher in all the regions with South-East Asia having 943 DALYs per 100,000 people and 38 deaths per 100, 000. Exposure to air pollution caused over 7.0 million deaths and 103.1 million lost years of healthy life in 2015, caused an estimated 7.6% of total global mortality in 2015 [121]. The WHO estimation report for Malaysia's Environmental burden of disease was published in 2009 [24]. Estimates were based on Comparative Risk Assessment, evidence synthesis and expert evaluation for regional exposure and WHO country health statistics 2004. Cardiovascular disease, respiratory infections, COPD, asthma and lung cancers were among the disease listed. The preliminary estimations for the diseases were 2.5, 1.6, 1.4, 1.2 and 0.5 DALYs/1000 capita, per year, respectively [120].

Author details

Muhammad Ikram Bin A Wahab
Environmental Health and Industrial Safety Programme, Center for Toxicology and Health Risk Studies, Faculty of Health Sciences, Universiti Kebangsaan Malaysia, Kuala Lumpur, Malaysia

*Address all correspondence to: ikram@ukm.edu.my

References

[1] Manisalidis, I., Stavropoulou, E., Stavropoulos, A., Bezirtzoglou, E. 2020. Environmental and Health Impacts of Air Pollution: A Review. Front.Public Health, 8:14.

[2] Brauer, M., Amann, M., Burnett, R.T., Cohen, A., Dentener, F., Ezzati. M., et al., 2011. Exposure assessment for estimation of the global burden of disease attributable to outdoor air pollution. *Environ Sci Tech*. 46: 652-660.

[3] Allen, A.G., Nemitz, E., Shi, J.P., Harrison, R.M. and Greenwood, J.C., 2001. Size Distributions of Trace Metals in Atmospheric Aerosols in the United Kingdom. *Atmospheric Environment*. 35: 4581-4591.

[4] Fernández Espinosa, A.J.F., Rodriguez, M.T., Rosa, F.J.B., Sanchez, J.C.J. 2001.Size distribution of metals in urban aerosols in Seville (Spain). Atmospheric Environment 35: 2595-2601.

[5] Voutsa, D., Samara, C. 2002. Labile and bioaccessible fractions of heavy metals in the airborne particulate matter from urban and industrial areas, Atmospheric Environment 36: 3583-3590.

[6] ATSDR, 2003. Toxicological profile information sheet. Division of Toxicology, Atlanta, GA. http:/www. Atsdr.cdc.gov./ toxiprofiles.

[7] Wang, X., X. Bi, G. Sheng and J. Fu. 2006. Hospital indoor PM10/PM2.5 and associated trace elements in Huangzhou, China. Sci. Total Environ. 366: 124-135.

[8] Benoff, S., A. Jacob and I.R. Hurley., 2000. Male infertility and environmental exposure to lead and cadmium. *Human Reproduction Update*. 6: 107-121.

[9] Hu, Y., Bai, Z., Zhang, L., Wang, X., Zhang, L., Yu, Q. and Zhu, T. 2007. Health Risk Assessment for Traffic Policemen Exposed to Polycyclic Aromatic Hydrocarbons (PAHs) in Tianjin, China. Sci. Total Environ. 382: 240– 250.

[10] Manalis, N., G. Grivas, V. Protonotarios, A. Moutsatsou, S. Samara and A. Chaloulakou. 2005. Toxic metal content of particulate matter (PM10), within the Greater Area of Athens. Chemosphere, 60: 557-566.

[11] Santos-Burgoa, C., C. Rios, L.A. Nercadi, R. Arecguga-Serrano, F. Cano- Vall and R.A. Eden-Wynter. 2001. Exposure to manganese; health effects on the general population, a pilot study in central Mexico. Environ. Res. Sect. A, 85: 90-104.

[12] ATSDR, 2002. Toxicological profile of copper. Division of Toxicology, Atlanta, GA.

[13] Dong, T.T.T., and Lee, B.K. 2009. Characteristics, toxicity, and source apportionment of polycylic aromatic hydrocarbons (PAHs) in road dust of Ulsan, Korea. Chemosphere 74: 1245-1253.

[14] Sharma, H., Jain, V.K., Khan, Z.H. 2008. Atmospheric polycyclic aromatic hydrocarbons (PAHs) in the urban air of Delhi during 2003. Environ Monit Assess147:43-55.

[15] Aiken A.C., DeCarlo P.F., Kroll J.H., Worsnop D.R., Huffman J.A., Docherty K.S., Ulbrich I.M., Mohr C., Kimmel J.R., Sueper D., Sun Y., Zhang Q., Trimborn A., Northway M., Ziemann P.J., Canagaratna M.R., Onasch T.B., Alfarra M.R., Prevot A.S.H., Dommen J., Duplissy J., Metzger A., Baltensperger U., Jimenez J.L., 2008. O/C and OM/OC Ratios of Primary, Secondary, and Ambient Organic Aerosols with High-Resolution Time-of-Flight Aerosol Mass

Spectrometry. *Environmental Science & Technology.* 42: 4478-4485.

[16] Heintzenberg, J., Raes, F., and Schwartz, S. 2003. Chapter 4, Tropospheric Aerosols, The Changing Atmosphere: An Integration and Synthesis of a Decade of Tropospheric Chemistry Research (Global Change– The Igbp Series,), Guy Brasseur, Ronald G. Prinn, Alexander A. P. Pszenny, Springer.

[17] CAFE Working Group on Particulate Matter; Second position paper on particulate matter; 2004 December.

[18] Easter, R.C., and Peters, L.K. 1994. Binary homogeneous nucleation: temperature and relative humidity fluctuations, non-linearity and aspects of new particles production in the atmosphere. J. Geophys. Res. J. Appl. Meteorol. 33:775-784.

[19] Hoppel, W., Frick, G. M., Larson, R.E., 1986. Effect of nonprecipitatin clouds on the aerosol size distribution in the marine boundary layer. Geophysical Research Letters, 13(2): 125-128.

[20] WHO, 2005. WHO air quality guidelines global update. 2005. World Health Organization, Regional Office for Europe, Bonn, Germany, Copenhagen, Denmark.

[21] Pope, C.A. and Dockery, D.W. 2006. Health effects of fine particulate air pollution: Lines that connect. Journal of the Air & Waste Management Association. 56(6): 709-742.

[22] Schlesinger, R.B., Kunzli, N., Hidy, G.M., Gotschi, T. and Jerrett, M., 2006. The health relevance of ambient particulate matter characteristics: Coherence of toxicological and epidemiological inferences. Inhalation Toxicology. 18(2): 95-125.

[23] WHO, 2006. WHO Air quality guidelines for particulate matter, ozone, nitrogen dioxide and sulfur dioxide, Geneva, Switzerland.

[24] World Health Organization. 2009. Country profile of Environmental Burden of Disease: Malaysia. Geneva.

[25] World Health Organization (WHO), 2004. Health aspects of air pollution: results from the WHO project "Systematic review of health aspects of air pollution in Europe", World Health Organization, Regional Office for Europe, Bonn, Germany, Copenhagen, Denmark.

[26] Nel A., Xia T., Mädler L., Li N. 2006. Toxic Potential of Materials at the Nanolevel. Science. 311: 622-627.

[27] Bostrom, C.E., Gerde, P., Hanberg, A., Jernstrom, B., Johansson, C., Kyrklund, T., Rannug, A., Tornqvist, M., Victorin, K., Westerholm, R., 2002. Cancer risk assessment, indicators, and guidelines for polycyclic aromatic hydrocarbons in the ambient air. *Environmental Health Perspectives* 110 Suppl 3, 451-488.

[28] Topinka J., Rossner P.Jr, Milcova A., Schmuczerova J., Svecova V., Sram R.J. 2011. DNA adducts and oxidative DNA damage induced by organic extracts from PM2.5 in an a cellular assay. Toxicology Letters. 202: 186-192.

[29] Amann, M., Bertok, I., Cofala, J., Gyarfas, F., Heyes, C., Klimont, Z., Schöpp, W. and Winiwarter, W., 2005. Baseline Scenarios for the Clean Air for Europe (CAFE) Programme
Final Report.

[30] Naeher, L., Brauer, M., Lipsett, M., Zelikoff, J., Simpson, C., Koenig, J. and Smith, K., 2007. Woodsmoke Health Effects: A Review. Inhalation Toxicology. 19(1): 67- 106.

[31] Boman, B.C., Forsberg, A.B. and Järvholm, B.G. , 2003. Adverse health effects from ambient air pollution in relation to residential wood combustion

in modern society. *Scandinavian journal of work, environment and health*. 29(4): 251-260.

[32] GBD 2016 Risk Factors Collaborators. Global, regional, and national comparative risk assessment of 79 behavioural, environmental and occupational, and metabolic risks or clusters of risks, 1990-2016: a systematic analysis for the Global Burden of Disease Study. *Lancet* 2017; 390:1343-1422.

[33] WHO 2016. Ambient air pollution: A global assessment of exposure and burden of disease. WHO Geneva, Switzerland.

[34] ASM 2016. Local and transboundary haze study. ASM Task Force on Haze, Academy of Sciences Malaysia.

[35] Koplitz SN, Mickley LJ, Marlier MM, Buonocore JJ, Kim PS, Liu T, Sulprizio MP, DeFries RS, DJ, Schwartz J, Pongsiri M, Myers SS. Public health impacts of the severe haze in Equatorial Asia in September–October 2015: demonstration of a new framework for informing fire Page **18** of **18** management strategies to reduce downwind smoke exposure. *Environmental Research Letters* 2016; 11: 094023.

[36] UNICEF 2017. Danger in the air: How air pollution can affect brain development in young children. UNICEF Division of Data, Research and Policy Working Paper, New York.

[37] WHO 2013a. Review of evidence on health aspects of air pollution – REVIHAAP Project Technical Report. WHO Regional Office for Europe, Copenhagen, Denmark.

[38] Brook RD, Rajagopalan S, Pope III CA, Brook JR, Bhatnagar A, et al. Particulate matter air pollution and cardiovascular disease: an update to the scientific statement from the American Heart Association. *Circulation* 2010; 121(21): 2331-2378.

[39] Shinggu, D. Y., Ogugbuaja, V. O., Barminas, J. T. & Toma, I. 2007. Analysis of street dust for heavy metal pollutants in Mubi, Adamawa State, Nigeria. International Journal of Physical Sciences. 2(11) :290-293.

[40] Adriaenssens, E., 2007. Focus on Spectroscopy: Analysis of Heavy Metals in Ambient Air. International Labmate. 15:1.

[41] Pacyna, Jozef. 2002. Heavy Metals in the Artie: Artie Monitoring and Assessment Plan. Amap Report. 98:8.

[42] Jacob, D. J., 1999. Introduction to Atmospheric Chemistry. Chapter 3: Simple Methods. Princeton University Press, Princeton New Jersey. Page 24.

[43] Qishlaqi, A. & Moore, F. 2007. Statistical Analysis of Accumulation and Sources of Heavy Metals Occurrence in Agricultural Soils of Khoshk River Banks, Shiraz, Iran. American-Eurasian Journal ofAgriculture & Environmental Science. 2 (5): 565-573.

[44] Abegaz, M. S., 2005. Investigation of input and distribution of polluting elements in Tinishu Akaki River, Ethiopia, based on the determination by ICP-MS. Laboratory of Analytical Chemistry Institutefor Nuclear Science (UMI No. AAT 471762).

[45] Hall, J. L. 2002. Cellular Mechanisms for Heavy Metal Detoxification. Journal of Experimental Botany. Vol 53, No. 366, pp 1-11.

[46] Agency for Toxic Substances and Disease Registry (ATSDR), 2008. Toxicological Profile for Chromium.

[47] Gidlow, D. A. 2004. In-Depth Review, Lead toxicity. Occupational Medicine. 54:76- 81.

[48] Onder, S., Dursun, S., Gezgin, S.& Demirbas, A. 2007. Determination of heavy metal pollution in grass of city centre green areas (Konya,Turkey). Polish Journal of Environmental Studies. 16 (1): 145-154.

[49] Selinus, O. 2004. Medical Geology: an emerging speciality. Terrae. 1(1): p. 8-15.

[50] Fellenberg, G. 2000. The Chemistry of Pollution, John Wiley & Sons, England.

[51] Cooper, R.G. 2008.Zinc toxicology following particulate inhalation. Indian J Occup Environ Med12(1): 10-13.

[52] Duruibe, J.O., M.O.C. Ogwuegbu, and J.N. Egwurugwu, 2007. Heavy metal pollution and human biotoxic effects. International Journal of Physical Sciences. 2(5): p. 112-118.

[53] Brzoska, M.M. and J. Moniuszko-Jakoniuk., 2001. Interactions between Cadmium and Zinc in the organism. *Food and Chemical Toxicology*. 967-980.

[54] Bergdahl, I.A. , Schütz, A. , Gerhardsson, L. , Jensen, A. , Skerfving, S., 1997. Lead concentrations in human plasma, urine and whole blood. *Scand J Work Environ Health*. 23: 359-363.

[55] Bocskay, K., Tang, D., Orjuela, M., Liu, X., Warburton, D., Perera, F., et al., 2005. Chromosomal aberrations in cord blood are associated with prenatal exposure to carcinogenic polycyclic aromatic hydrocarbons. *Cancer Epidemiol Biomarkers Prev*. 14(2): 506-511.

[56] Landrigan, P., Aykanat, B., Ertas, N., et al. 2004. Children's health and the environment: public health issues and challengers for risk assessment. Environ Health Perspect. 112(2): 257-265.

[57] O'Dell, B.L., and Sunde, R.A. 1997. Handbook of Nutritionally Essential Mineral Elements. Marcel Decker Inc. New York. 2:6.

[58] Marcè, R.M and Borrull, F. 2000. Solid Phase extraction of polycyclic aromatic compounds, Journal of Chromatography A. 885: 273-290.

[59] Baek S.O., Field R.A., Goldstone M.E., Kirk P.W., Lester J.N., Perry R., 1991. A review of atmospheric polycyclic aromatic hydrocarbons: sources, fate and behavior. *Water, Air and Soil Pollution*. 60: 279-300.

[60] Mastral A.M. and Callen M.S. 2000. A review on polycyclic aromatic hydrocarbon (PAH) emissions from energy generation, Environmental Science and Tecshnology, 34: 3051-3057.

[61] Hansen, L.D., and Eatough, D.J. 1991. Organic Chemistry of the atmosphere. Boca Raton. CRC Press.

[62] Golomb, D., Barry E., Fisher G., Varanusupakul P., Koleda M., Rooney T. 2001. Atmospheric deposition of polycyclic aromatic hydrocarbons near New England coastal waters, Atmospheric Environment, 35: 6245-6258.

[63] Fisher G.F. 2001. Source apportionment of polycyclic aromatic hydrocarbon wet and dry deposition at Massachusetts Bay. Ph.D. Thesis, Department of Chemistry, University of Massachusetts Lowell, USA.

[64] Kameda, T. 2011. Atmospheric chemistry of polycyclic aromatic hydrocarbons and related compounds. J Health Sc. 57(6): 504-11.

[65] Cecinato A., Marino F., Di Filippo P., Leopore L., Possanzini M., 1999. Distribution of n-Alkanes, polynuclear aromatic hydrocarbons and nitrated polynuclear aromatic hydrocarbons between the fine and coarse fractions of inhalable

atmospheric particulates. *Journal of Chromatography A*. 846: 255- 264.

[66] Ramesh, A., Walker, S.A., Hood, D.B., Guillen, M.D., Schneider, K., Weyand, E.H. 2004. Bioavailability and risk assessment of orally ingested polycyclic aromatic hydrocarbons. International Journal of Toxicology 23: 301-333.

[67] ATSDR, 1995. Toxicological Profile for Polycyclic Aromatic Hydrocarbons. sAtlanta: Agency for Toxic Substances and Disease Registry. Available: http://www.atsdr.cdc.gov/toxprofiles/tp69.html.

[68] Xue,W, and Warshawsky,D. 2005. Metabolic activation of polycyclic and heterocyclic aromatic hydrocarbon and DNA damage: a review. Toxicol Appl Pharmacol. 206: (1), 73-93.

[69] Jacob, J., Seidel, A. 2002. Biomonitoring of polycyclic aromatic hydrocarbons in human urine. Journal of Chromatography B - Analytical Technologies in the Biomedical and Life Sciences 778: 31-47.

[70] Angerer, J.; Ewers, U.; Wilhelm, M., 2007. Human biomonitoring: State of the art. Int. J. Hyg. *Environ. Health*. 210: 201-228.

[71] Gallo, V.; Khan, A.; Gonzales, C.; Phillips, D.H.; Schoket, B.; Gyorffy, E.; Anna, L.; Kovacs, K.; Moller, P.; Loft, S.; et al. 2008. Validation of biomarkers for the study of environmental carcinogens: A review. Biomarkers. 13: 505-534.

[72] Poirier, M.C.; Santella, R.M.; Weston, A. Carcinogen macromolecular adducts and their measurement. Carcinogenesis 2000, 21, 353-359.

[73] Poirier, M.C. and Weston, A. 1996. Human DNA adduct measurements state of the art. Environ. Health. Perspect.104 (Suppl 5): 883-893.

[74] Kafferlein, H.U., Marczynski, B., Mensing, T., Bruning, T. 2010. Albumin and hemoglobin adducts of benzo(a) pyrene in human-analyticalmethods, exposure assessment, and recommendations for future directions. Crtic. Rev. Toxicol. 40: 126-150.

[75] Divi, R.L.; Beland, F.A.; Fu, P.P.; Von Tungeln, L.S.; Schoket, B.; Camara, J.E.; Ghei, M.; Rothman, N.; Sinha, R.; Poirier, M.C. Highly sensitive chemiluminescence immunoassay for benzo[a]pyrene-DNA adducts: Validation by comparison with other methods, and use in human biomonitoring. Carcinogenesis 2002, 23, 2043-2049.

[76] Jacobson-Kram, D.; Albertini, R.J.; Branda, R.F.; Falta, M.T.; Iype, P.T.; Kolodner, K.; Liou, S.H.; McDiarmid, M.A.; Morris, M.; Nicklas, J.A.; et al. 1993. Measurement of chromosomal aberrations, sister chromatid exchange hprt mutations and DNA adducts in peripheral lymphocytes of human populations at increased risk for cancer. Environ. Health Perspect. 101: 121-125.

[77] Poirier, M.C.; Weston, A.; Schoket, B.; Shamkhani, H.; Pan, C.F.; McDiarmid, M.A.; Scott, B.G.; Deeter, D.P.; Heller, J.M.; Jacobson-Kram, D.; et al. Biomonitoring of United States Army soldiers serving in Kuwait in 1991. Cancer Epidemiol. Biomarkers Prev. 1998, 7, 545-551.

[78] Garcia-Suastegui, W.A.; Huerta-Chagoya, A.; Carrasco-Colın, K.L.; Pratt, M.M.; John, K.; Petrosyan, P.; Rubio, J.; Poirier, M.C.; Gonsebatt, M.E. 2011. Seasonal variations in the levels of PAH–DNA adducts in young adults living in Mexico City. Mutagenesis. 26: 385-391.

[79] Sahu, L., 2012. Volatile organic compounds and their measurements in the troposphere. Curr. Sci. 102, 1645-1649.

[80] Atkinson, R., Arey, J., 2003. Gas-phase tropospheric chemistry of biogenic volatile organic compounds: a review. Atmos. Environ. 37, 197-219.

[81] Montero-Montoya, R., Lopez-Vargas, R., Arellano-Aguilar, O., 2018. Volatile organic compounds in air: sources, distribution, exposure and associated illnesses in children. Ann. Glob. Health 84 (2), 225-238.

[82] Liao, K.-J., Hou, X., Baker, D.R., 2014. Impacts of interstate transport of pollutants on high ozone events over the Mid-Atlantic United States. Atmos. Environ. 84, 100-112.

[83] Marty, M., Spurlock, F., Barry, T., 2010. Chapter 19 - Volatile Organic Compounds from Pesticide Application and Contribution to Tropospheric Ozone, in: Krieger, R. (Ed.), Hayes' Handbook of Pesticide Toxicology (Third Edition). Academic Press, New York, pp. 571-585.

[84] Lee, S.C., Chiu, M.Y., Ho, K.F., Zou, S.C., Wang, X., 2002. Volatile organic compounds (VOCs) in urban atmosphere of Hong Kong. Chemosphere 48 (3), 375-382.

[85] Montzka, S.A., Reimann, S.C.L.A., O'Doherty, S., Engel, A., Krüger, K., Sturges, W.T., 2011. Ozone-depleting substances (ODSs) and related chemicals. World Meteorological Organization.

[86] Zhang, J.J., Wei, Y., Fang, Z., 2019. Ozone pollution: a major health hazard worldwide. Front. Immunol. 10, 2518.

[87] Dedoussi, I.C., Eastham, S.D., Monier, E., Barrett, S.R.H., 2020. Premature mortality related to United States cross-state air pollution. Nature 578 (7794), 261-265.

[88] Fann, N., Lamson, A.D., Anenberg, S.C., Wesson, K., Risley, D., Hubbell, B.J., 2012. Estimating the national public health burden associated with exposure to ambient PM2.5 and ozone. Risk Anal. 32 (1), 81-95.

[89] Jerrett, M., Burnett, R.T., Arden Pope Iii, C., Ito, K., Thurston, G., Krewski, D., Shi, Y., Calle, E., Thun, M., 2009. Long-term ozone exposure and mortality. N. Engl. J. Med. 360 (11), 1085-1095.

[90] Samet, J.M., Dominici, F., Curriero, F.C., Coursac, I., Zeger, S.L., 2000. Fine particulate air pollution and mortality in 20 U.S. cities, 1987-1994. N. Engl. J. Med. 343 (24), 1742-1749.

[91] Masekameni, M.D., Moolla, R., Gulumian, M., Brouwer, D., 2019. Risk assessment of benzene, toluene, ethyl benzene and xylene concentrations from the combustion of coal in a controlled laboratory environment. Int. J. Environ. Res. Public Health 16 (1), 95.

[92] De Bruinen Bruin, Y., Koistinen, K., Kephalopoulos, S., Geiss, O., Tirendi, S., Kotzias, D., 2008. Characterisation of urban inhalation exposures to benzene, formaldehyde and acetaldehyde in the European Union: comparison of measured and modelled exposure data. Environ. Sci. Pollut. Res. 15 (5), 417-430.

[93] Duerta-Davidson, R., Courage, C., Rushton, L., Levy, L., 2001. Benzene in the environment: an assessment of the potential risks to health of the population. Occupational and Environmental Medicine 58 (1), 2-13.

[94] Cruz, L.P.S., Alve, L.P., Santos, A.V.S., Esteves, M.B., Gomes, Í.V.S., Nunes, L.S.S., 2017. Assessment of BTEX concentrations in air ambient of gas stations using passive sampling and the health risks for workers. J. Environ. Prot. 8 (01), 12-25.

[95] Pierrehumbert, G., Oliver Droz, P., Tardif, R., Charest-Tardif, G., Truchon, G. 2002. Impact of human variability on

biological monitoring of exposure to toluiene, phenol, lead, and mercuty: II. Compartmental based toxicokinetic modelling. Toxicology Letters, 134-165-175.

[96] Cavalleri, A., Gobba, F., Nicali., E., Fiocchi, V. 2000. Dose-related color vision impairment in toluene exposed workers. Arch Environ Health, 55, 399-404.

[97] Paramei, G.V., Meyer-Baron, M., Seeber, A. 2004. Impairment of color vision induced by organic solvents: a meta-analysisstudy. NeuroToxicology, 25, 803-816.

[98] Campagna, D., Stengel., B., Mergler, D., Lamasset, J.C., Diebold, F., Michard, D., ET AL. 2001. Color vision and occupational toluene exposure. Neurotoxicology and Teratology, 23, 473-480.

[99] Muttray, A., Wolters, V., Jung, D., Konietzko, J., 1999. Effects of high doses of toluene on color vision. Neurotoxicology Teratology, 21, 41-45.

[100] IARC. 2000. Ethylbenzene 9100-41-4). IARC monographs on evaluation of carcinogenic risks to human. Volume 77: Some industrial chemical. WHO. International Agency fro Research on Cancer.

[101] Scott, W.T., Johnson, K.A., Bahnemann, R., Day., S.J., McGuirk, R.J. 2003. Carcinogenicity; Evaluation of potential modes of action of inhaled ethylbenzene in rats and mice. Toxicological Science, 71, 53-66.

[102] Adams, J.C., Dills, R.L., Morgam, M.S., Kalmaan, D.A., Pierce, C.H. 2005. A physiologically based toxicokinetic model of inhalation exposure, to xylene in Caucasian men. Regulatory Toxicology and Pharmacology, 43, 203-214.

[103] Kitwattanavong, M., Prueksasit, T., Morknoy, D., Tunsaringkarn, T., Siriwong, W., 2013. Health risk assessment of petrol station workers in the inner city of Bangkok, Thailand, to the exposure to BTEX and Carbonyl Compounds by inhalation. Hum. Ecol. Risk Assess. 19 (6), 1424-1439.

[104] RAIS, Toxicity Profile. 2020. https://rais.ornl.gov/tools/tox_profiles.html.

[105] Nuvolone D, Petri D, Voller F. 2018. The effects of ozone on human health. *Environ Sci Pollut Res*. 25:8074-88. 26.

[106] Koman PD, Mancuso P. 2017. Ozone exposure, cardiopulmonary health, and obesity: a substantive review. *Chem Res Toxicol*. 30:1384-95.

[107] Stieb D.M., Judek S. and Burnett R.T. 2002. Meta-analysis of time-series studies ofair pollution and mortality: Effects of gases and particles and the influence ofcause of death, age and season. *J. Air & Waste Manage. Assoc.*, **52**: 470-484.

[108] Burnett R.T., Cakmak S. and Brook J.R. 1998. The effect of the urban ambient air pollution mix on daily mortality rates in 11 Canadian cities. s*Can. J. Public Health*, **89**: 152-156.

[109] Dockery D.W., Pope C.A., Xu X., Spengler J.D., Ware J.H., Fay M.E., Ferris B.G.Jr. and Spezer F.E. 1993. An association between air pollution and mortality in six US cities. *The New Eng. J. Med*. **329**(24): 1753-1759.

[110] Cohen AJ, Ross Anderson H, Ostro B, Pandey KD, Krzyzanowski M, Künzli N, et al. 2005. The global burden of disease due to outdoor air pollution. J Toxicol Environ Health Part A., 68: 1301-7.

[111] Beelen R, Raaschou-Nielsen O, Stafoggia M, Andersen ZJ, Weinmayr G, Hoffmann B, et al. Effects of long-term exposure to air pollution on naturalcause mortality: an analysis of 22 European cohorts within the

multicentre ESCAPE project. Lancet. 2014; 383: 785-95.

[112] Engardt M, Bergstro¨m R, Andersson C. 2009. Climate and emission changes contributing to changes in near-surface ozone in Europe over the coming decades: results from model studies. Ambio, 38: 452-45.

[113] Analitis A, Katsouyanni K, Biggeri A, Baccini M, Forsberg B, Bisanti L, et al. Effects of cold weather on mortality: results from 15 European cities within the PHEWE project. Am J Epidemiol. 2008; 168(12): 1397-408.

[114] Dang, T.N., Honda, Y., Do, D.V., Pham, A.L.T., Chu, C., Huang, C., Phung, D. 2019. Effects of Extreme Temperetures on Mortality and Hospitalization in Ho Chi Minh City, Vietnam. Int. J.Environ.Res.Public Health, 16: 432.

[115] Gasparrini A, Armstrong B, Kovats S, Wilkinson P. The effect of high temperatures on cause-specific mortality in England and Wales. Occupational & Environmental Medicine. 2011;69(1):56.

[116] Basu R, Samet JM. Relation between elevated ambient temperature and mortality: a review of the epidemiologic evidence. 2002. Epidemiol Rev. 2002; 24(2):190-202.

[117] Ye X, Wolff R, Yu W, Vaneckova P, Pan X, Tong S. Ambient temperature and morbidity: a review of epidemiological evidence. Environ Health Perspect. 2012; 120(1): 19-28.

[118] Yiju Zhao, Zhao Huang, Shengyong Wang, Jianxiong Hu, Jianpeng Xiao, Xing Li, Tao Liu, Weilin Zeng, Lingchuan Guo, Qingfeng Du and Wenjun Ma. 2019. Morbidity burden of respiratory diseases attributable to ambient temperature: a case study in a subtropical city in China. Environmental Health, 18:89.

[119] Chung, S.C., Hae-Kwan Cheong, Jae-Hyun Park, Jong-Hun Kim, and Hyunjin Han. 2017. Current and Projected Burden of Disease From High Ambient Temperature in Korea. Epidemiology, 28. S98-S105.

[120] World Health Organization. 2012. *Burden of disease from Ambient Air Pollution for 2012.* World Health Organization.

[121] Babatola, S.S. 2018. Global burden of disease attributable to air pollution. Journal of Public Health in Africa, 9:813.

Chapter 6

Medellin Air Quality Initiative (MAUI)

Andres Yarce Botero, Olga Lucia Quintero Montoya, Santiago Lopez-Restrepo, Nicolás Pinel, Jhon Edinson Hinestroza, Elias David Niño-Ruiz, Jimmy Anderson Flórez, Angela María Rendón, Monica Lucia Alvarez-Laínez, Andres Felipe Zapata-Gonzalez, Jose Fernando Duque Trujillo, Elena Montilla, Andres Pareja, Jean Paul Delgado, Jose Ignacio Marulanda Bernal, Jaime Andres Betancur, Alejandro Vélez, Arjo Segers, Arnold Heemink, Juan Ernesto Soto, Bibiana Esperanza Boada Sanabria and Sara Lorduy

Abstract

This chapter book presents Medellín Air qUality Initiative or MAUI Project; it tells a brief story of this teamwork, their scientific and technological directions. The modeling work focuses on the ecosystems and human health impact due to the exposition of several pollutants transported from long-range places and deposited. For this objective, the WRF and LOTOS-EUROS were configurated and implemented over the región of interest previously updating some input conditions like land use and orography. By other side, a spinoff initiative named SimpleSpace was also born during this time, developing, through this instrumentation branch a very compact and modular low-cost sensor to deploy in new air quality networks over the study domain. For testing this instrument and find an alternative way to measure pollutants in the vertical layers, the Helicopter In-Situ Pollution Assessment Experiment HIPAE misión was developed to take data through the overflight of a helicopter over Medellín. From the data obtained from the Simple units and other experiments in the payload, a citogenotoxicity analysis quantify the cellular damage caused by the exposition of the pollutants.

Keywords: Chemical Transport Model, LOTOS-EUROS, contaminant deposition, airborne measurenments, cellular damage, SimpleSpace

1. Introduction

1.1 Medellin air qUality initiative

Cyber-physical systems are concepts that have been positioned in the area of systems and control, such as the way to integrate different types of information sources in schemes that allow human interaction with machines for proper decision-making. From this perspective, many aspects of man's interaction with the environment in the era of the fourth revolution become a subject of study that must approach in an interdisciplinary way. Although they have been widely discussed worldwide, air pollution problems increasingly are appearing in people's daily lives with higher frequency than before, what has made this problem a center of public attention during the recent years.

This work has a public policy background on human and ecosystem health and its impacts aims to be on the science-based decision-making processes. In the framework of cyber-physical systems, mathematical models require not only initial and boundary conditions to represent the physical and chemical phenomena but the data and convert it into useful information that is reliably transmitted to the computers where these models live. The reduction of uncertainty in the representation of these phenomena, with the aim of control or mitigation of human actions, makes Bayesian data assimilation techniques play a fundamental role. When trying to attack the problem of the ability to make decisions based on mathematical models that accurately predict atmospheric pollutants' behavior in sources, a general approach is necessary to methodically and systematically address each of the issues questions and formalize the research questions.

1.1.1 A brief history

The project named Data Assimilation Schemes in Colombian Geodynamics - Cooperative Research Plan for 2017–2020 Between Universidad EAFIT in Colombia and TUDelft in the Netherlands started in 2016 to reduce uncertainty and incorporating data in large-scale mathematical models. In particular, the modeling of Colombian geodynamics using a chemistry and transport model responds to the complexities of atmospheric dynamics in the country due to its global location and the rugged terrain determined by the Andes mountain range system.

As in [1] The peculiar topography (such as abrupt elevation changes, with narrow inter-mountain valleys over 1000 meters deep, and transitions from sea level to over 5000 meters in less than 200 km), the environmental conditions of the region (such as the bi-annual transit of the Inter-Tropical Convergence Zone, and the irregular behavior of the El Niño Southern Oscillation) generate atmospheric conditions that impoverish air quality. The Tropical Andean Region (TAR) geographical singularities present unprecedented challenges for modeling atmospheric chemical dynamics.

However, this is not just a chemical modeling problem. The chemistry and transport models used to determine air quality aspects require meteorological (short term) and climatic (long time) information that allows them to reproduce the atmosphere's physics state. These numerical weather prediction (NWP) models are also problematic in terms of phenomenology, susceptible to initial and boundary conditions, and the effects of the numerical solution methods of these equations. This, without considering the difficulty of modeling the microphysics parametrization in terrain as complex as Colombia and neighboring countries in the TAR.

In this scenario, Bayesian data assimilation techniques for incorporating information into models play a fundamental role improving this atmosphere states. It is required to use different source of information such as satellite information for example GOES16, radar, metar, SYNOP, sounding probes, measurements on-site, and airborne measurements to complete the meteorology scenario in Colombia to achieve models with greater prediction capacity of meteorological phenomena that impact the chemistry representation of the atmosphere. The former questions arise and the research project called Sensitivity and uncertainty sources in numerical modeling to forecast atmospheric systems: High-resolution WRF model simulations in urban valleys applied to air quality issues. Was written in 2017 an effort to focus on the meteorological fields for the complex TAR.

The estimation of states and parameters represents mathematical challenges that must approach from the estimation theory, the numerical analysis, and control theory. It is here where not only technological and computational aspects give rise to questions that allow quantifying the uncertainty for applications not only of air quality in human health and ecosystems but also in factors such as navigation and air safety.

Speaking of atmospheric knowledge, in the effort to know and adequately model the country's meteorological conditions, a series of institutions whose mission is aligned must be incorporated. In terms of scientific research, university researchers tend to answer more fundamental questions regularly. Numerical models such as the WRF (Weather Research Forecast) and Chemical transport Model (LOTOS EUROS) can be coupled to use meteorology to improve the representation of the reality. Institutions such as the Colombian Air Force (FAC) daily solve high-value scientific problems in their mission of guaranteeing the safety of air navigation (Air navigation direction - Dirección de navegación aerea DINAV and the Centre of Technological Development for Defense CETAD). The data assimilation techniques in meteorological prediction models such as WRF vary from the variational in three dimensions to approximations using particle filters and location. None of them is cheap computationally and requires an infrastructure not only for computing but also data, information, and knowledge. Techniques of fusion of meteorological data and 3D variational assimilation in the WRF model existing in the FAC allowed the universities' researchers to pave the way to determining sources of inertness and quantification of the sensitivity of the models to focus efforts on modeling and parameterization of atmospheric phenomena.

Medellin Air Quality Initiative: MAUI aims to bring together local, regional, and international experts on subjects related to air quality and its impact on human and ecosystem health to establish a knowledge network, gather the available tools to tackle the problem, and identify knowledge gaps to contribute to a deeper understanding of the local and regional scale of the impact of air pollution, motivate research collaborations, and inform public policy [2].

Until now, several universities and the state institutions efforts began to align in search of the Colombian atmosphere's representation. The particularities of the country's principal cities and the vulnerability of Colombian biodiversity offer a field of research where the questions and the rigor they must be answered give rise to interdisciplinarity spaces. In MAUI, larger-scale aspects such as numerical models and the integration of satellite information and data from various sources are integrated, and research results at the micro and nanometric level are incorporated.

Chemistry and transport models require emissions inventories within their initial and boundary conditions regularly updated by environmental authorities. This statement is entirely proper for Europe and the United States, but in Latin America, some ethnographic and political factors mean that the agencies in charge of them cannot give prompt answers. The uncertainties in dedicated emissions inventories generate problems of over and underestimating the models. The assimilation of data from the high, medium, and low-cost sensor networks can help the models improve the representation of dynamics in various magnitude orders. However, the information on the concentration of pollutants such as 10 and 2.5-micron particulate matter or tropospheric ozone and nitrogenous is not complete.

The use of European emissions inventory leads to revising the biogenic and anthropogenic emissions and revisiting the current national emissions inventory. Several experts joined the effort, and some preliminary and results are close to being published. However, it is a reality that with the rapid growth of cities in Latin America and the poor planning of land use, an inventory regularly does not end up being updated on the date of its launch. Some modeling experiments showed how large Colombian cities behave like volcanoes that expel pollutants to the country's protected areas.

Other potential forms of emission modeling appear when using traffic information from large cities to estimate density, flow, velocity, and behavior models that allow the short, medium, and long term to understand the dynamics of emissions from traffic. This question is not easy to solve either and in the project Scale-FreeBack from the Research at the National Center for Scientific Research (CNRS) in France [3] aims to understand when and how a complex, large-scale network system can be represented by a scale-free (low complexity) model usable for control design, understanding how the internal states of a scale-free network system are monitored and estimated using information originating from sources of various kinds. This project aims to know how it will be possible to design scale-free control algorithms and make them resilient to changes of scale, node/link failures/disconnections/attacks, and how can those ideas be applied to large-scale road urban networks. MAUI initiative used their ideas and joined their team for the emission modeling and estimation research.

Using available mobility information and merging it with previous knowledge of the inventories of static emissions and other sources incorporates the ability to control within the extensive cyber-physical system the number of pollutants emitted into the atmosphere city and eventually mitigate its effect. The mathematical and control problems are not easy to solve, but the results of the beginning to bear fruit in the chain of description and modeling of air chemistry within and outside cities.

In 2016, simultaneous work was being done on the magnetic characterization of the particulate material of Medellín analyzed atmos9070283 via optical spectroscopy to determine the presence of ferromagnetic material in the 10 and 2.5-micron particulate. This research naturally raised more questions about the possibility of using analogous methods of validation of the results of the models and the potential quantification of exposure levels. The biosensors' findings allowed incorporating more advanced passive sensing techniques and subsequent analysis with optical spectroscopy. The PARTICLE VISION group joined the MAUI capabilities.

Particle Vision is a highly qualified team with long-standing experience in particle analysis, aims to develop innovative analytical methods for the characterization and quantification of particles and apply them to a wide variety of questions. With its technology and support from FAC SEM/EDX instruments, the project made progress in knowing the city's chemical and morphological characteristics of Medellin's particulate material. These investigations have potential impacts on public health issues but are beyond the scope of this book chapter.

Thus, the knowledge at the microscopic level of urban particles' components allows establishing models of validation of numerical results of environmental chemistry. At the local level, nanotechnology has the technical-scientific capacity to develop electrospinning polymeric nanofibers to capture finer particles and offer an additional ability to the analytical aspects of environmental pollutants. This ability represents a leap in scale in orders of magnitude and value since it is possible to obtain particulate material in nanofibers and generate a solution from them that allows evaluating their damage capacity at smaller scales.

The work team is recently completed with experts in in-vitro and in-vivo biological models that, from the nanofibers' material, can analyze and quantify cellular and molecular damage using state-of-the-art techniques.

Our curiosity and technical capacity then led us to advance in different sampling techniques, and in 2019 the first airborne pollutant measurement mission called Helicopter Insitu Pollution Acquisition Experiment HIPAE was carried out. In this mission, the counting instruments for particulate matter, nanofilters, measurement units (inertial, geoposition, meteorological, and gasses measurements) build by the EAFIT spinoff SimpleSpace, and passive samplers attached to air pumps to simulate the expected breath air flux through them. The helicopter with the instruments traveled through the Aburra Valley at different heights. In this aircraft of the Colombian Air Force (FAC), not only gases and meteorological variables were measured, but contamination profiles were established, and the experimental model's evaluation was carried out.

The research groups, researchers, and students linked to this proposal compose an interdisciplinary group that has been working not only on the understanding of atmospheric phenomena, chemistry and transport of pollutants, assimilation of data, and artificial intelligence but also on studying the impact of anthropogenic activity on air pollution dynamics with effects on human health, ecosystems, and agriculture.

The result of this research history will be expanded a little more with the development of this chapter, sending the reader to the primary references of the works where they can obtain specific details of the principal methodologies, techniques, and findings of each area knowledge participating in MAUI.

Detailed review and explanation of some regional and worldwide initiatives similar to ours can be found in the critical review of QuinteroMontoya2020, and the recent paper of Advances in air quality modeling and forecasting BAKLANOV2020261. Some of the remarkably similar missions to HIPAE will be addressed in the following sections.

1.1.2 Scientific and technological directions

As a consequence of this interaction, much more specific goals are proposed in two directions: human health and ecosystem health. In the former direction, the research program called $ExPoR^2$ is proposed.

Exposure to Pollutants Regional Research $ExPoR^2$ aims to develop models of human exposure to air pollution in urban areas as a decision-making tool. Our purpose is to develop high-level scientific research that enables science-based decision-making and solutions to environmental and sustainability problems due to air pollution in human health, agriculture, and ecosystems. Three different research paths are going to be followed:

1. Ensemble of models to estimate human exposure to atmospheric pollutants. The General purpose is to develop an assembly of models to estimate human exposure to atmospheric pollutants in different areas of the Aburrá Valley. Specific objectives: Coupling the WRF model with the Open LOTOS-EUROS models at high resolution to simulate concentrations and dispersion of pollutants within the Aburrá Valley. Implement traffic models for the Aburrá Valley, to simulate the zonal contributions to atmospheric pollutant concentrations by the automotive fleet. Assimilate and merge data from various sources on concentrations of atmospheric pollutants. Coupling

high-resolution simulations of concentrations and dispersion of pollutants with data on human activity and occupation to estimate levels of human zonal exposure to air pollutants. Chemical and morphological profiles of the nano and microparticulate material present in the air of the Aburrá Valley. This research aims to establish the chemical and morphological profile of the nano and microparticulate material present in the Aburrá Valley air to establish its toxicological effect. Establish sampling areas and measurement periods to capture particulate matter of different sizes with the three filtration systems. Chemically and morphologically, characterize the particulate material present in the Aburrá Valley to identify the emission source and Manufacture nanofiber membranes from the selection of polymers with different surface energies to favor the capture of ultrafine particulate material.

2. Cytogenotoxicity of air pollutants. This research will particularly develop an in vitro evaluation model of the impact of particulate material on the health of the occupationally exposed population of the Aburrá Valley. Through this, build a zoned profile of the cytogenotoxic potential of particulate pollutants in the region. Specific objectives: To evaluate the sensitization, irritation and dermal corrosion of PM of different sizes collected with DRMPs, in 2D and 3D in vitro cell models. To evaluate the irritation and serious ocular damage of PM of different sizes collected with DRMPs, in 3D in vitro cell models. To evaluate PM's cytotoxicity and genotoxicity of different sizes collected in different city center areas in 2D and 3D in vitro cell models. To evaluate the mechanism of action of particulate matter in atmospheric pollutants on the distribution of cell cycle phases, mitochondrial activity, and the induction of senescence. Generate a Biosensor transgenic line by inserting the Hyper gene to detect intracellular ROS production after exposure to particulate matter from atmospheric pollutants. Relate the biological effects in vitro with the morphological and chemical composition of the particulate material collected in the different areas of the Aburrá Valley.

Focusing on the health of ecosystems and the various implications that cities can have on protected areas, the research program *3D + 1 Air Pollution Study: In situ, Surface, Remote Detection and Atmospheric Modeling Measurements" 4D Air - MISDAM*. The objectives of this research are oriented to develop remote laser and in situ detection technology to characterize pollutant particles and improve the precision of assimilation techniques in atmospheric modeling.

Some main goals of our current research directions are:

1. Design and implement a multi-channel laser remote detection system to obtain in real time the volumetric concentration of atmospheric aerosols.

2. Design and implement network sensors for in situ and surface monitoring of particulate matter, NO2 and O3 with aerospace technology.

3. Develop and implement the signal inversion algorithm backscattering of atmospheric aerosols.

4. Develop and implement the experimental model and the depolarization signal calibration algorithm for tropospheric aerosols.

5. Estimate the operating parameters of the HSI-DOAS instrument for its operation in the geomorphological and climatic conditions of northwestern South America.

6. Carry out joint measurement campaigns of remote and in situ detection systems on the surface to obtain the microphysical and radiative properties of the atmospheric components (gases and particles) in the Aburrá Valley.

7. Demonstrate the scientific relevance of determining the chemical and morphological characteristics of particulate matter as a differentiating element in determining sources of pollutants.

8. Compositionally and morphologically characterize the coarse particulate material (PM10) and (PM2.5) present in the city of Medellín through the use of SEMEDX techniques, in order to carry out a categorization of sources corresponding to the production of this particulate material.

9. Study the regional dynamics of emission and deposition of pollutants and their effect on protected areas, ecosystems and agriculture and their impact on biodiversity, its evolution and conservation.

10. Develop large-scale model evaluation techniques using robust functional data analysis techniques applied to modeling pollutant transport at high spatial resolutions within the Aburrá Valley.

11. Develop new and better data assimilation schemes in complex systems studying the problems of uncertainty, not Gaussianity, taking into account the problems of spatially dispersed data. With applications in the study of the dispersion of pollutants in the Aburrá Valley.

Additionally, MAUI's research path will allow obtaining results that extend not only to society in matters of public health but also to become antecedents of regional positioning in science and technology and aerospace technology's knowledge.

The confluence of public and private universities with the Colombian state through the Colombian Air Force and the Ministry of Sciences allows the work's speed to be doubled. The results also led to improved quality of life and the protection of biodiversity and economic growth.

At the national level, our purpose is to scale our research to the entire territory. Expand our goals to human and ecosystem health to guarantee our national resources and proper land use. To achieve this, it is evident that meteorological modeling and data incorporation to the models via assimilation must be carried out.

Colombian Air Force system for meteorological information (in Spanish, Sistema de Información meteorológica de la Fuerza Aerea Colombiana -SIMFAC) is the integrated and operational system for meteorological information for military forces and public state aviation. SIMFAC operates with the Weather Research and Forecasting (WRF) model version 4.1 under the 3DVar data assimilation scheme. Our multidisciplinary team seeks to review and test the performance of the current implementation of the data assimilation techniques on the WRF model, looking to understand the technique's results in 3DVar implementations for a known micro-physics. In order to achieve this, we use the micro-physics parameterization of the SIMFAC to generate a background on the WRF model and, via comparison against the 3DVar implementation using satellite, radar, and in-situ observations, Potential improvements of the 3DVar implementation were verified, looking for an accurate assessment and determining the potential risks for air navigation. Our main goal is to scale the operational set up to 4DVar implementation to enhance the Colombian Air Force's capabilities.

Observational and model weather records are a source of valuable data since hydro climatological variables like precipitation are recorded through time. Functional data analysis (FDA) encompasses different numerical and statistical methodologies that could help separate climate variability aspects and explore their consequences, improving the system understanding. These methodologies seek to represent the data in ways that aid additional analysis, highlighting different data characteristics. Further, a comparison between observational and model functional data analyses would serve as a model evaluation technique. FDA is used to compare two or more sets of data concerning certain types of variation, where two sets of data can contain different sets of replicates of the same functions or different functions for a standard set of replicates.

The FDA seeks to group multivariate functional data, identifying homogeneous groups without using any additional label to the model's records. The objective is to promote a classy and compare it with a natural and intuitive geographical classification offering a basis for the correction of biases.

With the achievement of a correct and precise representation of the national meteorology at different time scales, different objectives are achieved that are fundamental for studies of climate change, air pollution, and protection of Colombian biodiversity. Additionally, the national airspace's flight safety standards increase with information from different sources via data assimilation and deep learning techniques. Significant improvements on command and control for national security are part of the non-predicted results of our joint effort.

It is undeniable that thanks to the concern to attend to human health and ecosystems' relevant problems, it has been possible to articulate a network of knowledge that can also leverage Colombia's aerospace development. This is based on the development capacity of low-cost sensors with CanSat technology developed from the Spinoff SimpleSpace, which, due to their versatility in modularity and adaptive communications for different scenarios, allows the completion of the temporal space measurements, reducing the covariance of the state matrices and measurements in the assimilation of large scale models. Additionally, some preliminary results of the work team's ability to measure over the airspace with SimpleSpace units aboard Colombian Air Force aircraft show how data use can lead to precise knowledge and strategies for making environmental decisions matters.

Regarding human health, in a section, we will present some results obtained in the HIPAE mission where analysis of the particulate material captured with polymer nanofibers was carried out through cellular and molecular analyzes.

The Tropical Andean countries (Venezuela, Colombia, Ecuador, Peru, and Bolivia) face deteriorating air quality. When thinking about the assimilation of data and intelligent systems for data fusion or automatic learning, the quantity and quality of the sources begin to be a concern.

If a complete understanding of TAR's atmospheric dynamics is desired, satellite information on land use and chemical species combined with images of unmanned aerial vehicles is becoming a viable option for developing land-use models.

In the subsequent sections, the reader will find some scientific and technical details of the activities carried out by MAUI. In particular, we have selected some sections that are not published. In them, aspects of ecosystem health and the influence of the correct estimation of land use parameters to determine deposits of species such as nitrogenous are discussed. Also, the models with which our team has been advancing data assimilation techniques, artificial intelligence, and coupling are presented. We mention some hypothetical modeling scenarios where meteorological information can give relevant clues about the influence of cities' emissions and their impact on surrounding areas such as water sources.

Factors such as reducing uncertainty in the topographic representations and land cover in the tropicalization of models developed in other latitudes are briefly mentioned. The need to know and make this type of knowledge technologically independent has led to the emergence and potentiation of several technological developments such as SimpleSpace units and LIDAR 4D systems that are not specifically mentioned due to their current state design.

It is expected to advance in the development of more frequent measurement campaigns, but during 2020 they have been suspended as MAUI researchers overturn our ability to respond with science for the country in the situation of COVID-19. Our computing, nanomaterials, and cellular and molecular biology laboratories had to slow down the expected results. However, the design of additional sampling mechanisms means that in 2021 we announce a new airborne measurenment platform dedicated to meteorology and air quality.

Finally, concrete examples of the data assimilation for applying contamination by the particulate matter of 10 and 2.5 microns are presented in the second chapter of this book. These sections will be partially published in magazines on the subject in a short time.

We hope that our work experience, scientific findings, technological developments, and research directions allow the reader to define the strategy that best suits him for the correct knowledge of the surrounding systems, however, even more so for developing strategic decision-making tools in public health, environment, biodiversity, air safety, and national security. None of the above can be achieved without a network that involves actors from different sectors and researchers from many specialties.

2. Modeling pollution for health: human and ecosystems

2.1 Ecosystem health

This work presents the implementation of the LOTOS-EUROS regional atmospheric Chemical Transport Model (CTM) on Northwestern South America. The impact of land use and orography update in the model was analyzed to identify potential vulnerable natural areas by quantifying atmospheric deposition pollutants. CTMs allow simulating the physical dynamics of trace gasses and aerosols, including processes such as emission, chemical reactions, transport, and deposition. The deposition of atmospheric contaminants like nitrogen dioxide (NO_2) and ammonia (NH_3) induces chemical fluxes in natural ecosystems, with potential subsequent severe impacts like biodiversity loss. Due to the vast geographical diversity present in the study area, the LOTOS-EUROS model was updated for the land and topography inputs to simulate more representative conditions for the study region. Depositions were very sensitive for the change of land cover maps used in the model, and on the other side, topography update impacts more in the high layer of the model above harsh terrain. Additional simulations for the updated scenario using point sources were performed to identify the deposition area's spatial extent for the principal Colombian cities.

Through atmospheric transport and deposition, emitted pollutants from different anthropogenic and natural sources can alter remote ecosystems' dynamics [4, 5]. Atmospheric deposition is described as the mechanism that induces a flux of gasses and particles to the land surface due to meteorological, chemical, and biological phenomena [6] which is responsible for the balance concentration change of the different soil components.

Anthropogenic activities are significant sources of reactive nitrogen (N_r) to the atmosphere [7]. Photochemical reactions of NO_x and NH_3 create secondary inorganic aerosols [8] that can be transported over large distances [9]. Long-distance transport of secondary inorganic aerosol accounts for over 8% of the planet's reactive nitrogen flow in terrestrial ecosystems [7], and a significant source of Nr to the ocean [10]. Deposition of atmospheric N_r alters oligotrophic ecosystems [8, 11], affecting the distribution of communities of species [12–16] and ecosystem stability [17]. NO_2 is a gas emitted by anthropogenic and natural sources as part of the family of the nitrogen oxides $NO_x(NO+NO_2)$. NO_2 is emitted from anthropogenic (industrial activity, transport, and biomass burning) and natural (NO_x soil emissions and lighting) sources. Agricultural activities and livestock feedlot operations are the primary sources of atmospheric NH_3, followed by wood-burning (including forest fires) and to a lesser degree fossil fuel combustion. Acute exposures to NH_3 near its origin (4–5 km) can lead to substantial foliar damage to the plant, growth and productivity decrease [18].

Atmospheric ozone exposure can cause ecological and agricultural harm (e.g., [19] and references therein), so it also poses a risk to protected areas. The identification of natural and agricultural areas that may be excessively exposed to ground-level ozone will allow us to estimate the detrimental impact that urban pollution may be having on our protection and agricultural efforts.

Páramos are high-altitude ecosystems that serve as water supplies for a large part of Colombia and as suppliers of critical ecosystem services, they are protected under Colombian law. The following picture shows a quantification analysis of the paramos ecosystems (**Figure 1(a)**) and natural protected areas (**Figure 1(b)**) delimitation, quantifying the model output and the total deposition those areas are receiving.

The Paramo de Las Baldias was the paramo zone with greater exposure to atmospheric pollutants. It receives 14 kg/(ha yr), above the standard critical load of Nr, and above 60 kg/(ha yr) of ozone. The paramos of western Antioquia receive 5–6 kg/(ha yr), while the Sonson Paramos are identified as those with the lowest exposure to atmospheric pollutants (2.2–2.6 kg/(ha yr) Nr; 35–40 kg/(ha yr) ozone). The west of the Caribbean coast ecosystems receives the highest ozone load nationwide (50–100 kg/(ha yr)).

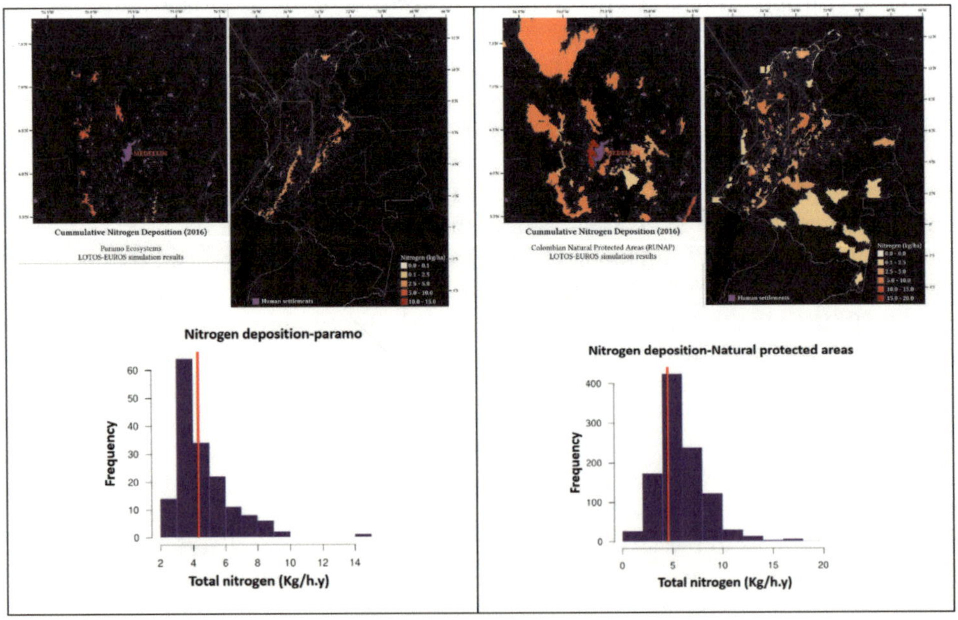

Figure 1.
Simulated calculation of annual (2016) deposited nitrogen budget for natural protected areas and paramo ecosystem.

Figure 1 shows a histogram quantification over the natural protected areas and paramo ecosystems to identify the critical deposition load of NO_2 in ecosystems and how many areas are at higher risk for the transport of pollutants. The critical load in a tropical ecosystem is not very well determine, and probably, due to the high biodiversity index in this region, the critical load should be less.

2.1.1 The WRF (weather research and forecasting) model

The WRF (Weather Research and Forecasting) model is a numerical weather prediction and atmospheric simulation system designed for research and operational applications. This model tries to advance the understanding and prediction of mesoscale weather and accelerate the transfer of research advances into operations. This model mainly is the result of the effort of the National Center for Atmospheric Research (NCAR) which is operated by the University Corporation for Atmospheric Research (UCAR), but, have contributions of many other sources, like Mesoscale and Microscale Meteorology (MMM) Division, the National Oceanic and Atmospheric Administration's (NOAA) National Centers for Environmental Prediction (NCEP) and Earth System Research Laboratory (ESRL), the Department of Defense's Air Force Weather Agency (AFWA) and Naval Research Laboratory (NRL), the Center for Analysis and Prediction of Storms (CAPS) at the University of Oklahoma, and the Federal Aviation Administration (FAA), so like contributions of researchers and Universities [20].

It's worth pointing out that for the operations of a model like WRF is necessary to set aspects like parameterizations, boundary and initial conditions. So, usually, if it desires to evaluate the performance of a model like WRF in a domain, different configurations are used, for example, changing parameters the initial and boundary conditions [21], doing online modeled [21, 22], modified the planetary boundary layer schemes, micro-physics, land-surface models, radiation schemes, sea surface temperature and 4-dimensional data assimilation [23] or to integrate new methodologies such as data assimilation [23, 24]. This aspect correspond to the physic options of the ARW, grouped in the micro-physics, cumulus parameterization, planetary boundary layer (PBL), land-surface model and radiation [20]. These options are important because they let obtaining a solution that according to the particularities of the work domain. It important to say that an important research for us is developed in Cauca river valley-Colombia, by [25], where the topography is similar to Aburrá Valley. This team try with convective parameterization schemes KF, BMJ, and G3D and show that the best is schemes KF.

With WRF model is possible to understand the behavior of a meteorological variable set through simulations, where the parameters of the model are configured in relation with the domain [21, 22, 26–29] of study like the Aburrá Valley that is our interest. However, here the intention is to understand initially the behavior of the model to alteration or modification in the information that received, particularity, the initial conditions.

One part from WRF model is its solver, ARW module or the Advanced Research WRF (ARW). Through this module is possible to producing a simulation [20]. The ARW is really important in the representations of the modeled in WRF. this component is composed by several programs for the idealized, and real simulations, as like as numerical integration program[1]. As ARW is the solver of WRF, this one have the mathematical representation of the physic and/or meteorological process, in

[1] http://www2.mmm.ucar.edu/wrf/users/docs/user_guide_V4/WRFUsersGuide.pdf, Consulted: September 10 2018.

general this is represented by a set of partial differential equations obtained of the flux-form Euler equations.

The WRF (Weather Research and Forecasting) model is a numerical weather prediction and atmospheric simulation system designed for research and operational applications. However, this model's behavior depends on domain configuration, physic parameterizations, boundary and initial conditions, and terrain characteristics. The model has to be configured according to the researchers' interest or particularities characteristic of the domain of interest and the information available for assimilation purposes. For this reason, one way to improve the model's performance is to use data assimilation for integrating additional information to understand the dynamics in an irregular territory like Colombian geography.

The accuracy of the numerical weather prediction is limited by the precision of the mathematical and physical representations of reality, bringing uncertainty from the initial and boundary conditions incorporated into a numerical model.

In WRF, the implementations of data assimilation cover since hybrid method as ensemble transform Kalman filter–three-dimensional variational data assimilation (ETKF–3DVAR) system [30], three dimensional variational data assimilation 3DVAR, four-dimensional variational data assimilation (4Dvar) [31–33], or ensemble methods as [34–38].

2.1.2 The LOTOS-EUROS model

The LOTOS-EUROS (LOng Term Ozone Simulation- EURopean Operational Smog model) [39] is a chemical transport model that models in three dimensions different species in the lower troposphere. This model was developed in 2004 by TNO and RIVM/MNP organizations, in The Netherlands, unifying the previously developed LOTOS and EUROS models. At the beginning it was developed as a model focused on ozone, but currently, the LOTOS-EUROS calculates concentrations of ozone, particulate matter, nitrogen dioxide, heavy metals and organic components [40]. LOTOS-EUROS has been widely used in different projects around the world, demonstrating the capacity of the model [41]. The dynamics of pollutants in LOTOS-EUROS is regulated by processes of chemical reactions, diffusion, drag, dry and wet deposition, emissions and aversion [42]. The LOTOS-EUROS dynamics are given by:

$$\begin{aligned}\frac{\partial C}{\partial t} + U\frac{\partial C}{\partial x} + V\frac{\partial C}{\partial y} + W\frac{\partial C}{\partial z} &= \frac{\partial}{\partial x}\left(K_h\frac{\partial C}{\partial x}\right) + \frac{\partial}{\partial y}\left(K_h\frac{\partial C}{\partial y}\right) \\ &\quad + \frac{\partial}{\partial z}\left(K_z\frac{\partial C}{\partial z}\right) + E + R + Q - D - W\end{aligned} \tag{1}$$

where C is the concentration of a specie; U, V and W, are wind components in West–East, South–North direction and vertical direction, respectively; K_h and K_z are horizontal and vertical coefficients of diffusion by turbulenc; E represents the entrainment or detrainment due to variations in layer height; R represents generation and consumption rates of pollutant by chemical reactions; Q is contribution by emissions and D and W are loss by dry and wet deposition process, respectively.

The main equation of LOTOS-EUROS dynamics is composed of different operators. Each one models different components of pollutants behavior. The operators that compose the LOTOS-EUROS are: i) the transport operator, ii) the chemistry operator, iii) the dry deposition operator, and iv) the wet deposition operator. Emissions and values related to meteorology are directly taken as input data. The transport operator consists of the dynamics of advection in three dimensions, diffusion and entrainment. The chemistry operator models everything related to the production and consumption of components by different chemical reactions in the atmosphere. The dry deposition is divided in two phases, the dry deposition of gases and the dry deposition of particles. The dry deposition of gases is modeled through the transfer of gases between the land surface and the atmosphere, result of the difference in concentrations and resistance between them. In the dry deposition of particles the scheme used depends on the given land use over the analysis region. The operator of wet deposition is modeled through the below cloud scavenging process. The below cloud scavenging process uses a sweep coefficient that describes the mass transfer speed of a pollutant from the air to the raindrops.

The LOTOS-EUROS model as other CTMs, is considered a high-uncertainty system due to the large number of uncertainty sources like meteorology, emissions, depositions parameters, among others. Additionally it is a high-dimensional system since the solution of Equation (1) is executed for different components and in each point belonging to a grid on the region of analysis. Due to this process, the state vector has a dimension in the order of millions.

The simulation of the deposition of atmospheric species is a challenging task highly dependent on whether the phenomena are modeled in the near, local or long-field range. Near-field deposition is dominated by larger particles. Local field events occur in the portion of the plume dominated by the wind-driven trajectory where the peak-to-mean concentration ratios are much smaller than closer to the source, resulting in more uniform deposition patterns. In long-range fields the larger particles have been removed so smaller particles are the main concern of the modeling [43]. The parameters that describe the deposition phenomena are highly uncertain, especially when land use and vegetation type data are low in detail or spatial resolution. Existing models of atmospheric deposition might not be appropriate for simulating deposition fluxes around urbanized environments, which present varied and rapidly changing forms of land use [44]. This study focuses on the impact of updating the land use data on the estimates of nitrogen deposition in Colombia. Whereas previous studies were limited by the use of rather coarse spatial resolutions [45, 46], here a resolution of 0.09° x 0.09° was used to increase the level of detail in the simulations. The sensitivity to input (land use and topography) modifications was studied to determine the impacts on the model deposition predictions. Additionally, point source experiments were conducted to start to understand the spatial deposition dynamics of N_r emanating from the principal urban areas of Country. This work contributes in assessing the performance of the LOTOS-EUROS CTM in NW South America.

In order to have a model that represents the principal meteorological, chemical and transport dynamics over a region is important first to improve as much as is possible the input information. Two input information were updated for the LOTOS-EUROS implementation over the northwest-southamerica region, the orography and the land use information.

The default elevation model for LOTOS-EUROS is obtained from the ECMWF meteorological data, which has a resolution of 0.07° (≈ 7 km). An updated elevation model used for the region was obtained from the Global Multi resolution Terrain Elevation Data (GMTED 2010) [47], with a resolution of 0.002° (≈ 220 m). **Figure 2** shows a comparison between the default (**Figure 2(A)**) and the updated elevation model (**Figure 2(A)**). The inset zoomed in around the Aburrá Valley (Medellín and neighboring munici-palities) demonstrates how the valley is entirely absent in the default elevation model (**Figure 2(A)**).

LOTOS-EUROS interpolates the input elevation data within each grid cell according to the resolution of the simulation. Changing the input elevation model can generate changes in the outcome of variables such as the temperature profiles of

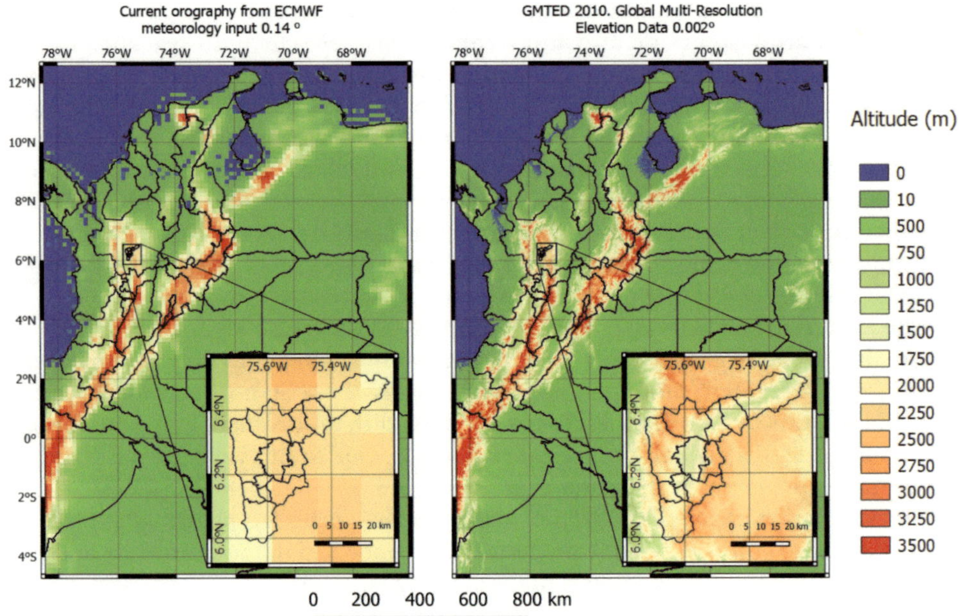

Figure 2.
(Comparison between the default elevation model for LOTOS-EUROS (A) and the updated elevation model derived from the GMTED 2010 data (B). The insets illustrate the differences in elevation representation for the Aburrá Valley.

the vertical layers. The effect of an updated elevation model depends on the desired simulation grid resolution. **Figure 3** shows a transverse cut at a latitude of 6.6°North for the simulation at a horizontal resolution of 0.09° x 0.09°, illustrating the impact of the change through the Aburrá Valley. The most significant changes in temperature occurred in the upper layers, reaching differences of up to 5°C degrees in top layers.

For modeling the deposition dynamics LOTOS-EUROS requires a map with deposition properties per grid cell. Land use characteristics are relevant for the deposition dynamics of a CTM because they define the parameters of the terrain roughness and canopy altitude of each category that determine the velocity at which the component is going to be deposited dependent on the vegetation type. The default land use/land cover (LU/LC) input data for LOTOS-EUROS were derived from the Global Land Cover (GLC2000) project [48], which includes 23 categories consistent with the Land Cover Classification System of the Food and Agriculture

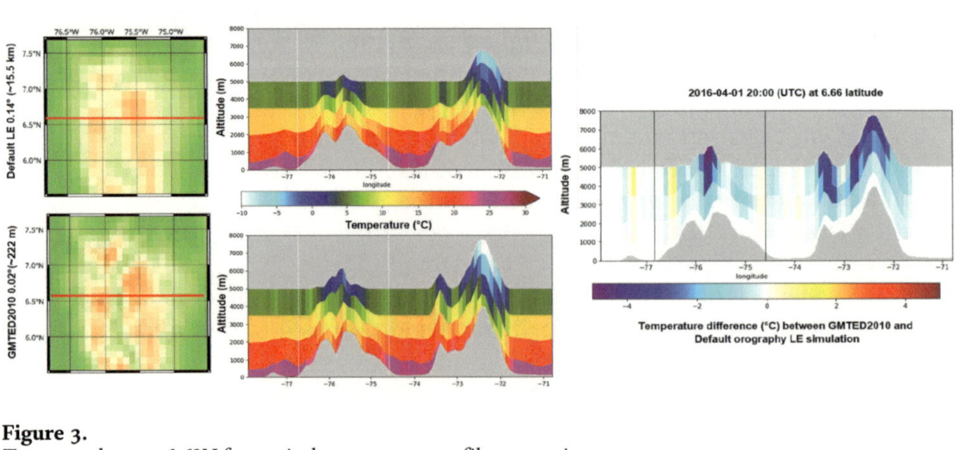

Figure 3.
Transversal cut at 6.6°N for vertical temperature profile comparison.

Organization (FAO) classification. For South America, the mapping of these categories at spatial resolutions of 1 km x 1 km was done in [49], with contributions from some regional experts based on multi-resolution satellite data. In this work, the LU/LC data was updated with the 2009 Land Cover Climate Change Initiative (CCI) dataset [50], with 38 categories at a horizontal resolution of 300 m x 300 m. **Figure 4** compares the default and updated LU/LC models for Aburrá Valley. The mapping of the 39 (CCI) and 23 (GLC) LU/LC categories to the 9 classes of the DEPAC deposition model is illustrated in **Figure 5**. The descriptions of each category are presented in **Table 1**. The mapping from GLC to DEPAC is the standard scheme constructed for LOTOS EUROS. The mapping from CCI to GLC

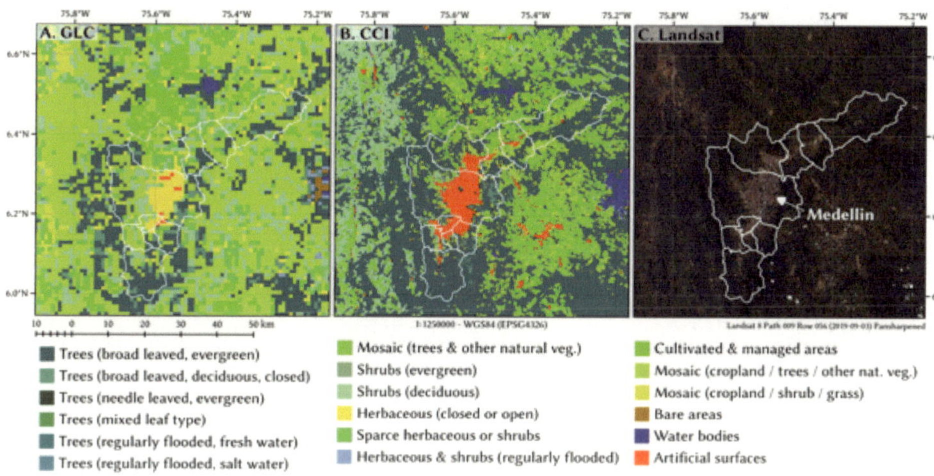

Figure 4.
Comparison between (A) the LOTOS-EUROS's original land cover model (global land cover, with resolution of 1 km x 1 km.) and (B) the updated land cover scheme (land cover from the climate change initiative, with resolution 0.3 km x 0.3 km). Real color Landsat cloudless imagery for the date 2019-09-03 is included in (C) as a reference for the artificial surfaces from the city infrastructure. The political boundaries correspond to the municipality of Medellín and the other nine municipalities that constitute the Aburrá Valley metropolitan area conurbation.

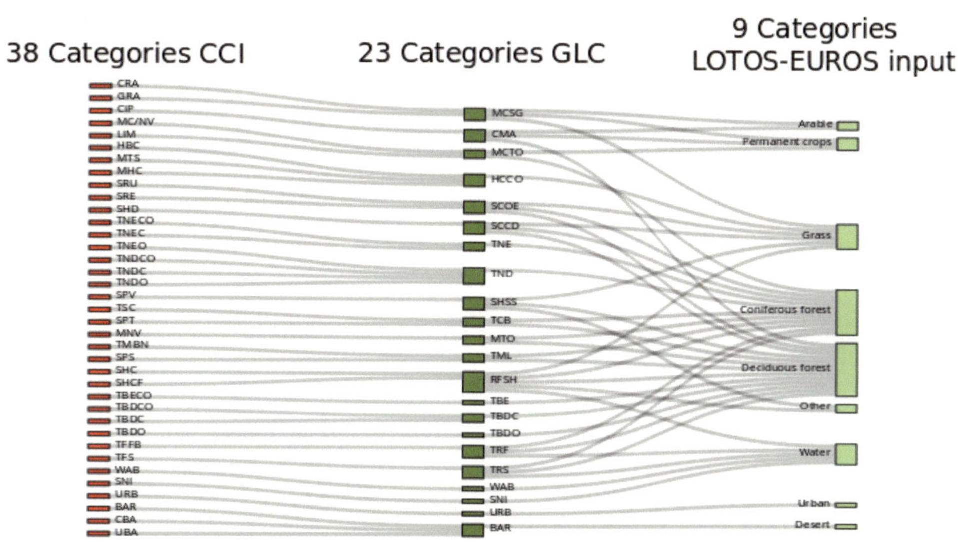

Figure 5.
*Climate change initiative (CCI) land cover data categories mapped to the global land cover (GLC). The GLC categories are in turned mapped to the model's equivalencies. The category's codes are as in **Table 1**.*

Number	Land Cover Climate Change Initiative (CCI) label	Code CCI	Number	Global Land Cover (GLC) label	Code CORINE
0	No data	ND	1	Tree Cover, broadleaved, evergreen	TBE
10	Cropland, rainfed	CRA	2	Tree Cover, broadleved, deciduous, closed	TBDC
11	Herbaceous cover	HBC	3	Tree Cover, broadleaved, deciduous, open	TBDO
12	Tree or shrub cover	TSC	4	Tree Cover, needle-leaved, evergreen	TNE
20	Cropland, irrigated or post-flooding	CIP	5	Tree Cover, needle-leaved, deciduous	TND
30	Mosaic cropland (>50%)/natural vegetation (tree, shrub, herbaceous cover) (<50%)	MC/NV	6	Tree Cover, mixed leaf type	TML
40	Mosaic natural vegetation (tree, shrub, herbaceous cover) (>50%)/ cropland (<50%)	MNV	7	Tree Cover, regularly flooded, fresh water	TRF
50	Tree cover, broadleaved, evergreen, closed to open (>15%)	TBECO	8	Tree Cover, regularly flooded, saline water	TRS
60	Tree cover, broadleaved, deciduous, closed to open (>15%)	TBDCO	9	Mosaic: Tree Cover/ Other natural vegetation	MTO
61	Tree cover, broadleaved, deciduous, closed (>40%)	TBDC	10	Tree Cover, burnt	TCB
62	Tree cover, broadleaved, deciduous, open (15–40%)	TBDO	11	Shrub Cover, closed-open, evergreen	SCOE
70	Tree cover, needleleaved, evergreen, closed to open (>15%)	TNECO	12	Shrub Cover, closed-open, deciduous	SCCD
71	Tree cover, needleleaved, evergreen, closed (>40%)	TNEC	13	Herbaceous Cover, closed-open	HCCO
72	Tree cover, needleleaved, evergreen, open (15–40%)	TNEO	14	Sparse herbaceous or sparse shrub cover	SHSS
80	Tree cover, needleleaved, deciduous, closed to open (>15%)	TNDCO	15	Regularly flooded shrub and/or herbaceous cover	RFSH
81	Tree cover, needleleaved, deciduous, closed (>40%)	TNDC	16	Cultivated and managed areas	CMA
82	Tree cover, needleleaved, deciduous, open (15–40%)	TNDO	17	Mosaic: Cropland/ Tree Cover / Other natural vege	MCTO
90	Tree cover, mixed leaf type (broadleaved and needleleaved)	TMBN	18	Mosaic: Cropland/ Shrub and/or grass cover	CSG
100	Mosaic tree and shrub (>50%)/ herbaceous cover (<50%)	MTS	19	Bare areas	BAR
110	Mosaic herbaceous cover (>50%)/ tree and shrub (<50%)	MHC	20	Water bodies	WAB

Number	Land Cover Climate Change Initiative (CCI) label	Code CCI	Number	Global Land Cover (GLC) label	Code CORINE
120	Shrubland	SRU	21	Snow an Ice	SNI
121	Shrubland evergreen	SRE	22	Artificial surfaces and associated areas	URB
122	Shrubland deciduous	SHD	23	No Data	ND
130	Grassland	GRA	**Number**	**LOTOS-EUROS fractional categories**	**LE-fract-code**
140	Lichens and mosses	LIM	1	Arable	ARA
150	Sparse vegetation (tree, shrub, herbaceous cover) (<15%)	SPV	2	Permanent crops	CRP
151	Sparse tree (<15%)	SPT	3	Grass	GRS
152	Sparse shrub (<15%)	SPS	4	Coniferous forest	CNF
153	Sparse herbaceous cover (<15%)	SHC	5	Decidious forest	DEC
160	Tree cover, flooded, fresh or brackish water	TFFB	6	Other	OTH
170	Tree cover, flooded, saline water	TFS	7	Water	WAT
180	Shrub or herbaceous cover, flooded, fresh/saline/brackish water	SHCF	8	Urban	NO2
190	Urban areas	URB	9	Desert	DSR
200	Bare areas	BAR			
201	Consolidated bare areas	CBA			
202	Unconsolidated bare areas	UBA			
210	Water bodies	WAB			
220	Permanent snow and ice	SNI			

Table 1.
Land use/land cover categories for the two sources of data used in this study and for the DEPAC module.

took into account the similar morphological characteristics between categories, and the aseasonality in this tropical region. The model defines for each grid cell the fraction covered by each one of the LU/LC classes used by the DEPAC module, and calculates the deposition over each fractions.

Deposition depends on land-use changes. It was essential to analyze the temporal behavior of the area of interest to identify zones with over and subs estimation comparing with the early reference. Natural protected areas with notable changes in deposition between the default and updated input data were identified, emphasizing the importance of using up-to-date and accurate land cover data in the simulation model. Vulnerable areas like natural protected areas and paramos ecosystems may require more than a local conservation effort for the preservation of their ecological functions.

To explore the fate of nitrogenous atmospheric species emitted from the main Colombian cities, the grid cells housing the centroids of the urban area for Bogotá, Medellín, Cali, and Barranquilla were assumed as artificial point sources of emissions. The simulations were conducted with the updated elevation model and updated LU/LC scheme detailed above, for a total of 10 days in four different times of the year: March 1–10, June 1–10, September 1–10, and December 1–10. After a

2-day model spin up, the point source was from 08:00–18:00 of day 3 of the simulation, emitting a total of 1000 kg/hour NO_2, which is the amount of daily NO_2 emissions reported for Medellń [51]. The artificial emissions were monitored during seven additional days, during which time all of the emitted species had either deposited or transformed. Similar simulations were conducted but without the activation of the point source in order to estimate the background deposition values for each grid cell.

A second experiment was conducted as above, but focusing on either Medellín or Rionegro. The latter city is located to the East of the Aburrá Valley, near the international airport that serves Medellín and the immediate region. Rionegro is the largest and fastest growing city in the Valley of San Nicolás. The urban growth in this region is being fueled in part by the migration of people from the Aburrá Valley. Many residents in the Valley of San Nicolás hold jobs inside the Aburrá Valley, and many schools that serve middle and upper income families from the Aburrá Valley are located in the Valley of San Nicolás. The second experiment was aimed at understanding the potential implications for regional Nr deposition derived from increase urban develpment in the Valley of San Nicolás.

The LOTOS-EUROS emission module explains the discharge of tracers and aerosols from various sources (anthropogenic, biogenic, marine, airborne dust, fires) that can be configured to define emissions in specific point sources to simulate scenarios. **Figure 6(A)** shows the simulation of the total deposition (dry and wet) for Nitrogen taking into account the emission of the four principal cities in Colombia. Cities here were assumed to be point emission sources, which works to determine the influence area of this city. **Figure 6(A)** shows the contours generated

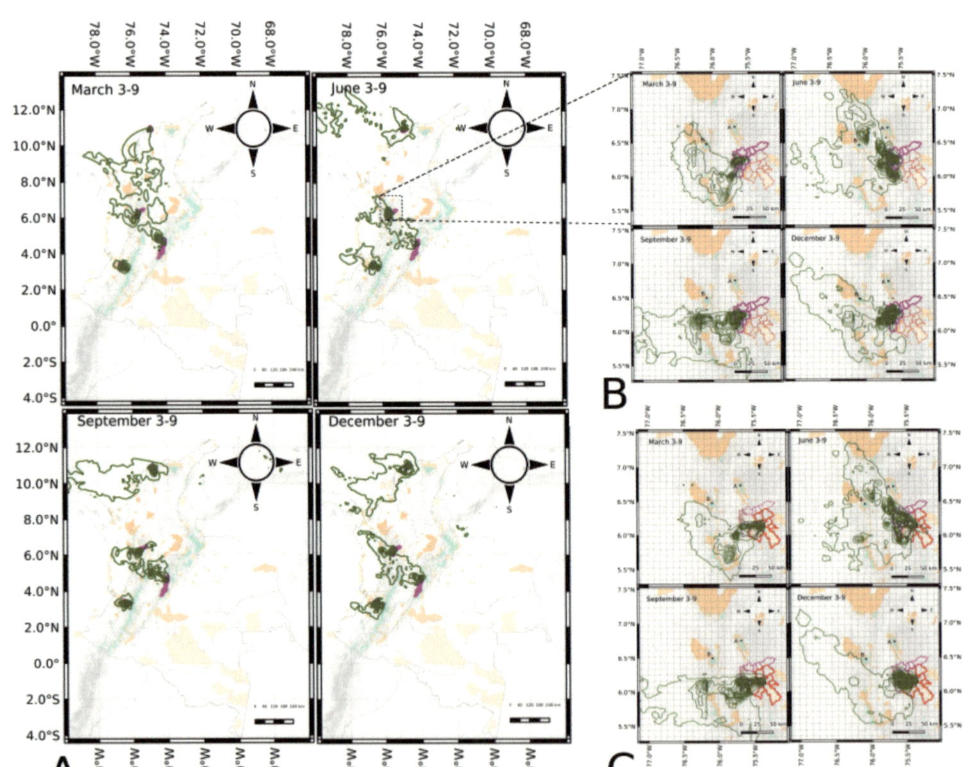

Figure 6.
Point source contour experiment for the 4 biggest cities in the country and contour limitation of the influence area.

with an increment of $5\,g/ha$ between contour level and bias correction of $+2\,g/ha$ to avoid negative values that appears as a numerical noise from the rest of the reference run minus the punctual perturbed emission run for the city of Medellín as a point source in the **Figure 6(B)** this same magnitude contour levels but in this case having the emission source located in Rionegro city. The more rounded contour delimits the impact zone for each of the cities.

It is possible to identify some trends in the wind direction for the four different times of the year, for which was performed simulations for the 1–9 days of March, June, September, and December. The influence of Barranquilla to faraway zones is perceivable due to the near location of this city to the Caribbean coast, where intense wind conditions exist as well as flat topography that drives the transport dynamics far away. More of the depositions from Barranquilla is going to the ocean direction and, in other time of the year to the southwest of the city reaching inclusive the other cities deposition areas. For the other cities, the impact area is more limited but with higher deposition values due to the roughness of the mountainous terrain and the less magnitude of the wind patterns presented. It is also interesting to see that the deposition from Bogotá could reach other cities as Medellín at some times of the year.

Figure 6(B) shows a simulation for what happens with Medellín for increase simulation resolution to 0.03x0.03°, conceiving the point source concentrated in a grid cell in the middle of the metropolitan area. The results of simulations indicate that the northwest-west area of the Aburra Valley (Medellín and municipalities of the metropolitan area) are the most affected, which could be seen in how paramos ecosystems located in the points a and B in the map receive nitrogen that was emitted from the cities. In the mosaic in the right part of **Figure 6(C)** a detailed simulation, conceiving the point source concentrated now in a grid cell in the middle of the metropolitan Rionegro area. It is important to notice during some parts of the year that Rionegro influences Medellín in terms of the deposition of its emissions. It means there is a transport dynamic of contaminants that should be studied from more detail because of the increasing Rionegro area for the industrial and urban setting.

Simulations with point sources identified the transport patterns in the territory, and showed the regional influence of the major cities in base of qualitative and quantitative results to understand the dynamics of emission and deposition of contaminants for the principal cities of Colombia which consist in an attractive information supply to start understanding the transport of atmospheric contaminants over this territory. The atmospheric transport and deposition of pollutants present ecosystem risk factors that require an evaluation of impacts directly in the field based on the reported results, and their inclusion in conservation strategies.

2.1.3 Technological development: SimpleSpace

A Simple module is a "CanSat" form factor measurement device with a standard for pico-satellites (0.1–1 kg). It has a mass of 0.4 kg and a volume of approximately 330 *ml*. It is a product of SimpleSpace, a spin-off program started at Universidad EAFIT since the early 2020 (**Figure** 7). Its aerospace design is versatile, and it does not represent a heavy load for any possible airborne campaign. As the module has already been measured at a stratospheric level in different High Altitude Balloon activities in the country, the sensors' robustness is guaranteed. Two simple modules inside and outside the aircraft provided an exciting framework for distributing pollutants outside and inside the urban area.

A Simple module is depicted in **Figure 8**, showing that it contains the necessary subsystems that a satellite architecture requires: an energy power supply, a

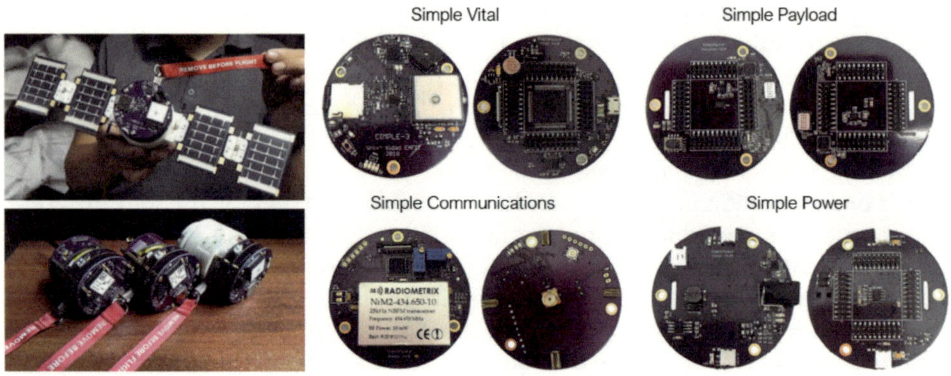

Figure 7.
Low-cost sensing solution from SimpleSpace. The simple unit is divided in subsystems which can be adapted into a vertical bus which allows the insertion of new pcb layers with new measurement instruments.

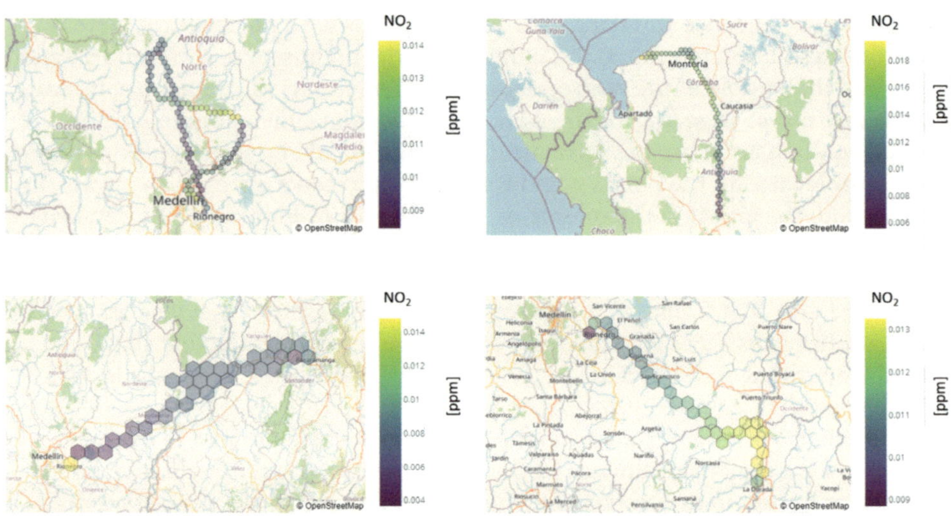

Figure 8.
Airborne measurements from the simple module for four days in our national territory for nitrogenous species taken from the airborne air quality platform (https://space-1516288940015.uc.r.appspot.com/).

communications subsystem, an on-board command data-handling subsystem, and a payload subsystem and a communication bus through all this subsystems.

The gases that were measured by the SimpleSpace modules for measurement in vertical layers in the HIPAE missio were: Carbon monoxide CO (1–1000 ppm), Nitrogen dioxide NO_2 (0.05–10 ppm), Ethanol C_2H_6OH (10–500 ppm), Hydrogen H_2 (1–1000 ppm), Ammonia NH_3 (1–500 ppm), Methane CH_4 (¿1000 ppm), Propane C_3H_8 (¿1000 ppm), Iso-butane C_4H_10 (¿1000 ppm), altitude, pressure and temperature using a CMOS based sensor. Additional sensor like the meteorology (Pressure, Humidity, Temperature, Wind direction and magnitude) and inertial sensor (accelerometers and magnetometers).

The previous figure shows the single units' observations onboard a non-pressurized aircraft of the Colombian Air Force for airspace protection. Valuable information is obtained in this research study regarding dynamics over the sensor devices and data assimilation experiments with the LOTOS-EUROS and WRF models. Scientists use numerical models that consider emissions and depositions of pollutants such as nitrogenous species and ozone in ecosystems to calculate these pollutants' transport and chemical reaction. The value of this lies in the possibility of defining courses of action to protect them because, in natural protected areas, the long-range transport pollution problem has not been extensively studying.

2.2 Human health

Behind many of the efforts of MAUI lies the intention to contribute to improving the environmental health in our region. Poor breathing air quality is amply recognized as a major environmental cause of morbidity and mortality [52], with global estimates suggesting nearly 9 million annual deaths being related to exposure to outdoor air pollution [53]. The National Institutes of Health in Colombia estimates that poor air quality is annually responsible for nearly 16,000 deaths representing the leading detriment to the national environmental health [54]. In Medellín and two other large Colombian cities, exposure of industrial air pollution has been associated with clusters of Acute Childhood Leukemia [55]. As has been repeatedly observed around the globe (e.g., [56]), urban exposure to air pollution has been found to be associated with increased rates of hospital visits related to respiratory and circulatory conditions in Medellín and other three large Colombian cities [57], and impact seemingly dominates by the effect of atmospheric NO_2 and its synergistic effect with other contaminants [58]. Atmospheric contaminants represent a threat to the health of Colombian people not only through their respiratory and mucosal exposure to polluted air, but also to surface-deposited contaminants [59].

The dynamics and spatial patterns of pollutant emissions in urban settings accentuate economic disparities with marked inequities in exposure to pollution [60]. Despite the reductions in air pollution resulting from total and partial lock downs in response to the COVID-19 pandemic [61], long-term exposure to poor air quality has been observed to exacerbate the manifestations of COVID-19 and increase its mortality rate [62]. Curiously, this effect was not apparent in Colombia, where sociodemographic differences appeared to be the leading factor influencing the dynamics of SARS-CoV-2 infections in the country [63]. Nevertheless, the cited authors suggest that ample errors affect the strength of the conclusions related to air pollution, and given the dearth of exposure models or assessments at sufficiently high spatial resolution available for Colombian cities, these results are indeed preliminary.

The development of mathematical models that could help us estimate the human exposure levels was one of the early goals at the outset of MAUI. With the research program $ExPoR^2$ we have been given the opportunity to finally pursue that goal.

2.2.1 HIPAE Mission

Research into nitrogenous species, ozone, and marine aerosols faces a common challenge posed by a scarcity of *in situ* measurements [8, 64]. Ideally, additional research infrastructure and sensors would be installed in the regions of interest. In the meantime, however, researchers will need to maximize the usefulness of satellite data, which is already available [11, 65] and alternative measurements sources as the airborne measurenment campaigns.

Airborne measurements constitute complementary *in situ* information for the vast array of measurements that may be collected from the atmosphere. To this end, HIPAE proves the viability of a measurement concept mission performed inside the

Colombian UH-60 L "Black Hawk" aircraft FAC4121, with the support of two groups within the Colombian Air Force: the fifth Air combat command -Comando aereo de combate N5 (CACOM5)- and Center for the technological development for defense -Centro de desarrollo tecnológico aeroespacial para la defensa (CETAD). This is the first time that this kind of measurement has been performed over Colombian territory.

Some references of airborne measuring campaigns come from the European project IAGOS (In-service Aircraft for a Global Observing System), a project that combines the expertise of two successful European research projects [66], MOZAIC (Measurements of OZone, water vapor, CO, NO_x by in service AIrbus airCraft) [67] and CARIBIC (Civil Aircraft for the Regular Investigation of the atmosphere Based on an Instrument Container) used in different scenarios to study the atmosphere composition [68]. The work of (Anthony et al., 2018) [69] used helicopter-borne electromagnetic measurements to derive a 3D resistivity model to study the permafrost carbon emission and compare a chemical transport model. One example from North America is the project from NOAA-ESRL named Tropospheric Aircraft Ozone Measurement Program focused on O_3 measurements [70]. Other project focused on aircraft measurements is called EMeRGe (Effect of Megacities on the transport and transformation of pollutants in the Regional and Global scales) which exploits the unique capabilities of the HALO (High Altitude and Long-Range aircraft) to investigate the impact of major population centers emissions on air pollution (Griffing et al., 2019).

In the mission HIPAE, we have the possibility of experimenting on topics such as nanomaterials, applied electromagnetism, atmospheric optics, environmental magnetism, biodiversity and conservation, mathematical modeling and meteorology [71]. This paper is the detailed extension of the work of [71]. The data assessed constitute a valuable source necessary for data assimilation experiments with the LOTOS EUROS to understand the dynamic of transport of contaminants in the region. The optical particle count and information retrieved from the 8 aerosol compounds of the Simple modules will allow valuable annotations to be used for the design of Lidar 4D systems; that will complement the TROPOMI ground and satellite measurements for the assimilation of data from the LOTOS-EUROS model at high and low resolution, increasing the impact from local to national in terms of the capacity to forecast the transport of pollutants (**Figure 9**).

The results of this experiment will not only validate the use of nanomaterials to capture particulate material, but will also offer valuable information on its composition, morphology, and chemistry via scanning electron microscopy (SEM) with energy-dispersive X-ray (EDX) analysis. These results are not going to be discussed in this paper and are considered under review on different journals. On the other hand, we address studies of cytogenotoxicity of pollutants in vertical columns of the atmosphere will provide valuable information to be used in human health risk assessments.

2.2.2 Preliminary results

From *in situ* aerosol measurements performed by HIPAE, a very clear picture emerged about the volumetric distribution of urban and rural pollutants in the Aburrá Valley. These data achieved good temporal and spatial resolutions. Direct analysis of numerous gases supported several decision-relevant findings, including spatial dynamics and source apportionment. This data will be assimilated into models in order to provide accurate dispersion scenarios and forecasting. Collected data by the OPC-N2 for PM 1.0, 2.5 and 10 match with those reported for the government services for the air quality monitoring. This demonstrates the reliability of the method.

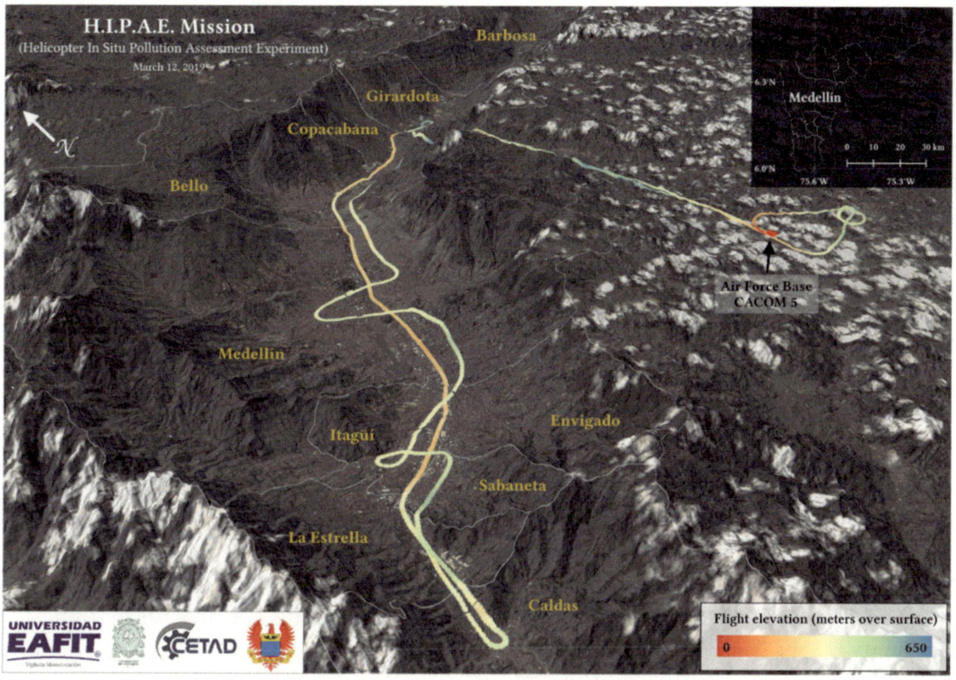

Figure 9.
Flight above the Aburrá Valley entering from the north and flying to the south and return at two different heights.

From a simple inspection of the comparison between the altimeter from the Simple device and the PM counters, several urban atmosphere dynamics were revealed. The concentrations of the two measures (PM10 and PM2.5) increased as the helicopter descended through the valley. Concentration data is useful for assimilation in a chemical transport models. The assimilated LE model was able to represent the concentrations of PM10 at low altitudes, as can be corroborated by ground stations; however, the model tends to underestimate concentrations proportional to the target altitude.

A remote laser detection device for research will be designed taking into account this information to define the physical design and its portability and of course, define the optimal channels to measure key particles i.e. wavelengths to detect, depolarization or Raman channels.

An automatized system for particulate matter collection is also needed to be designed, in order to have the possibility to capture particulate matter at different flying altitudes or city zones. This system also prevents to incorporate particle matter during take-off, landing and external load hooking maneuvers. Detailed information must be found in [71].

2.2.2.1 Polymer nanofibers for capture of particulate matter

In order to determine the capability of a novel nanofiber filter design, two types of filters were used in this study: one manufactured at Universidad EAFIT, based on PolyAcryloNitrile (PAN) nanofibers; and the other a commercial filter based on glass microfibers (GMF) [71].

The nanofilters were manufactured by an electrospinning technique. A high voltage was applied to a polymer solution in a syringe so that the polymer formed a small filament that traveled to a collection system. To manufacture the nanofilters, a concentration of 12 wt% of PAN, a voltage of 8 kV, and a 12-cm needle-collector were used.

Each fiber diameter was 560 +/− 130 nm, with a length of 50 +/− 5um.

The commercial filters were obtained from Electron Microscopy Sciences (Hatfield, PA). The assembly of the filters was done using air cassettes and GilAir 3 air sampling pumps (Sensidyne, USA).

The cassettes were connected by hoses to the pumps, which were programmed to intake 3 L/min. The filters were massed in a microbalance prior to installation (Mettler Toledo, USA) and were massed again at the end of the test [71]. The filtration efficiency of the particulate matter was determined in this way by gravimetry.

2.2.2.2 Celular and molecular damage quantification

2.2.2.2.1 Particulate matter extraction

The PAN and GMF filters were suspended in dichloro-methane (DCM) and stored at 20°C until extraction. The organic fraction of the PM was extracted with DCM in a sonicator for 45 min, broken into 15-min periods. The supernatant was rotavaporated at 35°C until the extract reached 0.5 mL. The extracts were then placed in amber bottles and left to dehydrate overnight. The PM extracts obtained from both filters were dissolved in 1 mL of dimethyl sulfoxide (DMSO).

2.2.2.2.2 Cell cultures

HaCaT cells were grown in Dulbecco's Modified Eagle's Medium (DMEM), which was supplemented with 5% fetal calf serum (Sigma, USA), 100 μg/mL penicillin, and 100 μg/mL streptomycin, in a humidified atmosphere of 5% CO2/95% air at 37°C. Exponentially growing cultures were used in all experiments.

2.2.2.2.3 Cytotoxicity assay

Cytotoxic evaluations of particulate matter captured with PAN and GMF filters were performed using MTT (3-(4,5-dimethylthiazol-2-yl)-2,5-diphenyltetrazolium bromide) assays, according to ISO/EN10993–5. Proliferating HaCaT cell lines were seeded in 96 well plates at a density of 10 cells/well in 100 mL of DMEM medium. Three concentrations of PM extract were tested: 25, 50, and 75 mg/mL. A positive control (methyl methanesulfonate, MMS, 100 mg/mL) and a negative control (DMSO, 1%) were implemented. Cells were incubated for 24 h with test extract concentrations, and all tests were carried out in triplicate. After the treatment, MTT (1 mg/mL) was added to each well and incubated for an additional 3 h. The performance of each well was quantified by measuring the spectral absorbance of the solution at 540 nm, using a microplate spectrophotometer.

2.2.2.2.4 Alkaline comet assay

Comet assays were performed with minor modifications, according to Sing et al. (1988). Briefly, HaCaT cells were seeded in each of 24 well plates, and exposed to 25, 50, 75 mg/mL PM solutions, and negative (DMSO 2%) and positive (50 mM H2O2) controls, for 1 hour. Cells were then trypsinized. Cell suspensions were mixed with low-melting-point agarose (0.75%), and immediately spread onto a microscope slide precoated with normal-melting-point agarose. Slides were then incubated in an ice-cold lysis solution (2.5 M NaCl, 10 mM Tris, 100 mM EDTA, 1%

Triton X-100, and 10% DMSO, at a pH of 10.0) at 4°C for 1 h. For the alkaline comet assay, slides were placed in an electrophoresis chamber containing freshly prepared alkaline buffer (300 mM NaOH and 1 mM EDTA, pH > 13.0) at 4°C for 20 min. After that a 300-mA and 25-V electric current was applied for 20 min to perform DNA electrophoresis. Slides were then neutralized (0.4 MTris, pH 7.5) and stained using 2% ethidium bromide solution. For the analysis, 200 randomly selected cells from each treatment were measured for DNA damage (% tail DNA). The tests were carried out in triplicate.

2.2.2.2.5 Cell cycle analysis

After 24 h of treatment, cell samples were fixed in 70% ethanol and subsequently incubated with 5 g/mL R5000 RNase (Sigma, USA), stained with 5 g/mL P4170 propidium iodide (Sigma, USA) for 30 min, and then analyzed for propidium iodide fluorescence using a BD LSRFortessa flow cytometer (BD Biosciences, USA). Percentages of cells in each phase were calculated using the FlowJo software package (FlowJo LLC, USA).

2.2.2.3 Cytogenotoxicity and genotoxicity analysis

2.2.2.3.1 Statistics on the data

As usual in biological sciences, a statistical analysis was carried out. Unless stated to the contrary, all data presented in this report represent results obtained from three independent experiments. Analysis of variance tests were performed with GraphPad Prism or Statview software, with post-hoc comparisons carried out by Fisher's protected least significant difference tests.

The cell viability examined by MTT assay showed that, compared with the unexposed control cells, there was statistically significant ($p < 0.05$) reduction in relative viability as the PM extract concentration increased from 25 to 50 to 75 mg/mL of PM. At the highest tested concentration, the cells reached a reduction of 25% of relative viability (**Figure 10**).

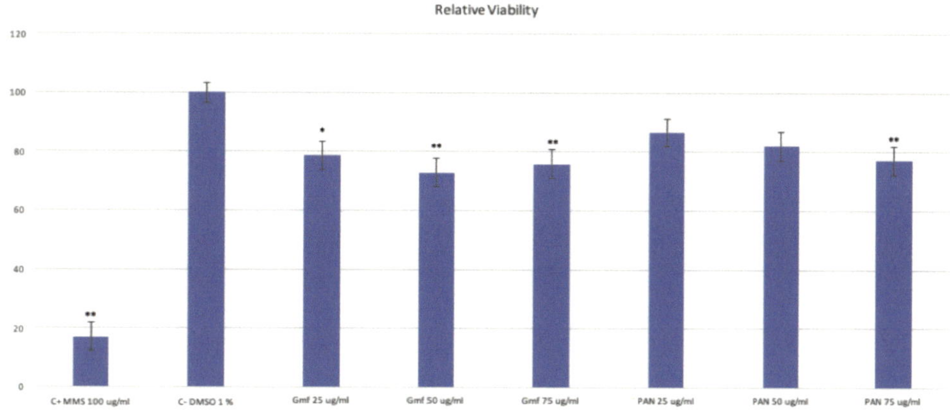

Figure 10.
*Cytotoxicity determined by relative viability, as measured by MTT assay of different concentrations of particulate matter extract obtained from GMF and PAN filters (25, 50, and 75 mg/mL). Positive control (C+; MMS 100 ug/mL) and negative control (C+; DMSO 1%) results are also depicted. The results are expressed as mean ± SE. Asterisks indicate statistical significance: * = (p ¡ 0.05), ** = (p < 0.01).*

2.2.2.3.2 Alkaline comet assay

As shown in **Figure 11**, the alkaline comet assay identified no significant statistical difference between the negative control (DMSO 2%) and the different concentrations of PM evaluated (25, 50, or 75 mg/mL).

2.2.2.3.3 Cell cycle analysis

Flow cytometry analysis indicated that, relative to control cells, the particulates captured by both GMF and PAN filters had no effect on the DNA content or the cell cycle phase distribution (**Figure 12**). The histograms in **Figure 13** show a slight population increase in the S phase, but without a significant statistical difference.

The cytotoxicity results show a clear reduction in the real viability related to the increase in the concentration of MP. The cytotoxicity results obtained in this work are in accordance with what was reported in other research papers that report cytotoxicity caused by PM after exposure to concentrations above 25 g/ml, for periods longer than 24 hours. (D. On the other hand, no evidence of genotoxicity or effects on the cell cycle of HaCaT cells was found at the concentrations evaluated. Which differs from the results reported by other researchers at similar concentrations of PM. These results may be related to the chemical composition of the PM at the sampling height.

Our analysis of the performance of PolyAcryloNitrile nanofibers in the capture and chemical and morphological analysis of particulate matter were promising. The designed nanomaterials outperformed the commercial filters at efficiently capturing particulate matter and in a way that was accessible for analysis.

Cell cycle analysis carried out on the particulate matter captured by the PAN nanofibers and GMF revealed lower rates of cellular damage compared with previous analysis on the surface. In addition, the comet assay identified no statistically significant difference between the negative control and the different concentrations of PM evaluated. The chemical composition of nano- or microscale particulate matter will determine the validity the hypothesis that cellular and genetic damage can be caused by human exposure to heavy particulate matter trapped near the surface in urban areas. Despite non-significative results on DNA damage and cell cycle analyses. We do strongly recommend taking into account these tests for future

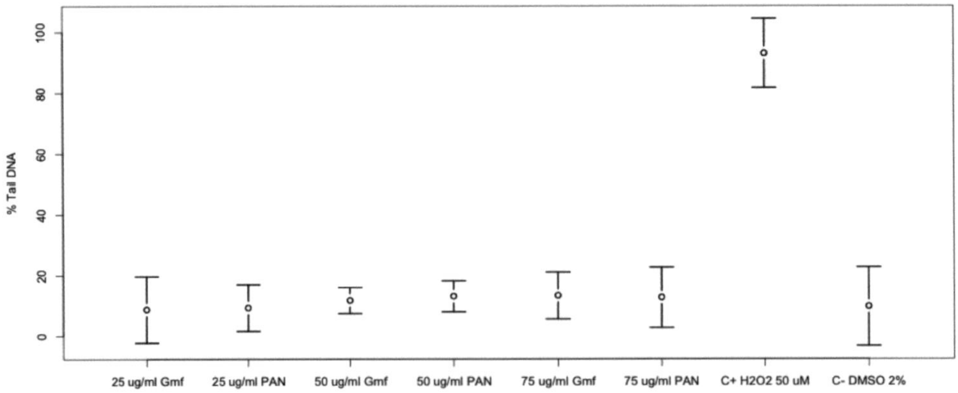

Figure 11.
Alkaline comet assay based on DNA breakage induced by different concentrations of PM extract obtained from GMF and PAN filters (25, 50, and 75 mg/mL). Positive control (C+; H2O2 50 mM) and negative control (C; DMSO 2%) results are also depicted. The results are expressed as mean ± SE.

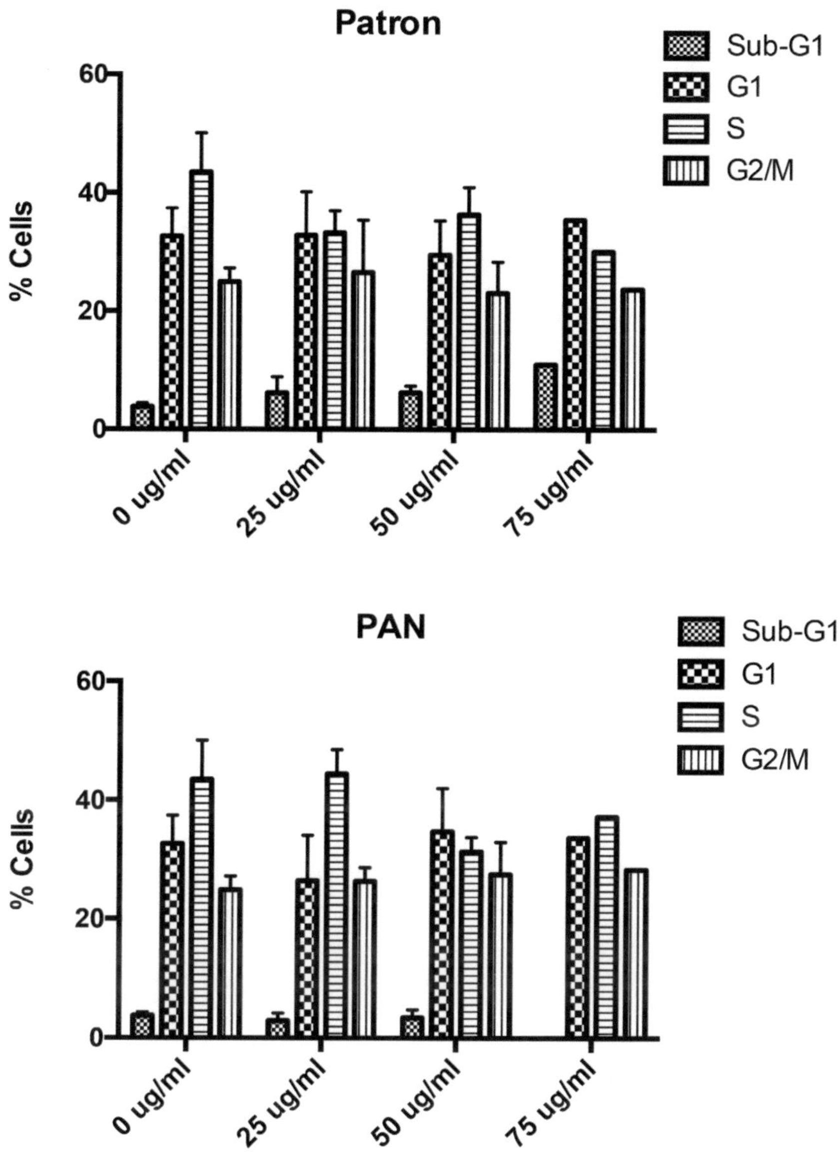

Figure 12.
Cell cycle distribution in HaCaT cells. Cells were treated with 25, 50 and 75 g/mL of GMF (A) and PAN (B) solutions. Values are expressed as mean ± SEM of three independent experiments for concentrations of 25 and 50 g/mL and one experiment for 75 g/mL.

analyses as a measure to monitor the evolution of the air polution and air quality in the valley.

Figure 14 depicts the Simple module measurements of ethanol and propane over the Aburrá Valley. The mission started in the military base near an urban area called Valle de San Nicolas. Flying north at a low altitude, the concentrations of C2H5OH were lower than when the aircraft entered the valley above Girardota. During the entire measurement campaign, and at several altitudes within the valley, there was no significant difference in ethanol concentrations in the vertical profiles, except for some portions in the south of the valley, where the land is not heavily urbanized. Human activities increase the concentration of ethanol, but when vegetation is present, it is mitigated. Similar behavior was detected by the measurements of

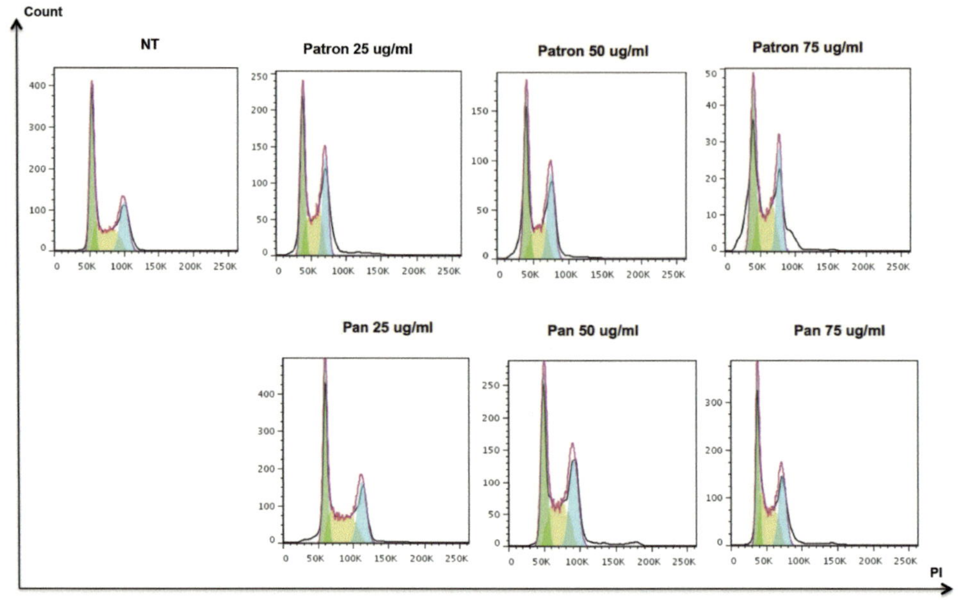

Figure 13.
Representative histograms of cell cycle distributions in HaCaT cells treated with GMF and PAN solutions.

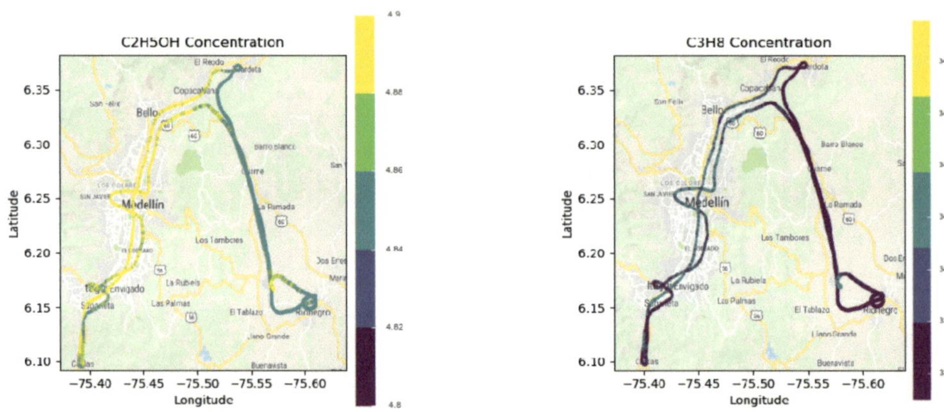

Figure 14.
Simple module measurements of C_2H_5OH and C3H8 over the Aburrá Valley.

propane, which reflect the presence of vegetation in urban areas with lower concentrations. Nevertheless, the highest urban concentration of propane was approximately 3410 ppm.

Simple module measurements of carbon monoxide and hydrogen over the Aburrá Valley are depicted in **Figure 15**. Carbon monoxide is toxic when encountered in concentrations above about 35 ppm. In the atmosphere, it is spatially variable and short-lived, having a role in the formation of ground-level ozone. Its presence indicates that tropospheric ozone is present in this urban area, having consequences for human health and nearby ecosystems. These measurements will enable several reaction schemes to be tested within a chemical transport modeling in the valley.

Simple module measurements of carbon monoxide and hydrogen over the Aburrá Valley are depicted in **Figure 15**. Carbon monoxide is toxic when encountered in concentrations above about 35 ppm. In the atmosphere, it is spatially

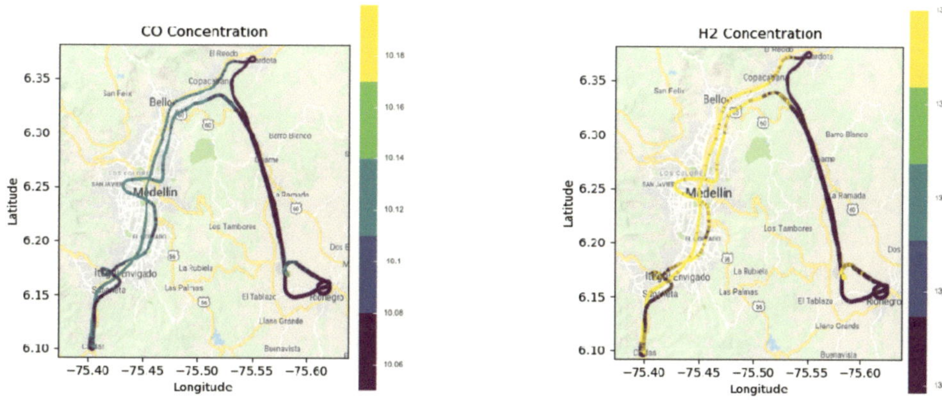

Figure 15.
Simple space measurements of CO and H_2 over the Aburra Valley.

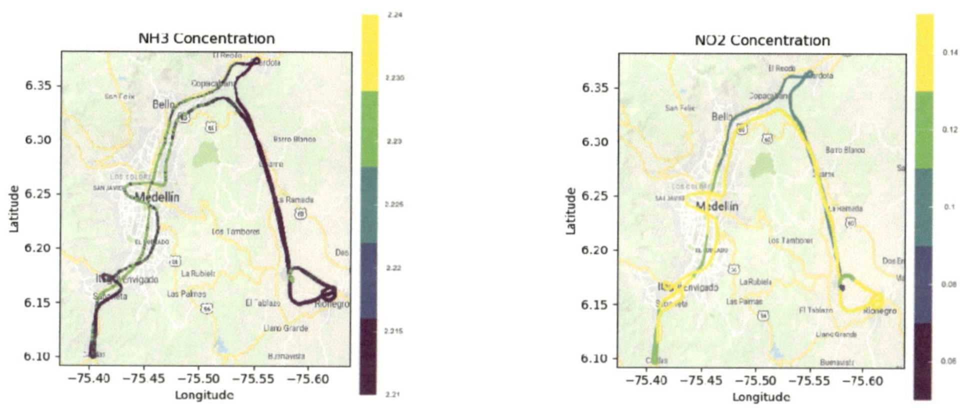

Figure 16.
Simple space measurements of NH3 and NO2 over the Aburra Valley.

variable and short-lived, having a role in the formation of ground-level ozone. Its presence indicates that tropospheric ozone is present in this urban area, having consequences for human health and nearby ecosystems. These measurements will enable several reaction schemes to be tested within modeling in the valley.

Simple module measurements of ammonia and nitrogen dioxide over the Aburrá Valley are depicted in **Figure 16**. The presence of elevated concentrations of nitrogenated species in the highest-altitude layers is deeply concerning. Ammonia is an irritant compound whose potency increases with its concentration. The permissible exposure limit is 25 ppm, and it becomes lethal above 500 ppm.

Author details

Andres Yarce Botero[1,2,3,4*], Olga Lucia Quintero Montoya[1,4], Santiago Lopez-Restrepo[1,2,4], Nicolás Pinel[2], Jhon Edinson Hinestroza[1], Elias David Niño-Ruiz[5], Jimmy Anderson Flórez[6], Angela María Rendón[7], Monica Lucia Alvarez-Laínez[8], Andres Felipe Zapata-Gonzalez[8], Jose Fernando Duque Trujillo[9], Elena Montilla[10], Andres Pareja[11], Jean Paul Delgado[12], Jose Ignacio Marulanda Bernal[13], Jaime Andres Betancur[6], Alejandro Vélez[6], Arjo Segers[14], Arnold Heemink[3], Juan Ernesto Soto[6], Bibiana Esperanza Boada Sanabria[1] and Sara Lorduy[1,6]

1 Mathematical Modelling Research Group, Universidad EAFIT, Medellin, Colombia

2 Department of Biological Sciences, Evolution and Conservation, Research Group on Biodiversity, Universidad EAFIT, Medellin, Colombia

3 Department of Applied Mathematics, TU Delft, Delft, The Netherlands

4 SimpleSpace, Universidad EAFIT, Medellin, Colombia

5 Applied Math and Computer Science Lab, Universidad del Norte, Colombia

6 Fuerza Aerea Colombiana, Colombia

7 Grupo de Ingenieria y Gestion Ambiental, Escuela Ambiental, Facultad de Ingeniería, Universidadde Antioquia, Medellín, Colombia

8 Design Engineering Research Group–GRID, Universidad EAFIT, Colombia

9 Departamento de Ciencias de la Tierra, Universidad EAFIT, Colombia

10 Applied Optics Research group, Universidad EAFIT, Colombia

11 Unidad de Toxicidad in vitro, Universidad CES, Colombia

12 Research Group of Genetics, Regeneration and Cancer, Institute of Biology, Universidad de Antioquia, Colombia

13 Applied Electromagnetism Reseacrh Group, Universidad EAFIT, Colombia

14 TNO Department of Climate, Air and Sustainability, Utrecht, The Netherlands

*Address all correspondence to: a.yarcebotero@tudelft.nl

References

[1] O. L. Quintero Montoya, Elías D. Niño-Ruiz, and Nicolás Pinel. On the mathematical modelling and data assimilation for air pollution assessment in the tropical Andes. Environmental Science and Pollution Research, 27(29): 35993–36012, 2020.

[2] Air Quality Index. World Air Quality Index. Technical Report, IQAir Air Visual, 2018.

[3] Muhammad Umar B Niazi, Diego Deplano, Carlos Canudas-de Wit, and Alain Y Kibangou. Scale-free estimation of the average state in large-scale systems. IEEE Control Systems Letters, 4(1):211–216, 2019.

[4] Shannon Mala Bard. Global transport of anthropogenic contaminants and the consequences for the arctic marine ecosystem. Marine Pollution Bulletin, 38 (5):356–379, 1999.

[5] David Fowler, J Neil Cape, Mhairi Coyle, Chris Flechard, Johan Kuylenstierna, Kevin Hicks, Dick Derwent, Colin Johnson, and David Stevenson. The global exposure of forests to air pollutants. Water, Air, and Soil Pollution, 116(1–2):5–32, 1999.

[6] M. Giardina and P. Buffa. A new approach for modeling dry deposition velocity of particles. Atmospheric Environment, 180(March):11–22, 2018.

[7] D Fowler, M Coyle, U Skiba, M A Sutton, J N Cape, S Reis, L J Sheppard, A Jenkins, B Grizzetti, J N Galloway, P Vitousek, A Leach, A F Bouwman, K Butterbach-Bahl, F Dentener, D Stevenson, M Amann, and M Voss. The global nitrogen cycle in the twenty-first century. Philos Trans R Soc Lond B Biol Sci, 368(1621):1–13, 2013.

[8] Jan Willem Erisman, Alex Vermeulen, Arjan Hensen, Chris Flechard, Ulrich Dämmgen, David Fowler, Mark Sutton, Ludger Grünhage, and Juha Pekka Tuovinen. Monitoring and modelling of biosphere/atmosphere exchange of gases and aerosols in Europe. Environmental Pollution, 133 (3):403–413, 2005.

[9] Yanlong Jia, Guirui Yu, Yanni Gao, Nianpeng He, Qiufeng Wang, Cuicui Jiao, and Yao Zuo. Global inorganic nitrogen dry deposition inferred from ground- and space-based measurements. Scientific Reports, 6: 19810, 2016.

[10] R. A. Duce, J. LaRoche, K. Altieri, K. R. Arrigo, A. R. Baker, D. G. Capone, S. Cornell, F. Dentener, J. Galloway, R. S. Ganeshram, R. J. Geider, T. Jickells, M. M. Kuypers, R. Langlois, P. S. Liss, S. M. Liu, J. J. Middelburg, C. M. Moore, S. Nickovic, A. Oschlies, T. Pedersen, J. Prospero, R. Schlitzer, S. Seitzinger, L. L. Sorensen, M. Uematsu, O. Ulloa, M. Voss, B. Ward, and L. Zamora. Impacts of atmospheric anthropogenic nitrogen on the open ocean. Science, 320(5878): 893–897, 2008.

[11] Jan Willem Erisman, James N Galloway, Sybil Seitzinger, Albert Bleeker, Nancy B Dise, A M Roxana Petrescu, Allison M Leach, and Wim de Vries. Consequences of human modification of the global nitrogen cycle. Philosophical transactions of the Royal Society of London. Series B, Biological sciences, 368(1621):20130116, 2013.

[12] R Bobbink, K Hicks, J Galloway, T Spranger, R Alkemade, M Ashmore, M Bustamante, S Cinderby, E Davidson, F Dentener, B Emmett, J-W Erisman, M Fenn, F Gilliam, A Nordin, L Pardo, and W De Vries. Global assessment of nitrogen deposition effects on terrestrial plant diversity: A synthesis. Ecological applications : a publication of the Ecological Society of America, 20(1):30–59, 2010.

[13] Emily C. Farrer and Katharine N. Suding. Teasing apart plant community responses to N enrichment: The roles of resource limitation, competition and soil microbes. Ecology letters, 19(10):1287–1296, 2016.

[14] Lindsay C. Maskell, Simon M. Smart, James M. Bullock, Ken Thompson, and Carly J. Stevens. Nitrogen deposition causes widespread loss of species richness in British habitats. Global Change Biology, 16(2): 671–679, 2010.

[15] Samuel M. Simkin, Edith B. Allen, William D. Bowman, Christopher M. Clark, Jayne Belnap, Matthew L. Brooks, Brian S. Cade, Scott L. Collins, Linda H. Geiser, Frank S. Gilliam, Sarah E. Jovan, Linda H. Pardo, Bethany K. Schulz, Carly J. Stevens, Katharine N. Suding, Heather L. Throop, and Donald M. Waller. Conditional vulnerability of plant diversity to atmospheric nitrogen deposition across the United States. Proceedings of the National Academy of Sciences, 113(15):4086–4091, 2016.

[16] Carly J Stevens, Nancy B. Dise, J. Owen Mountford, and David J. Gowing. Impact of nitrogen deposition grasslands. Science, 303(March):1876–1880, 2004.

[17] Sally E. Koerner, Meghan L. Avolio, Kimberly J. La Pierre, Kevin R. Wilcox, Melinda D. Smith, and Scott L. Collins. Nutrient additions cause divergence of tallgrass prairie plant communities resulting in loss of ecosystem stability. Journal of Ecology, 104(5):1478–1487, 2016.

[18] SV Krupa. Effects of atmospheric ammonia (nh3) on terrestrial vegetation: A review. Environmental pollution, 124(2):179–221, 2003.

[19] Rita Van Dingenen, Frank J. Dentener, Frank Raes, Maarten C. Krol, Lisa Emberson, and Janusz Cofala. The global impact of ozone on agricultural crop yields under current and future air quality legislation. Atmospheric Environment, 43(3):604–618, 2009.

[20] W. C. Skamarock, J. B. Klemp, J. Dudhi, D. O. Gill, D. M. Barker, M. G. Duda, X.-Y. Huang, W. Wang, and J. G. Powers. A Description of the Advanced Research WRF Version 3. *Technical Report*, (June):113, 2008.

[21] Chris Misenis and Yang Zhang. An examination of sensitivity of WRF/Chem predictions to physical parameterizations, horizontal grid spacing, and nesting options. Atmospheric Research, 97(3): 315–334, 2010.

[22] Anikender Kumar, Rodrigo Jiménez, Luis Carlos Belalcázar, and Néstor Y. Rojas. Application of WRF-Chem model to simulate PM10 concentration over Bogotá Aerosol and Air Quality Research, 16(5):1206–1221, 2016.

[23] Rafael Borge, Vassil Alexandrov, Juan José del Vas, Julio Lumbreras, and Encarnacion Rodríguez. A comprehensive sensitivity analysis of the WRF model for air quality applications over the Iberian Peninsula. Atmospheric Environment, 42(37): 8560–8574, 2008.

[24] Gregory R. Carmichael, Adrian Sandu, Tianfeng Chai, Dacian N. Daescu, Emil M. Constantinescu, and Youhua Tang. Predicting air quality: Improvements through advanced methods to integrate models and measurements. Journal of Computational Physics, 227(7):3540–3571, 2008.

[25] José A. Posada-Marín, Angela M. Rendón, Juan F. Salazar, John F. Mejía, and Juan Camilo Villegas. WRF downscaling improves ERAInterim representation of precipitation around a tropical Andean valley during El Niño: Implications for GCM-scale simulation of precipitation over complex terrain. Climate Dynamics, 2018.

[26] David Carvalho, Alfredo Rocha, Moncho Gómez-Gesteira, and Carlos Santos. A sensitivity study of the WRF model in wind simulation for an area of high wind energy. Environmental Modelling and Software, 33(December 2017):23–34, 2012.

[27] Paolo Tuccella, Gabriele Curci, Guido Visconti, Bertrand Bessagnet, Laurent Menut, and Rokjin J. Park. Modeling of gas and aerosol with WRF/Chem over Europe: Evaluation and sensitivity study. Journal of Geophysical Research Atmospheres, 117(3):1–15, 2012.

[28] Xiao Ming Hu, Petra M. Klein, and Ming Xue. Evaluation of the updated YSU planetary boundary layer scheme within WRF for wind resource and air quality assessments. Journal of Geophysical Research Atmospheres, 118 (18):10490–10505, 2013.

[29] María E. Dillon, Yanina García Skabar, Juan Ruiz, Eugenia Kalnay, Estela A. Collini, Pablo Echevarría, Marcos Saucedo, Takemasa Miyoshi, and Masaru Kunii. Application of the WRF-LETKF data assimilation system over southern South America: Sensitivity to model physics. Weather and Forecasting, 31(1):217–236, 2016.

[30] Xuguang Wang, Dale M. Barker, Chris Snyder, and Thomas M Hamill. A hybrid etkf–3dvar data assimilation scheme for the wrf model. Part i: Observing system simulation experiment. Monthly Weather Review, 136(12):5116–5131, 2008.

[31] Takuya Kawabata, Tohru Kuroda, Hiromu Seko, and Kazuo Saito. A cloud-resolving 4DVAR assimilation experiment for a local heavy rainfall event in the Tokyo metropolitan area. Monthly Weather Review, 139(6):1911–1931, 2011.

[32] Hongli Wang, Juanzhen Sun, Xin Zhang, Xiang-Yu Huang, and Thomas Auligné. Radar data assimilation with WRF 4D-Var. part I: System development and preliminary testing. Monthly Weather Review, 141(7):2224–2244, 2013.

[33] J Liu, M Bray, and D Han. A study on WRF radar data assimilation for hydrological rainfall prediction. Hydrol. Earth Syst. Sci, 17:3095–3110, 2013.

[34] Chengsi Liu, Qingnong Xiao, and Bin Wang. An ensemble-based four-dimensional Variational data assimilation scheme. Part II: Observing system simulation experiments with advanced research WRF (ARW). Monthly Weather Review, 137(5):1687–1704, 2009.

[35] Ryan D. Torn. Performance of a mesoscale ensemble Kalman filter (EnKF) during the NOAA high-resolution hurricane test. Monthly Weather Review, 138(12):4375–4392, 2010.

[36] Fuqing Zhang, Zhang Meng, and Jonathan Poterjoy. E3DVar: Coupling an ensemble Kalman filter with three-dimensional Variational data assimilation in a limited-area weather prediction model and comparison to E4DVar. Monthly Weather Review, 141: 900–917, 2013.

[37] Sara Q. Zhang, Milija Zupanski, Arthur Y. Hou, Xin Lin, and Samson H. Cheung. Assimilation of precipitation-affected radiances in a cloud-resolving WRF Ensemble data assimilation system. Monthly Weather Review, 141 (2):754–772, 2013.

[38] Juanzhen Sun and Hongli Wang. Radar data assimilation with WRF 4D-Var. part II: Comparison with 3D-Var for a squall line over the U.S. Great Plains. Monthly Weather Review, 141 (7):2245–2264, 2013.

[39] A. Mues, J. Kuenen, C. Hendriks, A. Manders, A. Segers, Y. Scholz, C.

Hueglin, P. Builtjes, and M. Schaap. Sensitivity of air pollution simulations with LOTOS-EUROS to the temporal distribution of anthropogenic emissions. Atmospheric Chemistry and Physics, 14 (2):939–955, 2014.

[40] Ferd Sauter, Eric Van der Swaluw, Astrid Manders-groot, Roy Wichink Kruit, Arjo Segers, and Henk Eskes. TNO Report TNO-060-UT-2012- 01451. Technical Report, TNO, Utrecht, Netherlands, 2012.

[41] Astrid M M Manders, Peter J H Builtjes, Lyana Curier, Hugo A C Denier Van Der Gon, Carlijn Hendriks, Sander Jonkers, Richard Kranenburg, Jeroen J P Kuenen, Arjo J Segers, Renske M A Timmermans, Antoon J H Visschedijk, Roy J Wichink Kruit, W Addo, J Van Pul, Ferd J Sauter, Eric Van Der Swaluw, Daan P J Swart, John Douros, Henk Eskes, Erik Van Meijgaard, Bert Van Ulft, Peter Van Velthoven, Sabine Banzhaf, Andrea C Mues, Rainer Stern, Guangliang Fu, Sha Lu, Arnold Heemink, Nils Van Velzen, and Martijn Schaap. Curriculum vitae of the LOTOS–EUROS (v2.0) chemistry transport model. Geosci. Model Dev, 10: 4145–4173, 2017.

[42] M. Van Loon, P. J. H. Builtjes, and A. J. Segers. Data assimilation of ozone in the atmospheric transport chemistry model LOTOS. Environmental Modelling and Software, 15(6–7 SPEC. ISS):603–609, 2000.

[43] James G Droppo. Improved Formulations for Air-Surface Exchanges Related to National Security Needs: Dry Deposition Models. Technical Report, Pacific Northwest National Lab. (PNNL), Richland, WA (United States), 2006.

[44] N. Chérin, Y. Roustan, L. Musson-Genon, and C. Seigneur. Modelling atmospheric dry deposition in urban areas using an urban canopy approach. Geoscientific Model Development, 8(3): 893–910, 2015. Revue OA, lien vers le full-text: http://www.geosci-model-dev.net/8/893/2015/gmd-8-893-2015.pdf.

[45] John Freddy Grajales and Astrid Baquero-Bernal. Inference of surface concentrations of nitrogen dioxide (no2) in Colombia from tropospheric columns of the ozone measurement instrument (omi). Atmósfera, 27(2): 193–214, 2014.

[46] J. G. M. Barten, L. N. Ganzeveld, A. J. Visser, R. Jiménez, and M. C. Krol. Evaluation of nitrogen oxides sources and sinks and ozone production in Colombia and surrounding areas. Atmospheric Chemistry and Physics Discussions, 2019:1–30, 2019.

[47] JJ Danielson and DB Gesch. Global Multi-resolution Terrain Elevation Data 2010(GMTED2010). U.S. Geological Survey Open-File Report 2011–1073, 2010:26, 2011.

[48] Steffen Fritz, Etienne Bartholomé, Alan Belward, Andrew Hartley, Hugh Eva, Philippe Mayaux, Sergey Bartalev, Rasim Latifovic, Partha Sarathi Roy, Shefali Agrawal, Wu Bingfang, Xu Wenting, Jean-francois Pekel, Chandra Giri, Sander Mücher, Erik De Badts, Ryutaro Tateishi, Jeanlouis Champeaux, and Pierre Defourny. Harmonisation, Mosaicing and Production of the Global Land Cover 2000 Database. page 41, 2003.

[49] H.D Eva, E. Miranda, and C. Di Bella. Vegetation Map of South America, volume 148. 2002.

[50] P. Defourny, I. Moreau, and S. Bontemps. P Roduct U Ser G Uide. Technical Report, UCL, Gamma Remote Sensing, 2017.

[51] UPB and AMVA. Inventario de Emisiones Atmosféricas del Valle de Aburrá - actualización 2015. Technical report, Universidad Pontificia Bolivariana - Grupo de Investigaciones

Ambientales, Area Metropolitana del Valle de Aburra, Medellín, 2017.

[52] Aaron J. Cohen, Michael Brauer, Richard Burnett, H. Ross Anderson, Joseph Frostad, Kara Estep, Kalpana Balakrishnan, Bert Brunekreef, Lalit Dandona, Rakhi Dandona, Valery Feigin, Greg Freedman, Bryan Hubbell, Amelia Jobling, Haidong Kan, Luke Knibbs, Yang Liu, Randall Martin, Lidia Morawska, C. Arden Pope, Hwashin Shin, Kurt Straif, Gavin Shaddick, Matthew Thomas, Rita van Dingenen, Aaron van Donkelaar, Theo Vos, Christopher J.L. Murray, and Mohammad H. Forouzanfar. Estimates and 25-year trends of the global burden of disease attributable to ambient air pollution: An analysis of data from the global burden of diseases study 2015. The Lancet, 389(10082):1907–1918, 2017.

[53] Richard Burnett, Hong Chen, Mieczyslaw Szyszkowicz, Neal Fann, Bryan Hubbell, C. Arden Pope, Joshua S. Apte, Michael Brauer, Aaron Cohen, Scott Weichenthal, Jay Coggins, Qian Di, Bert Brunekreef, Joseph Frostad, Stephen S. Lim, Haidong Kan, Katherine D. Walker, George D. Thurston, Richard B. Hayes, Chris C. Lim, Michelle C. Turner, Michael Jerrett, Daniel Krewski, Susan M. Gapstur, W. Ryan Diver, Bart Ostro, Debbie Goldberg, Daniel L. Crouse, Randall V. Martin, Paul Peters, Lauren Pinault, Michael Tjepkema, Aaron Van Donkelaar, Paul J. Villeneuve, Anthony B. Miller, Peng Yin, Maigeng Zhou, Lijun Wang, Nicole A.H. Janssen, Marten Marra, Richard W. Atkinson, Hilda Tsang, Thuan Quoc Thach, John B. Cannon, Ryan T. Allen, Jaime E. Hart, Francine Laden, Giulia Cesaroni, Francesco Forastiere, Gudrun Weinmayr, Andrea Jaensch, Gabriele Nagel, Hans Concin, and Joseph V. Spadaro. Global estimates of mortality associated with longterm exposure to outdoor fine particulate matter. Proceedings of the National Academy of Sciences of the United States of America, 115(38):9592–9597, 2018.

[54] Instituto Nacional de Salud. Carga de enfermedad ambiental en Colombia. Technical report, 2018.

[55] Laura Andrea Rodríguez-Villamizar, Feisar Enrique Moreno-Corzo, Ana María Valbuena-García, Claudia Janeth Uribe Pérez, Mary Ruth Brome Bohórquez, Héctor Iván García García, Luis Eduardo Bravo, Rafael Gustavo Ortiz Martínez, Jürg Niederbacher Velásquez, and Alvaro R. Osornio-Vargas. Childhood leukemia in small geographical areas and proximity to industrial sources of air pollutants in three Colombian cities. International Journal of Environmental Research and Public Health, 17(21):7925, 2020.

[56] R. W. Atkinson, S. Kang, H. R. Anderson, I. C. Mills, and H. A. Walton. Epidemiological time series studies of $PM_{2.5}$ and daily mortality and hospital admissions: A systematic review and meta-analysis. Thorax, 69(7): 660–665, 2014.

[57] Laura Andrea Rodríguez-Villamizar, Néstor Yezid Rojas-Roa, Luis Camilo Blanco-Becerra, Víctor Mauricio Herrera-Galindo, and Julián Alfredo Fernández-Niño. Short-term effects of air pollution on respiratory and circulatory morbidity in Colombia 2011–2014: A multi-city, time-series analysis. International Journal of Environmental Research and Public Health, 15(8), 2018.

[58] Laura Andrea Rodríguez-Villamizar, Néstor Yezid Rojas-Roa, and Julián Alfredo Fernández-Niño. Short-term joint effects of ambient air pollutants on emergency department visits for respiratory and circulatory diseases in Colombia, 2011–2014. Environmental Pollution, 248:380–387, 2019.

[59] Erika P. Donado, Marcos L.S. Oliveira, Janaína O. Gonçalves,

Guilherme L. Dotto, and Luis F.O. Silva. Soil contamination in Colombian playgrounds: Effects of vehicles, construction, and traffic. Environmental Science and Pollution Research, 28:166–176, 2021.

[60] Christopher W Tessum, Joshua S Apte, Andrew L Goodkind, Nicholas Z Muller, Kimberley A Mullins, David A Paolella, Stephen Polasky, Nathaniel P Springer, Sumil K Thakrar, Julian D Marshall, and Jason D Hill. Inequity in consumption of goods and services adds to racial–ethnic disparities in air pollution exposure. Proceedings of the National Academy of Sciences, 116(13): 6001 – 6006, 2019.

[61] Juan F. Méndez-Espinosa, Néstor Y. Rojas, Jorge Vargas, Jorge E. Pachón, Luis C. Belalcázar, and Omar Ramírez. Air quality variations in northern South America during the COVID-19 lockdown. Science of the Total Environment, 749(2):141621, 2020.

[62] Andrea Pozzer, Francesca Dominici, Andy Haines, Christian Witt, Thomas Münzel, and Jos Lelieveld. Regional and global contributions of air pollution to risk of death from COVID-19. Cardiovascular Research, 116(14):2247–2253, 2020.

[63] Laura A. Rodriguez-Villamizar, Luis Carlos Belalcázar-Ceron, Julián Alfredo Fernández-Niño, Diana Marcela Marín-Pineda, Oscar Alberto Rojas-Sánchez, Lizbeth Alexandra Acuña-Merchán, Nathaly Ramírez-García, Sonia Cecilia Mangones-Matos, Jorge Mario Vargas-González, Julián Herrera-Torres, Dayana Milena Agudelo-Castañeda, Juan Gabriel Piñeros Jiménez, Néstor Y. Rojas-Roa, and Victor Mauricio Herrera-Galindo. Air pollution, sociodemographic and health conditions effects on COVID-19 mortality in Colombia: An ecological study. Science of the Total Environment, 756:144020, 2021.

[64] Shelley C. Van Der Graaf, Enrico Dammers, Martijn Schaap, and Jan Willem Erisman. Technical Note: How Are NH 3 Dry Deposition Estimates Affected by Combining the LOTOS-EUROS Model with IASI-NH 3 Satellite Observations? :1–36, 2018.

[65] J. W. Erisman and M. Schaap. The need for ammonia abatement with respect to secondary PM reductions in Europe. Environmental Pollution, 129 (1):159–163, 2004.

[66] F. Berkes, N. Houben, U. Bundke, H. Franke, H.-W. Paetz, F. Rohrer, A. Wahner, and A. Petzold. The iagos nox instrument - design, operation and first results from deployment aboard passenger aircraft. Atmospheric Measurement Techniques, 11(6):3737–3757, 2018.

[67] Andreas Stohl, Paul James, Caroline Forster, Nicole Spichtinger, Alain Marenco, V Thouret, and Herman Smit. An extension of measurement of ozone and water vapour by airbus in-service aircraft (mozaic) ozone climatologies using trajectory statistics. Journal of Geophysical Research, 106:27757–27768, 2001.

[68] Carl Brenninkmeijer, P Crutzen, F Boumard, T Dauer, B Dix, Ralf Ebinghaus, D Filippi, H Fischer, Harald Franke, U Frieß, Jost Heintzenberg, F Helleis, Martyna Hermann, H.H. Kock, Claus Koeppel, J Lelieveld, M Leuenberger, Bengt Martinsson, S Miemczyk, and Helmut Ziereis. Civil aircraft for the regular investigation of the atmosphere based on an instrumented container: The new caribic system. Atmospheric Chemistry and Physics, 7:4953–4976, 2007.

[69] Katey Walter Anthony, Thomas Schneider von Deimling, Ingmar Nitze, Steve Frolking, Abraham Emond, Ronald Daanen, Peter Anthony, Prajna Lindgren, Benjamin Jones, and Guido Grosse. 21st-century modeled permafrost carbon emissions accelerated by abrupt thaw beneath

lakes. Nature Communications, 9, 12 2018.

[70] M. Bocquet, H. Elbern, H. Eskes, M. Hirtl, R. Aabkar, G. R. Carmichael, J. Flemming, A. Inness, M. Pagowski, J. L. Pérez Camaño, P. E. Saide, R. San Jose, M. Sofiev, J. Vira, A. Baklanov, C. Carnevale, G. Grell, and C. Seigneur. Data assimilation in atmospheric chemistry models: Current status and future prospects for coupled chemistry meteorology models. Atmospheric Chemistry and Physics, 15(10):5325–5358, 2015.

[71] Juan Camilo et al. In Proceedings of the 4th CMAS South America Air Quality Conference, pages 313–315, Vitoria, 2019. Fundação Espírito Santense de Tecnologia.

Chapter 7

Data Assimilation as a Tool to Improve Chemical Transport Models Performance in Developing Countries

Santiago Lopez-Restrepo, Andrés Yarce Botero, Olga Lucia Quintero, Nicolás Pinel, Jhon Edinson Hinestroza, Elias David Niño-Ruiz, Jimmy Anderson Flórez, Angela Maíra Rendón, Monica Lucia Alvarez-Laínez, Andres Felipe Zapata-Gonzalez, Jose Fernando Duque Trujillo, Elena Montilla, Andres Pareja, Jean Paul Delgado, Jose Ignacio Marulanda Bernal, Bibiana Boada, Juan Ernesto Soto, Sara Lorduy, Jaime Andres Betancur, Arjo Segers and Arnold Heemink

Abstract

Particulate matter (PM) is one of the most problematic pollutants in urban air. The effects of PM on human health, associated especially with PM of $\leq 2.5\mu m$ in diameter, include asthma, lung cancer and cardiovascular disease. Consequently, major urban centers commonly monitor $PM_{2.5}$ as part of their air quality management strategies. The Chemical Transport models allow for a permanent monitoring and prediction of pollutant behavior for all the regions of interest, different to the sensor network where the concentration is just available in specific points. In this chapter a data assimilation system for the LOTOS-EUROS chemical transport model has been implemented to improve the simulation and forecast of Particulate Matter in a densely populated urban valley of the tropical Andes. The Aburrá Valley in Colombia was used as a case study, given data availability and current environmental issues related to population expansion. Using different experiments and observations sources, we shown how the Data Assimilation can improve the model representation of pollutants.

Keywords: chemical transport model, air quality, data assimilation, LOTOS-EUROS, low-cost networks

1. Introduction

Air pollution is defined as the presence of solid, liquid or gaseous components in the atmosphere that can cause risk and troubles for living beings or goods in general. Air pollution is one of the major environmental problem in modern human history [1]. Environmental pollution can be produced by natural or human actions. Natural sources include forest fires, volcanic emissions, dust, sand, vegetation (as pollen) and wildlife (as methane). The main human sources of air pollution are industry, power generation, transportation, deforestation and cattle raising [2].

The current exponential growth in world population heightens the importance of public health issues related to air quality [3, 4]. In developing countries, decision makers must cope with the environmental demands of expanding and overpopulated urban centers. Short term air quality forecasts and long term mitigation strategies for these centers are usually based on specialized assessments of particulate matter dynamics [5, 6]. The Aburrá Valley houses the city of Medellín and neighboring municipalities. It is the second most populous urban agglomeration in Colombia, and the third densest in the world. The valley traces the course of the Medellín River along 60 km of a deep mountain canyon that ranges in width between 3 and 10 km, and with a height difference of up to 1800 m. Air quality conditions deteriorate severely within the valley twice a year around the time of the arrival of the Intertropical Convergence Zone (March–April, and with lower intensity in October–November), when the atmospheric inversion layer persists throughout the day below the rim of the canyon, thus trapping all of the urban atmospheric contaminants within the lower atmosphere [7]. During these periods, the concentrations of particulate matter below $10\mu m$ (PM_{10}) and $2.510\mu m$ (PM2.5) remain at levels considered hazardous for vulnerable populations and even for the general population (**Figure 1**).

Due to the large stress on human health induced by this air pollution, efforts have been made to monitor, reduce, and prevent episodes in which concentrations of pollutants reach hazard levels. Before measures for reducing air pollution can be implemented it is important to know the actual concentration levels and how these evolve in time over the area of interest. This could be done using a Chemical Transport Model (CTM) to simulate concentrations of trace gasses and particulate matter [8, 9]. In the last 20 years, CTMs have seen a huge growth and development; in consequence a diversity of models exists, differing in their complexity, size of the region of study, and methods used for their development. CTMs can be broken down in four categories according to their dynamic behavior: i) Gaussian, ii)

Figure 1.
Perspective of the air quality in the city of Medellín. (August 26, 2016, www.elmundo.com).

statistic, iii) Lagrangian and iv) Eulerian [8]. Eulerian models are the most widely used and reported for monitoring and predicting the pollution behavior and define the air quality in bigger areas [9]. So, these are frequently used in areas with sizes like countries or continents and have been less used in areas like cities.

Data assimilation (DA) is a mathematical process that provides integration between measured values (observations) and a dynamic model, to improve the operation of the model. With DA, the output value provided by the model has a smaller error than the output value provided by the model without observations. DA has two key objectives: to improve the operation in predictions of model states; and estimate unknown parameters of the model [10]. DA has been tested in different science fields such as oceanography, climatology, CTMs, and reservoirs characterization [11]. DA allows integrating models and observations out different scales of size and temporal sampling [12]. When two sources of information are combined, DA assumes that both the model and the measurements are subject to errors. These errors are impossible to know with accuracy and need to be specified in statistical and probabilistic terms. DA is not only looking to reduce the model error in space or time with observations; its mission is to digest the observation based on the laws given by the model and to determine the dynamic evolution of the model state that represents better measurements [13].

Large-scale model uncertainty, especially in CTM, is a very complicated issue. Increasing the accuracy of initial conditions, such as accurate land cover representations or updated emissions inventories, or using observations and DA, may reduce uncertainty. Data assimilation offers an alternative that is dynamically driven to reduce the lack of knowledge about the behavior of air pollution. The addition of surface, satellite, in situ, and laser-based remote sensing data to a model will enhance the understanding of proper scenario simulation and online decision-making. A bounty promise lies in the incorporation of the DA, not only for its contribution to the reduction of uncertainty, but also for opening the door to air quality forecasting in atmospheric pollution modeling. CTM forecasting presents us with interesting and complex challenges associated with the uncertainty of weather forecasting, the lack of precise inventory of emissions, and the scarcity and sparsity of monitoring networks for air quality. Such challenges require creative solutions; these challenges are opportunities for knowledge advancement. Due to the scarcity of data and high uncertainty in the model inputs, a mathematical, analytical, and computational effort is needed to push the frontiers of knowledge in the field.

Public air quality monitoring networks often consist of fixed measuring stations equipped with expensive sensors and maintained under rigorous operational and calibration regimes in order to provide high quality data. The high costs associated with establishing and maintaining such stations means that not all cities in developing countries can afford monitoring networks of sufficient spatial coverage [14]. Even in large cities in developed countries, the official air quality monitoring networks do not always provide information at the spatial and temporal resolution required to assess the impact of pollution sources on health [15], as the cost of the equipment makes the necessary density prohibitive. In the metropolitan region of Medellín (Colombia) and its con-urban municipalities for example, there are 21 main PM2.5 monitoring stations, at an average density of 8.25 km2 over the entire area of the 10 municipalities. This has motivated the expansion and improvement of low-cost systems and programs to measure PM [16]. The limited number of studies that have evaluated newer generations of low-cost $PM_{2.5}$ sensors have shown that the most widely used low-cost sensors attain high accuracy when compared to standard monitoring stations (R^2 value ranging from 0.93 to 0.95) [17]. The data provided by these sensors can complement those generated by conventional systems, increasing the data resolution and allowing studies of exposure at the human level [15, 18]. By data assimilation, the incorporation of air pollution data into CTM increases the ability to grasp local and regional patterns and fill spatial coverage gaps. Additionally, the combination of different sources of information and knowledge (data and model) increases the robustness and reliability of low-cost observations [12, 19].

2. The ensemble Kalman filter

The Ensemble-Based DA is a family of methods that uses an ensemble to model the statistics of the first guess (background). In each assimilation step, a forecast from the previous model simulation is used as a first guess, using the available observation this forecast is modified in better agreement with these observations. Due to it is easily implemented, it is relatively low in computational costs (compared with other DA techniques), and has a very general statistical formulation it is one of the most widely used approaches for tackling real-time forecasting problems [20].

The Ensemble Kalman filter (EnKF) is the main Ensemble-based DA method [21]. Based on the Kalman Filter (KF) [22], EnKF is an alternative for nonlinear, high-dimensional systems. EnKF essentially is a Monte Carlo Ensemble-based method, based on the representation of the probability density of the state by an ensemble of N model realizations. Each ensemble member is assumed to be a single sample out of a distribution of the true state [23]. In the first step, a Monte Carlo ensemble of the initial condition is generated to represent the uncertainty in the initial condition. After that, and in the same way that the KF, the EnKF propagates each ensemble using the state-space operator, this step is called forecast step. When observations are available, the EnKF uses them to update each forecast ensemble members and obtain the analysis ensemble, this step is named analysis step. The update is proportional to the differences between the observations and the model outputs, by a gain called Kalman Gain. **Figure 2** shows a graphic representation and a comparison between the KF and EnKF.

3. Forecasting PM10. And PM2.5. in the Aburrá Valley (Medellín, Colombia) via EnKF based data assimilation

Understanding local and regional atmospheric particulate matter transport patterns becomes a top priority for urban valleys in the northern Andes. This work will help establish accurate air quality forecasting systems for the Aburr'a Valley (and other similar areas) and improve decision-making. Chemical Transport Models (CTM) are valuable resources for understanding atmospheric pollutants' dynamics and have thus been widely used in air quality monitoring [8, 9].

Here we use simulations of the LOTOS-EUROS (LE) chemistry transport model (CTM) to investigate the atmospheric contaminant dynamics in the Aburr'a valley, which spans ten municipalities, including Medellín city. The *Sistema de Alerta Temprana del Valle de Aburrá* (SIATA), a ground-based sensor network with stations throughout the valley, can provide particulate material observations. A preliminary exercise is carried out to assimilate these findings into the simulations and assess the system's forecast capacity. Due to the various sources of uncertainty present, this implementation poses a challenge from a scientific standpoint. The topography and scale of the valley and the physical conditions of the area of interest necessitate an extra effort to perform a regional high-resolution model simulation.

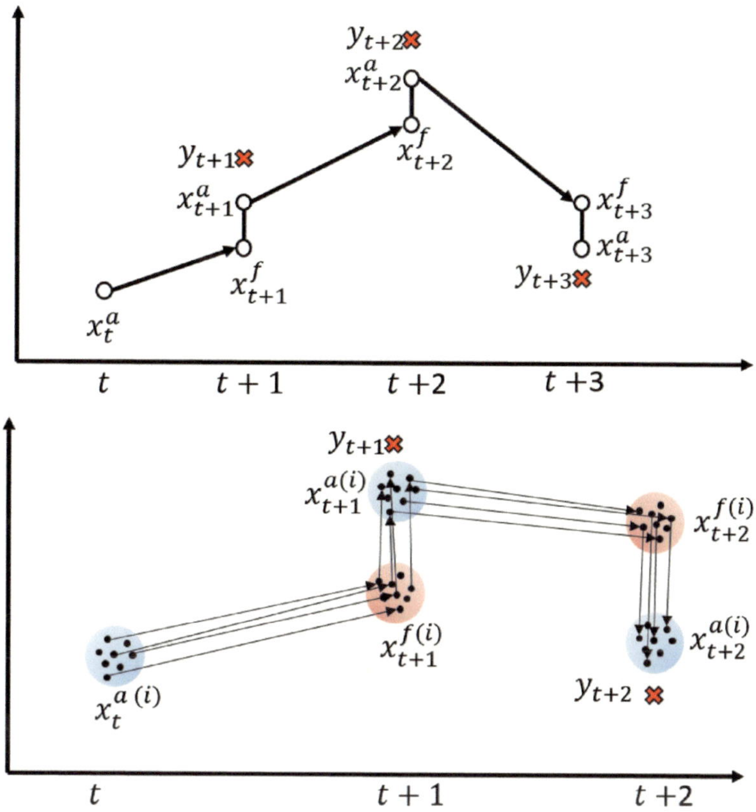

Figure 2.
Representation of Kalman filter (upper) and ensemble Kalman filter (lower).

Model inputs (emission inventory and meteorology) are not readily available with the desired resolution and accuracy, which adds to the experiment's uncertainty.

3.1 Material and methods

A data assimilation method for the LOTOS-EUROS chemical transport model has been introduced to boost the PM_{10} and $PM_{2.5}$ forecasts. The system uses an Ensemble Kalman filter with covariance localization, which is based on the specification of emissions uncertainties. The data was gathered from a surface network for the months of March and April 2016, during one of the region's worst air quality crises in recent memory. The SIATA is spread around the five most populous municipalities in the Aburrá Valley, with the bulk of the measuring stations in Medellín. **Figure 3** represents the distribution of observation sites.

Measurements for one station for each species (represented with a star in **Figure 3**) were used for validation, taking two stations with a considerable distance between them to obtain a acceptable spatial representation.

In a first series of experiments, the spatial length scale of the covariance localization and the temporal length scale of the stochastic model for the emission uncertainty were calibrated to optimize the assimilation system. The calibrated system was then used in a series of assimilation experiments. The summarized experimental setup is presented in the **Figure 4**.

Simulations were conducted with the LE model, adopting a nested domain configuration as depicted in **Figure 5** and detailed in **Table 1**. The data sets used in the model are summarized in **Table 2**.

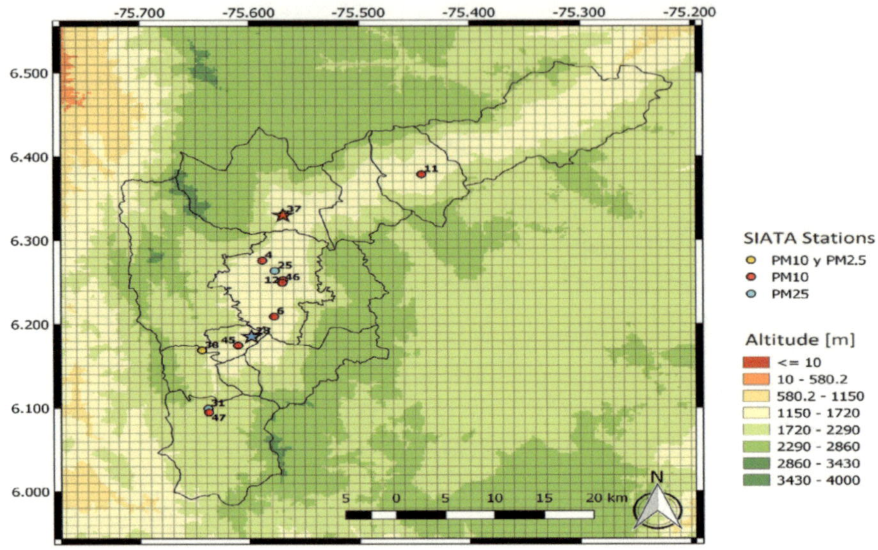

Figure 3.
SIATA sensor network for PM_{10} and $PM_{2.5}$. The stars represent observation points for validation and the circles represent observations points for assimilation. Taken from [24].

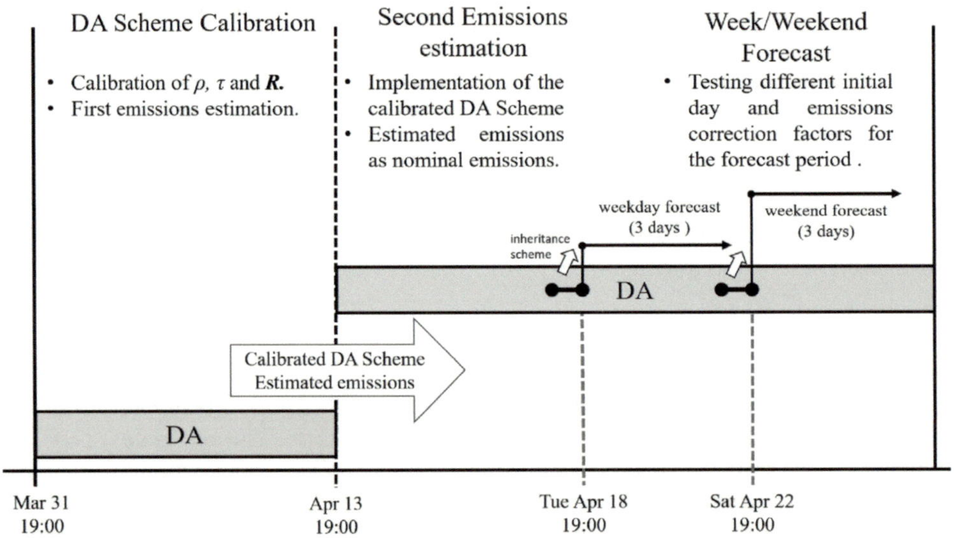

Figure 4.
Graphic representation of the experimental setup. Taken from [24].

3.2 Results

Estimated PM_{10} emissions and EDGAR nominal emissions are shown in **Figure 6**. The emissions hot-spots occur in rural zones with limited human activity in the EDGAR database. The estimated emissions attempt to remedy this behavior by projecting the most of the pollution into the metropolitan region of the valley (**Figure 6**).

The assimilated PM_{10} concentration match closely those measurements at the Universidad San Buenaventura (center of the valley) from April at 19:00 UTC-5 through April 25 at 11:00 UTC-5 (see **Figure 7**). The peak around 18:00 (and usually all day up to that hour) may be unreliable, which may be because of EDGAR's

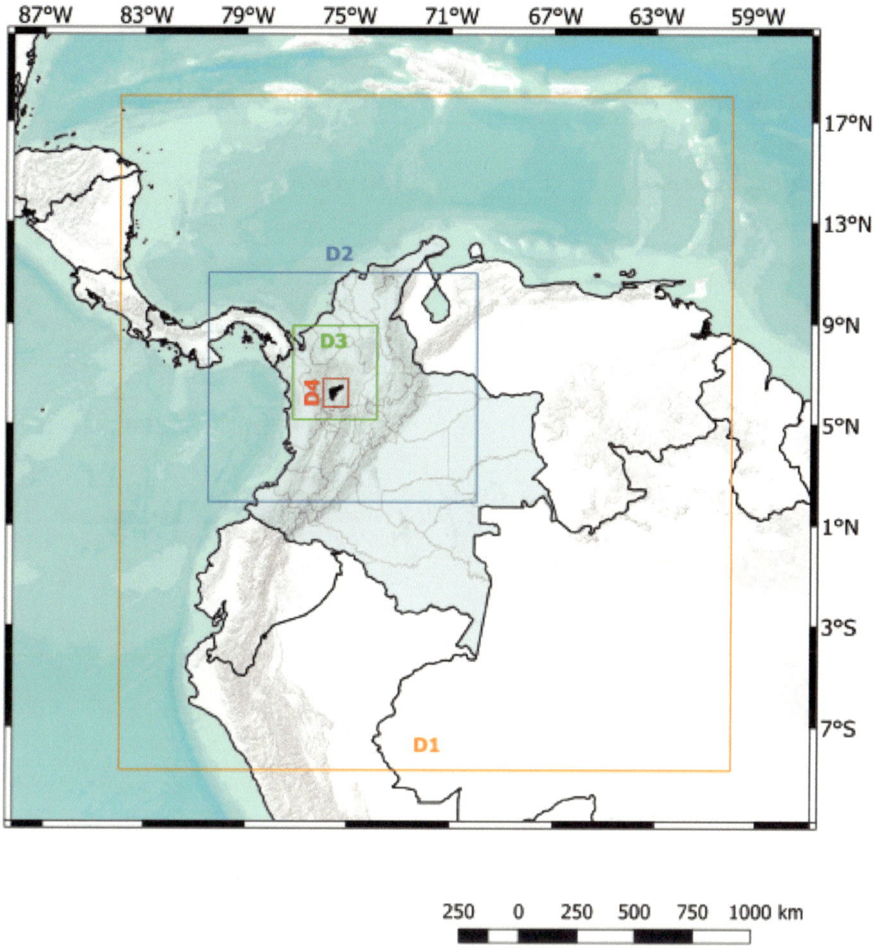

Figure 5.
Four nested domains for metropolitan area of Aburrá Valley assesment. Taken from [24].

Domain	Longitude	Latitude	Cell size
D1	84°W-60°W	8.5°S-18°N	0.27° × 0.27°
D2	80.5°W-70°W	2°N-11°N	0.09° × 0.09°
D3	77.2°W-73.9°W	5.2°N-8.9°N	0.03° × 0.03°
D4	76°W-75°W	5.7°N-6.8°N	0.01° × 0.01°

Table 1.
Nested domain specifications.

temporal emissions factors. Additionally, concentrations can be increased by the meteorological fields. Note that the daily cycle for the assimilated model remains closer to the observations than the model without assimilation.

Figure 8 shows a similar comparison for the $PM_{2.5}$ station. The model in a free run tends to over estimate the $PM_{2.5}$ concentrations (see peaks in 15 April at 23:00 UTC-5, 24 April at 22:00 and 25 April at 23:00 UTC-5). The results of the assimilation process offer a better average estimation. The daily cycle of $PM_{2.5}$ within the Aburrá valley is related to the industrial and mobile sources emissions profile and the meteorological conditions inside the valley.

Period	**31-March-2016 to 25-April-2016**
Metereology	ECMWF; Temp.res: 3 h; spat.res: 0.07° × 0.07°
Initial and boundary	LOTOS-EUROS (D3). Temp.res: 1 h.
conditions	Spat.Res: 0.03° × 0.03°
Anthropogenic emissions	EDGAR v4.2. Spat.res:10 km × 10 km
Biogenic emissions	MEGAN Spat.res:10 km × 10 km
Fire emissions	MACC/CAMS GFAS Spat.res:10 km × 10 km
Landuse	GLC2000. Spat.res:1 km × 1 km
Orography	GMTED2010. Spat.res: 0.002° × 0.002°

Table 2.
Data set used in the D4 domain.

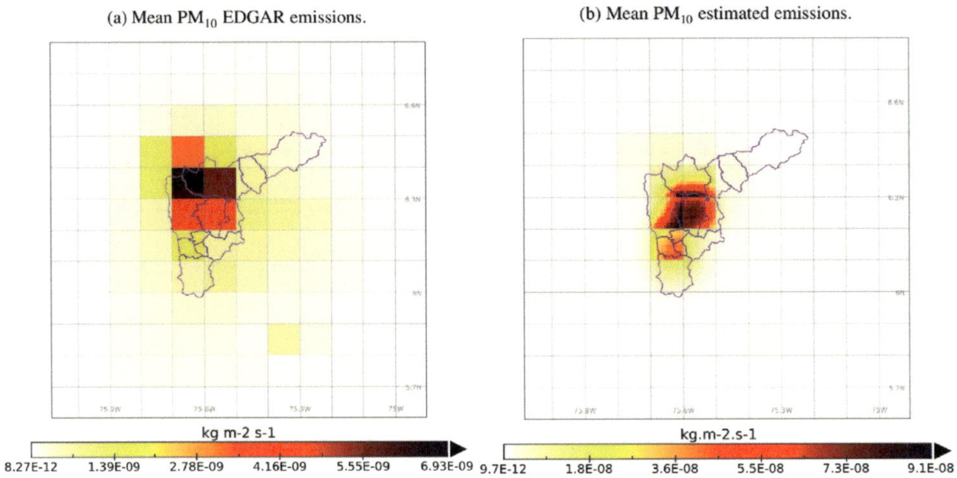

Figure 6.
Comparison between EDGAR PM_{10} and estimated PM_{10} emissions. Taken from [24].

3.3 Conclusions

Poor air quality is a current environmental problem in several Colombian cities. To be prepared for air quality degradation requires accurate and reliable data for decision-making in South America. This study shows that the LOTOS-EUROS model can function in areas with more complex topography, such as the Abura Valley, and encourages the development of fine-tuned weather forecasting systems to support the target. The use of regional, ground-based pollutant data from the SIATA sensor network, in the assimilation of the LOTOS-EUROS model, enhanced the PM_{10} and $PM_{2.5}$ representation.

4. Urban air quality modeling using low-cost sensor network and data assimilation in the Aburrá Valley, Colombia

Public air quality monitoring networks frequently consist of fixed measuring stations equipped with expensive sensors and maintained under strict operational and calibration regimes. Because of the high costs of setting up and maintaining

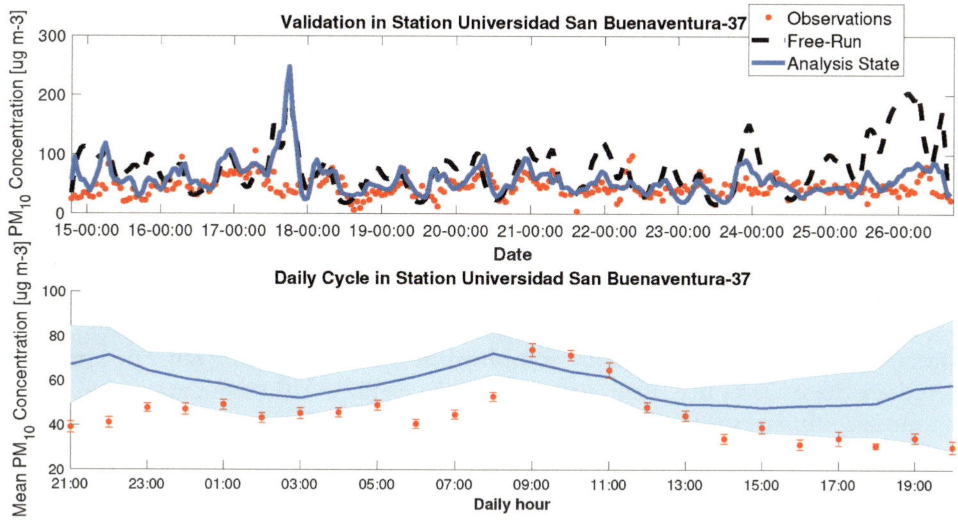

Figure 7.
PM_{10} validation for the second DA iteration. Estimated emissions were used as nominal emissions, the estimated observation error covariance is used in the assimilation step. Red points are observations, solid black line is the free run model and the solid blue line is the analysis step for the assimilated model. The diurnal cycles were obtained from 13 samples for each hour. The shadows and the bars represent the standard deviation of the 13 samples. The time axis corresponds with the local time zone UTC-5. Taken from [24].

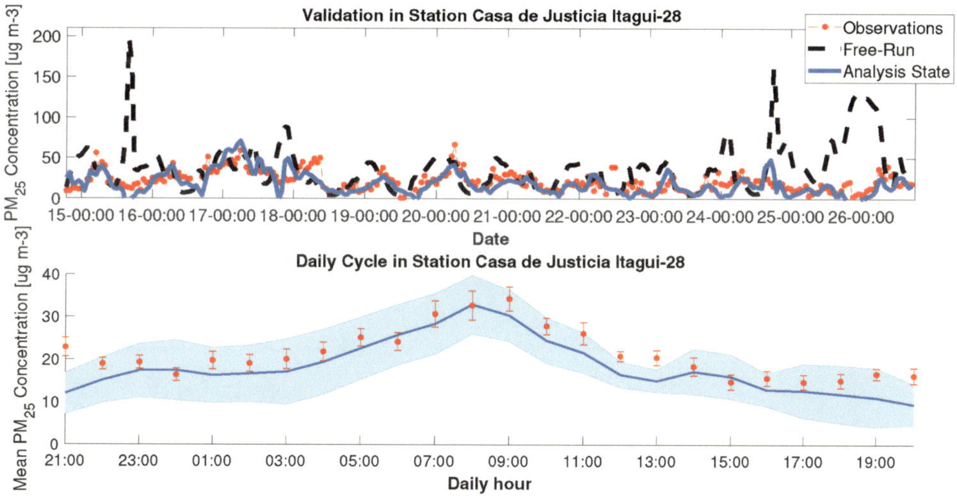

Figure 8.
$PM_{2.5}$ validation for the second DA iteration. Estimated emissions were used as nominal emissions, the estimated observation error covariance was used in the assimilation step. Red points are observations, solid black line the free run model and solid blue line the analysis step for the assimilated model. The diurnal cycles were obtained from 13 samples for each hour. The shadows and the bars represent the standard deviation of the 13 samples. The time axis corresponds with the local time zone UTC-5. Taken from [24].

such stations, not all cities in developing countries can afford monitoring networks with sufficient spatial coverage [14]. Even in developed cities, official air quality monitoring networks do not always provide information at the spatial and temporal resolution required to assess the impact of pollution sources on health, [15], due to the equipment's high cost. This has prompted the development and improvement of low-cost PM measurement systems and programs. According to [17], a small number of studies evaluating newer generations of low-cost PM2.5 sensors have found that the most widely used low-cost sensors achieve high accuracy when compared to standard monitoring stations (R^2 values ranging from 0.93 to 0.95). The data collected by these sensors can be used to supplement that collected by traditional systems, increasing data resolution and allowing studies of human exposure [15, 18].

Using techniques like data fusion or data assimilation to integrate observations from dense networks of low-cost sensors into mathematical models allows for a spatially continuous representation of concentration fields with significantly reduced bias citeLahoz2014. By spatially interpolating between monitoring locations and constraining the model with observations, these techniques add value to the sensor observations while also adding value to the model [17, 18, 25]. Both sources of information can thus be combined in a mathematically objective manner to reduce the uncertainty inherent in both sources [12]. Although data assimilation is a more complex family of methods than data fusion or interpolation techniques, it is the most versatile and robust of these approaches. The goal of evaluating the data from the low-cost sensor network as an alternative to monitoring PM2.5 concentrations in developing countries is to see if it is viable.

4.1 Material and methods

The SIATA project operates the official high-end air quality monitoring network (henceforth *official network*, and a hyper-dense, low-cost air quality network developed within the Citizen Scientist program (henceforth *low-cost network*).

The low-cost network was created with the aim of engaging the community in issues surrounding air quality, and as an extension of the official network. The low-cost network consists of 255 real-time $PM_{2.5}$ (**Figure 9**, panel b).The measuring equipment was developed by SIATA based on the well-known low-cost Shinyei PPD42NS, NOVA SDS011, and Bjhike HK-A5 sensors [27]. Each low-cost sensor is calibrated individually against BAM-1020 measurements [27]. The calibration process showed the measurements of 91% of the low-cost sensors with correlation values above 0.6 against the official measurements, and 67% with values above 0.8.

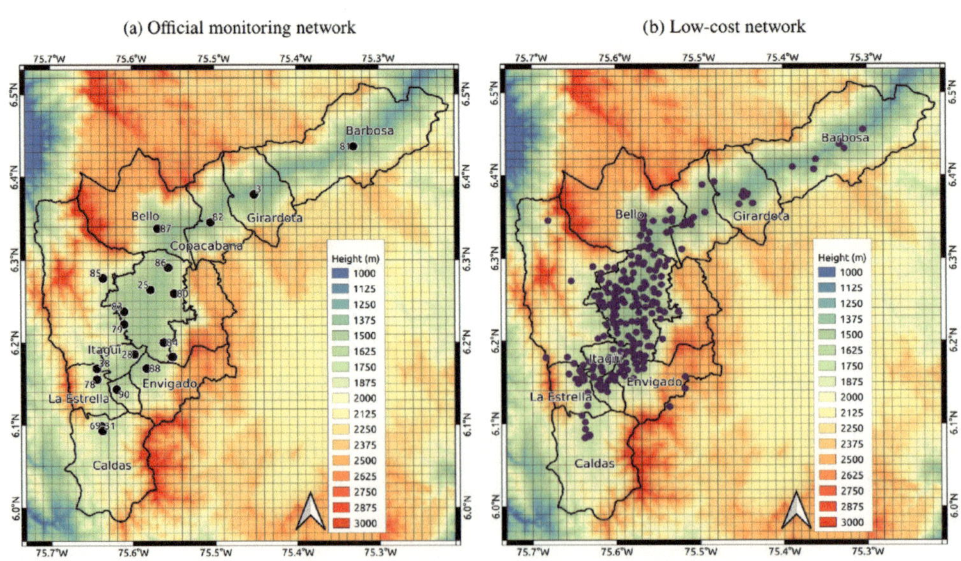

Figure 9.
Spatial distribution of the hyper-dense low-cost network citizen scientist and official monitoring air-quality network for $PM_{2.5}$. The gray raster represent the LOTOS-EUROS model grid. Taken from [26].

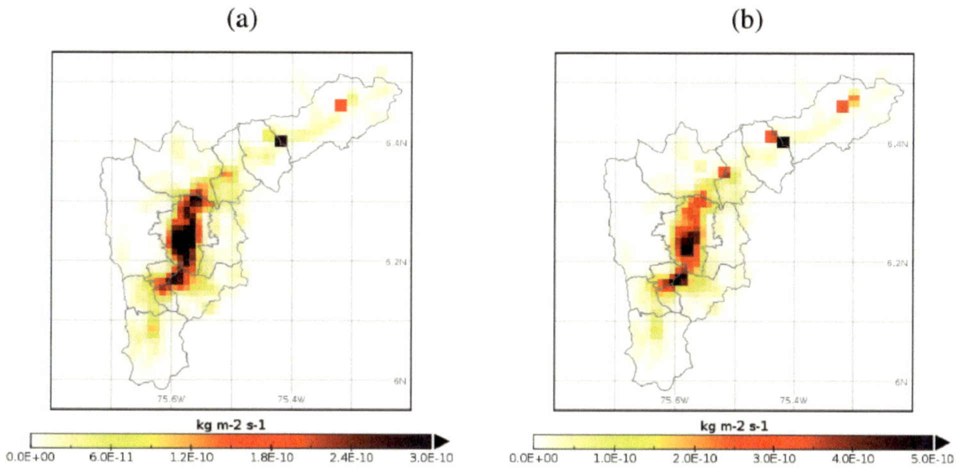

Figure 10.
Local particulate matter emission inventories for the Aburrá Valley: (a) $PM_{2.5}$, and (b) PM_{10}. The values correspond with the estimated annual emissions. Taken from [26].

The median of the root mean square error showed a value of 6.2 $\mu g/m^3$, with a tendency to decrease for higher concentrations [27]. The low-cost network thus represents satisfactorily the dynamics of $PM_{2.5}$ concentrations in the Valley's atmosphere.

An anthropogenic urban emissions inventory for 2016 specific to Medellín and the other nine municipalities of the Aburrá Valley was used for the simulations on the D4 domain. The construction of the inventory followed a bottom-up methodology, combining activity data (traffic intensities, industrial production) with emission factors. Only traffic and industrial point sources were considered, without accounting for neither household nor commercial emissions [28].

The emission inventory was disaggregated over the Aburrá Valley (76°W-75°W and 5.7°N-6.8°N) at a resolution of $0.01^o \times 0.01^o$ (approximately 1 km × 1 km), using a method based on road density as in [29]. The road network map was obtained from the OpenStreetMap database [30], and simplified by removing segments classified as residential, as recommended in [31, 32]. The simplification of the road network can reduce errors in the spatial disaggregation since residential roads correspond to a high portion of the road network length but carry a low percentage of total vehicular traffic. The point-source emissions were distributed on the grid using their known location [28]. **Figure 10** shows the resulting emissions maps for $PM_{2.5}$ and PM_{10}.

Two sets of low-cost sensors data were assembled: The first one included 255 sensors from the low-cost network that had a station from the official network within a 2-km radius. The second, higher quality one consisted of a subset of the previous set, including only those sensors whose data showed an R value equal or greater than 0.8 when evaluated against the official network.

We performed four different LOTOS-EUROS simulations:

1. a LOTOS-EUROS model simulation without data assimilation (henceforth *LE*);

2. a simulation with assimilation of data (observations) from the 14 stations of the official network (henceforth *LE-official*;

3. a simulation with assimilation of the data from the entire low-cost network (henceforth *LE-lowcost*)

4. a simulation with assimilation only of high-quality data from the low-cost network (henceforth *LE-lowcost-HQ*).

4.2 Results

The concentration fields were evaluated using seven of the official monitoring stations (*validation stations*. **Figure 11** shows the temporal series for the simulated and observed $PM_{2.5}$ concentrations at four of the validation stations. The four selected stations represent downtown Medellín (station 25), residential areas (station 86), areas with high vehicular flow (station 88), and a peri-urban area in the outskirts of the city (station 85). Those stations summarize the behavior of all seven validation stations. The LE simulation consistently underestimated the concentrations observed at stations 85 and 88. At stations 25 and 86, the LE simulation results were close in magnitude between February 24 and March 3 and March 10 to March 15; between March 3 and March 10, the model presented values much lower than those observed. The day-to-day variability was reduced for this same period, as seen in stations 85 and 86. This inconsistent behavior suggests a poor representation of the meteorological dynamics that govern the dispersion and accumulation of $PM_{2.5}$ within the valley. Simulations using data assimilation showed noisier behaviors than the LE simulation. This process is commonly observed when applying the EnKF and obeys the stochastic nature and the handling of uncertainty inherent to the method [21]. However, those simulations managed to correct the large discrepancies present

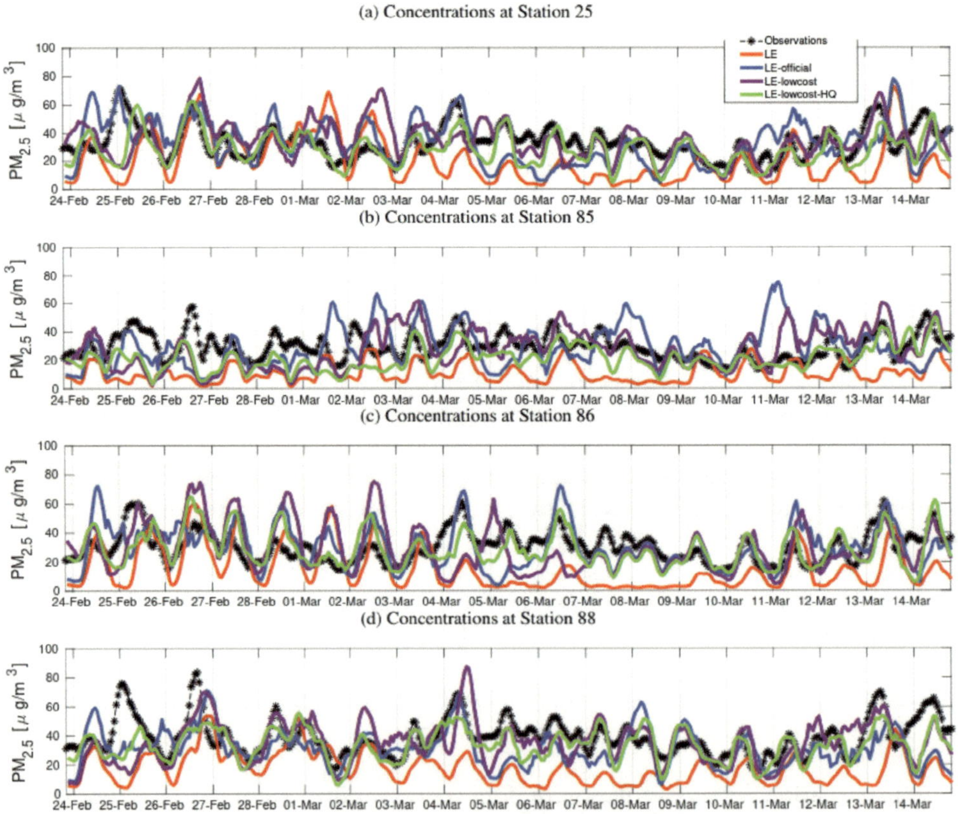

Figure 11.
Temporal series of $PM_{2.5}$ concentrations from selected validation stations of the official network, LOTOS-EUROS without assimilation, LE-official, LE-lowcost and LE-lowcost-HQ. Time stamps are valid for local time (UTC-5). A spin-up of 5 previous days was taken for each simulation. Taken from [26].

in the LE simulation. Both LE-official, LE-lowcost, and LE-lowcost-HQ represented more accurately the day-to-day variability of the observations than LE. In general terms, there was no evidence of a sizeable and persistent difference among the simulations with data assimilation throughout the entire period. Nevertheless, the LE-lowcost-HQ simulation reproduced with greater accuracy the concentrations observed in different periods, such as between February 26 and March 4 in station 25, between March 9 and March 14 in stations 85 and 86.

Figure 12 shows the diurnal cycles during the simulation period in the four selected validations stations. The diurnal cycle of the LE simulation differed from the observations in both magnitude and temporal behavior. The highest concentration peak that appears around 09:00 in all the stations is mainly due to traffic dynamics. In stations 25 and 88, the LE morning peak corresponded in time but not in magnitude with the observations; in stations 85 and 86, said peak appeared later in the simulations than in the observations. This time lag suggests a poor spatial representation of mobile emissions by the emissions inventory; or a deficiency it the wind fields in reproducing the valley dynamics, showing a late transport of the particulate material to these areas. The LE simulation did not capture the evening peak shown by the observations around 21:00 hours. The simulations using data assimilation presented diurnal cycles closer to the observations than did the LE simulation. The LE-official simulation captured the time and magnitude of the morning peak in stations 85 and 86. In station 88, LE-official corrected the time lag in the morning peak seen in LE, and improved the estimated magnitudes albeit still falling short of the observed values. A different behavior was seen for station 25,

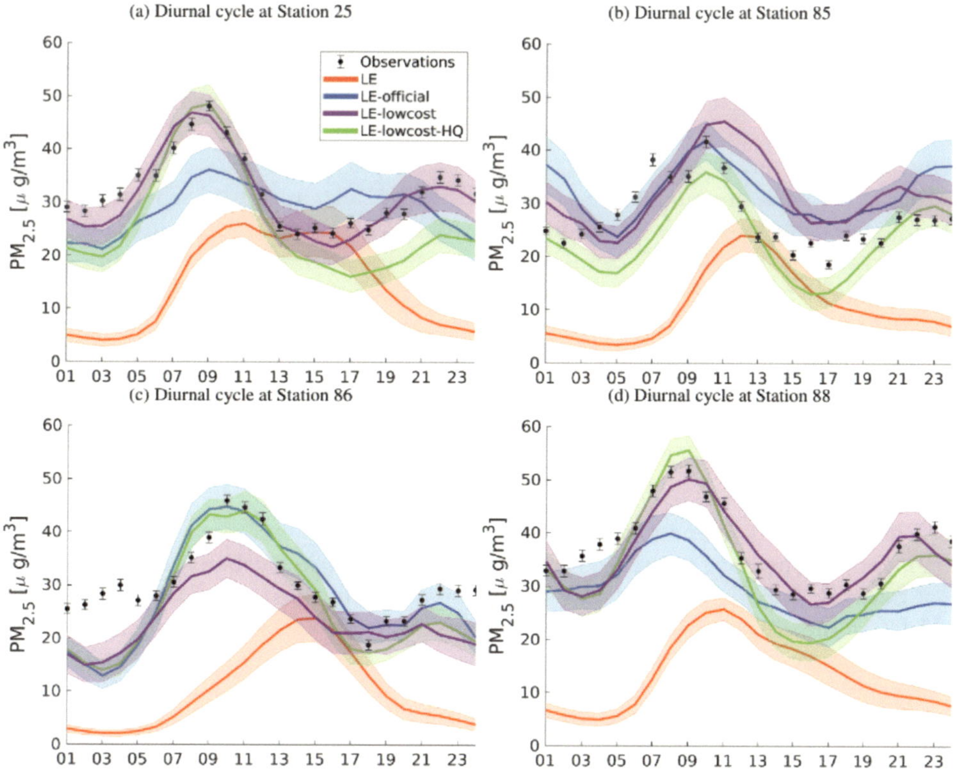

Figure 12.
Diurnal cycle of $PM_{2.5}$ concentrations from selection stations of the official network, LOTOS-EUROS without assimilation, LE-official, LE-lowcost and LE-lowcost-HQ. The bars and the shadows represent the standard deviation over the simulation period. The time stamps are valid for local time (UTC-5). Taken from [26].

	MFB	RMSE	*R*
LE	−0.65	27.38	0.42
LE-official	−0.07	20.69	0.64
LE-lowcost	0.08	18.39	0.76
LE-lowcost-HQ	0.06	17.46	0.82

Table 3.
Mean fractional bias, root mean square error and Pearson correlation coefficient for simulated $PM_{2.5}$. Values are averaged over all the validation stations for the simulation period.

where LE-official had low diurnal variability, with a slight underestimation in the morning, and an overestimation in the afternoon. The LE-lowcost and LE-lowcost-HQ simulations results resembled closely the diurnal behavior of the observations, especially the temporal component. In all the stations, both the morning and the evening peaks matched the observations. The observed concentrations for stations 25 and 88 fell inside the standard deviation range for the LE-lowcost simulation; the same simulation overestimated the concentrations between 11:00 and 19:00 for station 85, and underestimated the concentrations between 01:00 and 13:00 for station 86. The LE-lowcost-HQ simulation results were overall the closest to observations.

The averaged evaluation statistics among all the validation station are shown in **Table 3**. The simulation results without data assimilation (LE) underestimated the observed concentrations in all the validation stations. This was also seen in previous related works [24, 33]. The RMSE value reflected a low correspondence between the observed and simulated concentrations when using the model without data assimilation. The correlation coefficient was low, meaning that the model was not able to capture the variations in diurnal and day-to-day concentrations. In contrast, the three simulations using data assimilation had MFB values close to 0, without a significant difference among them. The data assimilation was thus effective in reducing between the model and reality. The RMSE also improved when using data assimilation, decreasing by 24.4% in the LE-official, 32.8% in the LE-lowcost, and 36.2% in the LE-lowcost-HQ simulations relative to the RMSE of the LE simulation. The *R* values were all above the criteria of good performance according with [34] **Table 2**, and based in [35, 36]. Assimilation of either data set from the low-cost network resulted in improved error statistics when compared to the LE-official simulation.

4.3 Conclusions

We present a data assimilation application of a hyper-dense low-cost PM network and the chemical transport model LOTOS-EUROS in a urban setting. The low-cost network provided high quality data comparable to those provided by the official monitoring network. The performance of the model with assimilation of the spatially-dense data from the low-cost network improved both in terms of its representation of the observed dynamics, as well as in its forecast capabilities, highlighting its value as an air-quality management tool. Our results support the idea than with the current advances in the low-cost sensors, it is possible to use low-cost networks and data assimilation to model and predict air quality in urban areas.

Jointly with previous work [15, 18, 25, 37–39], our results can support and motivate the development of future low-cost networks and their integration in data fusion applications. According to the literature, North America, Europe, and China concentrate most of the current low-cost implementations, with experimental, citizen, and data dissemination purposes [14, 40]. In developing countries, a low-cost network, together with a CTM and data assimilation can provide a valuable first approach to monitoring PM without the high cost of an official air quality network.

Although one of the main advantages of a low-cost networks is the possibility of implemented hyper-dense networks with relative low costs, it is recommended to prioritize in the quality of the data (sensor quality, calibration, maintenance) and the study of optimal localization. High quality and the correct number and localization of sensors improve the data assimilation process and minimizes operational and computational costs.

Acknowledgements

The authors acknowledge the supercomputing resources made available by the Centro de Computación Científica Apolo at Universidad EAFIT (http://www.eafit.edu.co/apolo) to conduct this work.

Conflict of interest

The authors declare no conflict of interest.

Author details

Santiago Lopez-Restrepo[1,2,3]*, Andrés Yarce Botero[1,2,3], Olga Lucia Quintero[2], Nicolás Pinel[4], Jhon Edinson Hinestroza[1], Elias David Niño-Ruiz[5], Jimmy Anderson Flórez[6], Angela Maíra Rendón[7], Monica Lucia Alvarez-Laínez[8], Andres Felipe Zapata-Gonzalez[8], Jose Fernando Duque Trujillo[9], Elena Montilla[10], Andres Pareja[11], Jean Paul Delgado[12], Jose Ignacio Marulanda Bernal[13], Bibiana Boada[1], Juan Ernesto Soto[6], Sara Lorduy[6], Jaime Andres Betancur[12], Arjo Segers[14] and Arnold Heemink[2]

1 Mathematical Modelling Research Group, Universidad EAFIT, Medellín, Colombia

2 Department of Applied Mathematics, Delft University of Technology, The Netherlands

3 SimpleSpace, Medelliín, Colombia

4 Department of Biological Sciences, Research Group on Biodiversity, Evolution and Conservation at Universidad EAFIT, Medellín, Colombia

5 Applied Math and Computer Science Laboratory, Department of Computer Science, Universidad del Norte, Colombia

6 Comando Aereo de Combate N 5 CACOM 5, Centro Tecnologico Aeroespacial para la Defensa CETAD, Fuerza Aérea Colombiana, Rionegro, Colombia

7 Grupo GIGA, Universidad de Antioquia, Medellín, Colombia

8 GRID Nano-Fiber Group, Universidad EAFIT, Medellín, Colombia

9 Environmental Magnetism, Universidad EAFIT, Medellín, Colombia

10 Applied Optics Research Group, Universidad EAFIT, Medellín, Colombia

11 Unidad de Toxicidad in vitro, Universidad CES, Sabaneta, Colombia

12 Grupo, genérica, regeneración y cáncer, Universidad de Antioquia, Medellín, Colombia

13 GEMA, Universidad EAFIT, Medellín, Colombia

14 Department of Climate, Air and Sustainability, TNO, The Netherlands

*Address all correspondence to: slopezr2@eafit.edu.co

References

[1] J. Green and S. Sánchez, "Air Quality in Latin America: An Overview," tech. rep., Clean air Institute, Washington D. C., USA, 2012.

[2] C. Borrego, M. Coutinho, a. M. Costa, J. Ginja, C. Ribeiro, a. Monteiro, I. Ribeiro, J. Valente, J. H. Amorim, H. Martins, D. Lopes, and a. I. Miranda, "Challenges for a New Air Quality Directive: The role of monitoring and modeling techniques," *Urban Climate*, vol. 14, pp. 328–341, 2015.

[3] H. Akimoto, "Global air quality and pollution," *Science*, vol. 302, no. 5651, pp. 1716–1719, 2003.

[4] B. Gurjar, T. Butler, M. Lawrence, and J. Lelieveld, "Evaluation of emissions and air quality in megacities," *Atmospheric Environment*, vol. 42, no. 7, pp. 1593–1606, 2008.

[5] M. L. Bell, L. A. Cifuentes, D. L. Davis, E. Cushing, A. G. Telles, and N. Gouveia, "Environmental health indicators and a case study of air pollution in latin american cities," *Environmental Research*, vol. 111, no. 1, pp. 57–66, 2011.

[6] J. F. Sallis, F. Bull, R. Burdett, L. D. Frank, P. Griffiths, B. Giles-Corti, and M. Stevenson, "Use of science to guide city planning policy and practice: how to achieve healthy and sustainable future cities," *The Lancet*, vol. 388, no. 10062, pp. 2936–2947, 2016.

[7] J. F. Jiménez, *Altura de la Capa de Mezcla en un área urbana montañosa y tropical. Caso de estudio: Valle de Aburrá (Colombia)*. Doctoral thesis, Universidad de Antioquia, Medellín, 2016.

[8] P. Thunis, *A. Miranda*, J. M. Baldasano, N. Blond, J. Douros, A. Graff, S. Janssen, K. Juda-Rezler, N. Karvosenoja, G. Maffeis, A. Martilli, M. Rasoloharimahefa, E. Real, P. Viaene, M. Volta, and L. White, "Overview of current regional and local scale air quality modeling practices: Assessment and planning tools in the EU," *Environmental Science & Policy*, vol. 65, pp. 13–21, 2016.

[9] M. Lateb, R. Meroney, M. Yataghene, H. Fellouah, F. Saleh, and M. Boufadel, "On the use of numerical modelling for near-field pollutant dispersion in urban environments: A review," *Environmental Pollution*, vol. 208, pp. 271–283, 2016.

[10] M. Berardi, A. Andrisani, *L. Lopez*, and M. Vurro, "A new data assimilation technique based on ensemble Kalman filter and Brownian bridges: An application to Richards' equation," *Computer Physics Communications*, vol. 208, pp. 43–53, 2016.

[11] M. Van Loon, P. J. H. Builtjes, and a. J. Segers, "Data assimilation of ozone in the atmospheric transport chemistry model LOTOS," *Environmental Modelling and Software*, vol. 15, no. 6–7 SPEC. ISS, pp. 603–609, 2000.

[12] W. A. Lahoz and P. Schneider, "Data assimilation: Making sense of Earth Observation," *Frontiers in Environmental Science*, vol. 2, no. MAY, pp. 1–28, 2014.

[13] M. Bocquet, H. Elbern, H. Eskes, M. Hirtl, R. Aabkar, G. R. Carmichael, J. Flemming, A. Inness, M. Pagowski, J. L. Pérez Camaño, P. E. Saide, R. San Jose, M. Sofiev, J. Vira, A. Baklanov, C. Carnevale, G. Grell, and C. Seigneur, "Data assimilation in atmospheric chemistry models: Current status and future prospects for coupled chemistry meteorology models," *Atmospheric Chemistry and Physics*, vol. 15, pp. 5325–5358, may 2015.

[14] A. Kumar and B. R. Gurjar, "Low-Cost Sensors for Air Quality Monitoring

in Developing Countries -A Critical View," *Asian Journal of Water, Environment and Pollution*, vol. 16, no. 2, pp. 65–70, 2019.

[15] F. E. Ahangar, F. R. Freedman, and A. Venkatram, "Using low-cost air quality sensor networks to improve the spatial and temporal resolution of concentration maps," *International Journal of Environmental Research and Public Health*, vol. 16, no. 7, 2019.

[16] P. Kumar, L. Morawska, C. Martani, G. Biskos, M. Neophytou, S. Di Sabatino, *M. bell*, L. Norford, and R. Britter, "The rise of low-cost sensing for managing air pollution in cities," *Environment International*, vol. 75, pp. 199–205, 2015.

[17] H. Y. Liu, P. Schneider, R. Haugen, and M. Vogt, "Performance assessment of a low-cost PM 2.5 sensor for a near four-month period in Oslo, Norway," *Atmosphere*, vol. 10, no. 2, 2019.

[18] P. Schneider, N. Castell, M. Vogt, F. R. Dauge, W. A. Lahoz, and A. Bartonova, "Mapping urban air quality in near real-time using observations from low-cost sensors and model information," *Environment International*, vol. 106, no. June, pp. 234–247, 2017.

[19] N. Castell, F. R. Dauge, P. Schneider, M. Vogt, U. Lerner, B. Fishbain, D. Broday, and A. Bartonova, "Can commercial low-cost sensor platforms contribute to air quality monitoring and exposure estimates?," *Environment International*, vol. 99, pp. 293–302, 2017.

[20] G. Fu, F. Prata, H. Xiang Lin, A. Heemink, A. Segers, and S. Lu, "Data assimilation for volcanic ash plumes using a satellite observational operator: A case study on the 2010 Eyjafjallajökull volcanic eruption," *Atmospheric Chemistry and Physics*, vol. 17, no. 2, pp. 1187–1205, 2017.

[21] G. Evensen, "The Ensemble Kalman Filter: Theoretical formulation and practical implementation," *Ocean Dynamics*, vol. 53, no. 4, pp. 343–367, 2003.

[22] R. E. Kalman, "A new approach to linear filtering and prediction problems," *Journal of Basic Engineering*, vol. 82, no. 1, pp. 35–45, 1960.

[23] G. Fu, *Improving volcanic ash forecasts with ensemble-based data assimilation*. PhD thesis, Delf University of Technology, 2017.

[24] S. Lopez-Restrepo, A. Yarce, N. Pinel, O. L. Quintero, A. Segers, and A. W. Heemink, "Forecasting PM10 and PM2.5 in the Aburrá Valley (Medellín, Colombia) via EnKF based data assimilation," *Atmospheric Environment*, vol. 232, no. April, p. 117507, 2020.

[25] O. A. Popoola, D. Carruthers, C. Lad, V. B. Bright, M. I. Mead, M. E. Stettler, J. R. Saffell, and R. L. Jones, "Use of networks of low cost air quality sensors to quantify air quality in urban settings," *Atmospheric Environment*, vol. 194, no. February, pp. 58–70, 2018.

[26] S. Lopez-restrepo, A. Yarce, N. Pinel, O. Quintero, A. Segers, and A. W. Heemink, "Urban Air Quality Modeling Using Low-Cost Sensor Network and Data Assimilation in the Aburrá Valley, Colombia," *Atmosphere*, vol. 12, no. 91, pp. 1–19, 2021.

[27] C. D. Hoyos, L. Herrera-Mejía, N. Roldán-Henao, and A. Isaza, "Effects of fireworks on particulate matter concentration in a narrow valley: the case of the medellín metropolitan area," *Environmental Monitoring and Assessment,* vol. 192, p. 6, Dec 2019.

[28] UPB and AMVA, "Inventario de Emisiones Atmosféricas del Valle de Aburrá - actualización 2015," tech. rep., Universidad Pontificia Bolivariana - Grupo de Investigaciones Ambientales,

Area Metropolitana del Valle de Aburra, Medellín, 2017.

[29] M. Ossés de Eicker, R. Zah, R. Triviño, and H. Hurni, "Spatial accuracy of a simplified disaggregation method for traffic emissions applied in seven mid-sized Chilean cities," *Atmospheric Environment*, vol. 42, no. 7, pp. 1491–1502, 2008.

[30] M. Haklay and P. Weber, "Openstreetmap: User-generated street maps," *IEEE Pervasive Computing*, vol. 7, no. 4, pp. 12–18, 2008.

[31] D. Tuia, M. Ossés de Eicker, R. Zah, M. Osses, *E. Zarate*, and A. Clappier, "Evaluation of a simplified top-down model for the spatial assessment of hot traffic emissions in mid-sized cities," *Atmospheric Environment*, vol. 41, pp. 3658–3671, 2007.

[32] C. D. Gómez, C. M. González, M. Osses, and B. H. Aristizábal, "Spatial and temporal disaggregation of the on-road vehicle emission inventory in a medium-sized Andean city. Comparison of GIS-based top-down methodologies," *Atmospheric Environment*, vol. 179, no. February, pp. 142–155, 2018.

[33] J. J. Henao, J. F. Mejía, A. M. Rendón, and J. F. Salazar, "Sub-kilometer dispersion simulation of a CO tracer for an inter-Andean urban valley," *Atmospheric Pollution Research*, no. January, pp. 0–1, 2020.

[34] C. Mogollón-sotelo, L. Belalcazar, and S. Vidal, "A support vector machine model to forecast ground-level PM 2. 5 in a highly populated city with a complex terrain," *Air Quality, Atmosphere & Health*, 2020.

[35] EPA, "Meteorological Monitoring Guidance for Regulatory Modeling Applications," tech. rep., U.S. ENVIRONMENTAL PROTECTION AGENCY, 2000.

[36] J. W. Boylan and A. G. Russell, "Pm and light extinction model performance metrics, goals, and criteria for three-dimensional air quality models," *Atmospheric Environment*, vol. 40, no. 26, pp. 4946–4959, 2006. Special issue on Model Evaluation: Evaluation of Urban and Regional Eulerian Air Quality Models.

[37] S. J. Johnston, P. J. Basford, F. M. Bulot, M. Apetroaie-Cristea, N. H. Easton, C. Davenport, G. L. Foster, M. Loxham, A. K. Morris, and S. J. Cox, "City scale particulate matter monitoring using LoRaWAN based air quality IoT devices," *Sensors (Switzerland)*, vol. 19, no. 1, pp. 1–20, 2019.

[38] V. Isakov, S. Arunachalam, R. Baldauf, M. Breen, P. Deshmukh, A. Hawkins, S. Kimbrough, S. Krabbe, B. Naess, M. Serre, and A. Valencia, "Combining dispersion modeling and monitoring data for community-scale air quality characterization," *Atmosphere*, vol. 10, no. 10, 2019.

[39] S. Moltchanov, I. Levy, Y. Etzion, U. Lerner, D. M. Broday, and B. Fishbain, "On the feasibility of measuring urban air pollution by wireless distributed sensor networks," *Science of the Total Environment*, vol. 502, pp. 537–547, 2015.

[40] L. Morawska, P. K. Thai, X. Liu, A. Asumadu-Sakyi, G. Ayoko, A. Bartonova, A. Bedini, F. Chai, B. Christensen, M. Dunbabin, J. Gao, G. S. Hagler, R. Jayaratne, P. Kumar, A. K. Lau, P. K. Louie, M. Mazaheri, Z. Ning, N. Motta, B. Mullins, M. M. Rahman, Z. Ristovski, M. Shafiei, D. Tjondronegoro, D. Westerdahl, and R. Williams, "Applications of low-cost sensing technologies for air quality monitoring and exposure assessment: How far have they gone?," *Environment International*, vol. 116, no. April, pp. 286–299, 2018.

Chapter 8

A Synthesis of the Information Given by Temporal Data Series: The Representative Day

Tiziano Tirabassi and Daniela Buske

Abstract

The recording of air pollution concentration values involves the measurement of a large volume of data. Generally, automatic selectors and explicators are provided by statistics. The use of the Representative Day allows the compilation of large amounts of data in a compact format that will supply meaningful information on the whole data set. The Representative Day (RD) is a real day that best represents (in the meaning of the least squares technique) the set of daily trends of the considered time series. The Least Representative Day (LRD), on the contrary, it is a real day that worst represents (in the meaning of the least squares technique) the set of daily trends of the same time series. The identification of RD and LRD can prove to be a very important tool for identifying both anomalous and standard behaviors of pollutants within the selected period and establishing measures of prevention, limitation and control. Two application examples, in two different areas, are presented related to meteorological and SO_2 and O_3 concentration data sets.

Keywords: air pollution, daily trends, data set, temporal series, air pollution management, representative day

1. Introduction

In recent years, environmental management and a suitable development have assumed great importance [1, 2]. Air quality management and protection presuppose knowledge of the state of the environment. Such knowledge involves a properly cognitive and interpretative ability.

Local or regional air pollution control is usually achieved through air quality monitoring networks. These networks are a useful tool for the protection of human health and the environment, and allow both to evaluate the benefit of remediation actions and to prepare specific interventions in case of exceeding the threshold levels considered dangerous. For economic and managerial reasons, the number of measuring points in a network is limited and, especially if their arrangement has not been carefully studied, the detection units risk being unrepresentative of the entire territory that is to be monitored. In this regard, the mathematical models that simulate the transport and diffusion of pollutants in the atmosphere constitute a valid integration to the measurements, allowing to have estimates of concentrations over the entire territory for which it is interesting to know the evolution of concentrations. Once the good quality of the answers provided by a model has been ascertained, it allows us to trace the contribution of the different sources to the overall pollution, and therefore to correctly address any actions to limit emissions. Furthermore, only with the models is it possible to make forecasts or simulate concentration scenarios in connection with emission limitation policies as part of the preparation of recovery plans. The analyzers network, together with the inventory of emissions sources, is of fundamental importance for the construction of the cognitive framework, but not the interpretive one. In reality, air quality control requires an instrument interpretative capable of extrapolating in space and time the values measured in the of the analyzers, while the improvement of the air quality can be obtained only with plans that reduce emissions and then with instruments (such as the air pollution mathematical model) capable of linking the cause (source) of pollution with the effect (the concentration of the pollutant) [3].

The introduction of mathematical modeling produces a qualitative leap in the management of atmospheric pollution compared to that possible through measurements alone, because the models allow functions that are not accessible to the latter [4].

Mathematical models are capable of:

- describe and interpret the experimental data;
- control in real time and/or analyze air quality;
- manage accidental releases and assess risk areas;
- identify pollution sources;
- evaluate the contribution of a single source to the pollution load;
- manage and plan the territory.

For the above consideration, they turn out to be a technical instrument indispensable for environmental management. In fact, it is of considerable importance the description of the processes that govern the transport and diffusion of pollutants. They are generally represented by meteorological preprocessors able to describe the transport operated by the wind and the variables useful for the different models to calculate the diffusion of pollutants [3]. They are extremely useful when local phenomena such as land-sea and/or upstream-downstream breezes have to be described.

While the complete dataset of the measured data is useful (also from a legal point of view) for an environmental control and the data can also be considered as alarm signals for particularly dangerous situations, on the contrary, for the use of the models it is generally necessary to limit the number of the data to simulate. To this regard it is useful to have techniques that allow to feel the data or identify subsets. Generally, in order to summarize information automatic selectors and explicators are used, many of which are provided by statistics as probability density function, mean standard deviation, median, quantiles. Moreover, time-series trends, spectral analysis, principal component analysis and cluster analysis are usually used. In support of good decision-making, the use of statistic has become widespread in air pollution assessment. In addition, collected data are used to define specific typical periods that may be of particular interest in a study of pollutant diffusion. We can also mention, for example, a typical working day, a typical holiday, a typical seasonal day, etc. The purpose of a such typifying is that of outlining characteristic scenarios for a given period under investigation. Afterwards, mathematical models make it possible to attempt simulation of a typical period trend.

It is very useful to identify from the set of data periods that can represent the peculiarity of the area under control and at the same time extraordinary events, in particular if they represent critical situations from the point of view of environmental pollution, so as to be able to study which meteorological situations and emissions conditions that cause them.

To meet these needs, it is proposed to use a methodology capable of identifying the most representative day of a series of daily data so that the simulation of that day allows to understand the diffusion and pollution situation typical of the study area. At the same time, the methodology should also identify the most anomalous day or the most anomalous days that correspond to situations of major pollution on the ground. Naturally, the methodology must identify real days, not fictitious as the typical day [5] constructed with time series of data composed of the average values of the concentration averaged (at the same time) over the whole dataset.

Identifying real days is also important because only in this way can be identified the meteorological and emission data that characterized that day and the measured ground level concentrations. This methodology will be presented below.

2. The representative day methodology

What we want to select in an annual, seasonal or monthly dataset of daily time series is the one that best represents the set of those stored. This can be achieved with the RD technique.

What we call RD technique is a daily data set, actually recorded at a field station, which is characterized by the minimal differences with respect to all the daily measurements series of that station's temporal series: that is, the daily series whose sum of the squared differences over one day turns out to be the smallest compared to all the other days of the period under consideration [6].

That is, if A is the matrix:

$$A_{ij} = \sum_{k=1}^{24} \left(c_{ki} - c_{kj}\right)^2, \quad i,j = 1, 2, \dots, N \tag{1}$$

where N is the number of days in the time period for which the representative day is calculated, and c_{ki} is the pollutant concentration of the $i-th$ day at the $k-th$ time period.

We adopt A_i to indicate the sum of all the squared residuals of the $i-th$ line (or column, A_{ij} being a symmetrical matrix with all zeros in the diagonal):

$$A_i = \left(\sum_{j=1}^{N} A_{ij}\right) \tag{2}$$

The RD is the one with the lowest sum, i.e. the $i-th$ day where A_i is the smallest of the quantities obtained. The purpose of such typifying is that of outlining characteristic scenarios for a given period under investigation. That is, identify the day of the data series that is closest (in the meaning of the least squares technique) to all the time series included in the data set under examination.

2.1 The least representative day

The shown above approach also allows the identification of the "least representative day" (LRD), i.e. the daily series that maximizes the mean sum of squared residuals. The LRD identifies an anomalous situation of pollutant dispersion.

As LRD is a real day, it allows us to identify the meteorological characteristics and air pollution emissions related to that day and, therefore, giving us the possibility to study the phenomena and conditions that contributed to the realization of that air pollution diffusion and that distribution situation of air pollution concentrations on the ground.

Of course, by eliminating the data of that day from the original series and repeating the procedure on the remaining data, it is possible to highlight the second less representative day. Proceeding in the same way, the third, fourth, etc., LRD can be highlighted.

2.2 Results normalization

To compare the degree of representativity of the most or least representative days with that obtained for other time periods and/or at other measurement stations or stations in different areas, a normalization is required in order to make the day independent of the length of the measurement series, sampling period and characteristics of the area under study.

The "representativity" of an RD can be quantified by introducing the adimensional index DI:

$$DI = \frac{\sum_{i=1}^{N}\sum_{k=1}^{24}(\Gamma_k - c_{ik})^2}{\sum_{i=1}^{N}\sum_{k=1}^{24}(\bar{c}_k - c_{ik})^2} \qquad (3)$$

where $\bar{c}_k$ is the time mean concentration of the RD at the $k - th$ data, and Γ_k is the time mean concentration at the $k - th$ data calculated over the period under consideration, i.e.:

$$\bar{c}_k = \frac{1}{N}\sum_{i=1}^{N} c_{ik}, \quad k = 1, 2, \dots, 24 \qquad (4)$$

N is the number of days making up the time interval and c_{ik} is the time mean concentration of the pollutant of the $i - th$ day at the $k - th$ data of the daily sequence.

DI is an adimensional quantity greater than or equal to 1, which is closer to unit the more the RD is representative of the period under consideration.

The LRD can also be normalized in the same way: one simply substitutes in Eq. (3) the time mean concentration of the RD (Γ_k) with the least representative ones. In this case, the value of DI will always be greater than one, providing an indication of the low degree of representativity of the day obtained; the more DI is greater than 1, the more the LRD is "anomalous", compared to the trend of RD.

The normalization procedure described above it is independent of the size of the measured concentrations and of the number of days (N) included in the time period considered.

2.3 A fictional day: the typical day

We introduce the typical day (TD), that could be defined as a "fictional" day, whose concentrations (the time concentrations that form the daily sequence) are given by the concentration means, calculated, time by time over all the days of the period of study. The daily sequences can belong to periods of a month, a season, a year, or grouping of particular days that share the features one wishes to study. This form of data representation is widely used in Italy [5].

The TD, with the notation already introduced, can be mathematically expressed by:

$$\overline{C}_k = \frac{1}{N}\sum_{i=1}^{N} c_{ik}, \quad k = 1, 2, \ldots, 24 \tag{5}$$

However, since the TD is not a real day, this form of evaluation provides a presentation which cannot take account of the variations characterizing the actual behavior of the quantity under examination. Furthermore, since it is not a real day, it cannot be associated with any meteorological or emission parameters and therefore is of little interest for a possible air pollution diffusion model applied in the area.

The TD can be considered as an extreme case: for an infinite series of data, the RD tends toward the TD. Therefore, the TD can be considered an asymptotic limit for the RD.

3. Application examples

We applied the method to SO_2 concentrations measurements recorded in the Ravenna area (Italy). The town of Ravenna is 10 *Km* from the sea, while his industrial area is situated between the sea and the town. The climate is basically continental but more temperate by the proximity of the coast. The entire area is subject to a series of weak local wind circulation, frequent inversion phenomena and high relative humidity [7, 8]. The methodology was applied to a time series of hourly SO_2 concentration data measured by a station of the automatic monitoring network located in the city of Ravenna. In **Figure 1** it is shown the RD and LRD of hourly SO_2 record. There is a big difference between the RD and LRD from 10 *am* onwards.

Such behavior can be explained by the fact that SO_2 is mainly emitted from point sources located in the industrial area (located between the sea and the town) and therefore the measurements are influenced by wind direction. The LRD correspond to the SO_2 concentration recorded on the January 3 where, as it is shown in **Figure 2**, at 10 *am* wind change direction with a corresponding doubling in the speed.

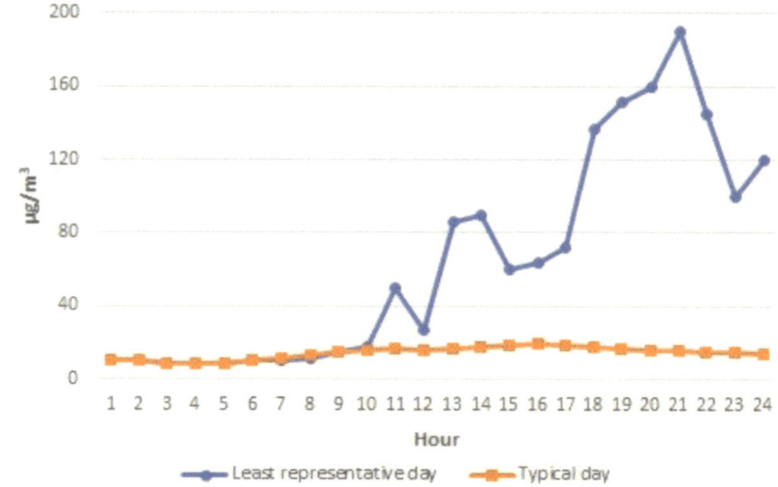

Figure 1.
Representative day (RD) and least representative day (LRD) of SO_2.

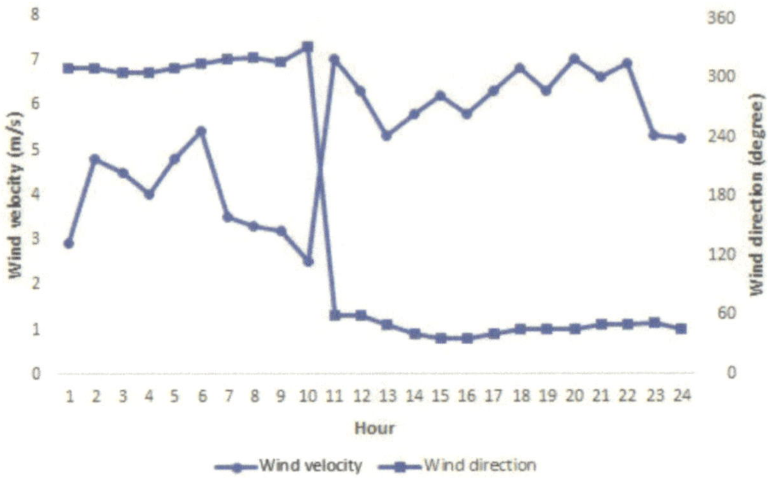

Figure 2.
Wind speed and direction during the day to which LRD of ***Figure 1*** *refers.*

Another interesting example of application of the methodology is that of concentration time series detected in the Falconara industrial area, where we have a high ozone production due to the characteristics of the local conditions and the emission sources [9]. In addition, there is an automatic monitoring network in the area which has collected a large amount of data.

Falconara is an urban center located north of Ancona on the Adriatic coastline of Italy. The Falconara area faces air pollution problems, mainly during summertime. Preliminary studies [10] showed that the most important factor contributing to urban air pollution in Falconara and its surroundings is the amount of emissions from mobile sources and industries.

The Falconara area can be roughly divided into two parts: a coastal area and an inland area. The coastal area is characterisd by the presence of a large oil-refinery. The inland area comprises the main urban area surrounded by hills. The description of the microclimate and landform of this area can be found in [9].

The accumulation of photochemically produced Ozone depends strongly on the prevailing meteorological conditions. In fact, meteorological conditions observed on days with high Ozone mixing ratios are often quite different from those when Ozone concentrations are low [11, 12].

In the first example we present, the Ozone shows the same trend on the RD and LRD, although with lower concentration values in the second part of the day.

Regarding **Figure 3**, the explanation can be found by analyzing the solar energy data for the day corresponding the LRD, June 21st., **Figure 4** shows lower solar radiation values in the middle of the day. The explanation for this case is very obvious because of the direct correlation between Ozone and solar radiation [13].

Another example, where on the contrary the Ozone shows the same trend in the RD and LRD, although with higher concentration values in the second part of the day (see example **Figure 5**). In this case the main wind direction during the last afternoon and the evening is North-East, indicating that the wind came from the sea (see **Figure 6**). Land breeze/sea breeze phenomenon seems to predominate in the Italian Adriatic area, so that the air pollution produced over the urban-industrial coast is transported as plumes over the sea and, subsequently, due to sea breezes, transported back to the coastal areas. The evening sea breezes transport the masses

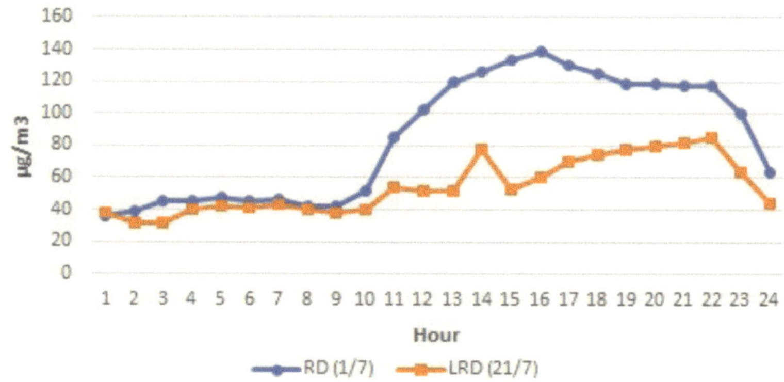

Figure 3.
Representative day (RD), least representative day (LRD) of O_3.

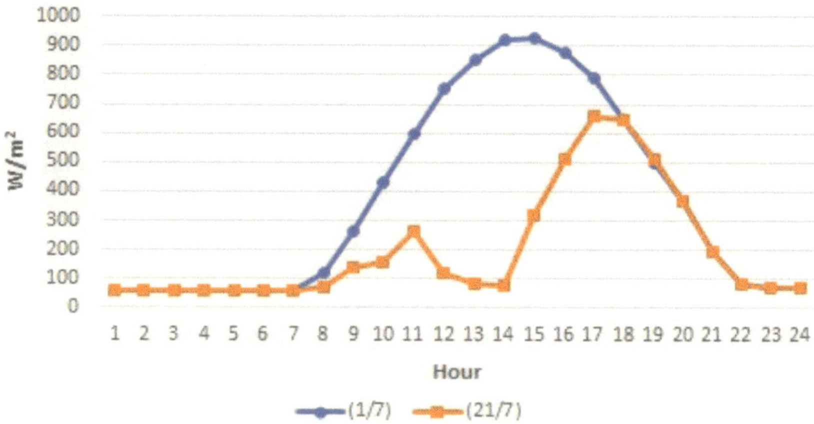

Figure 4.
Solar radiation measured during the RD and LRD of ***Figure 3****.*

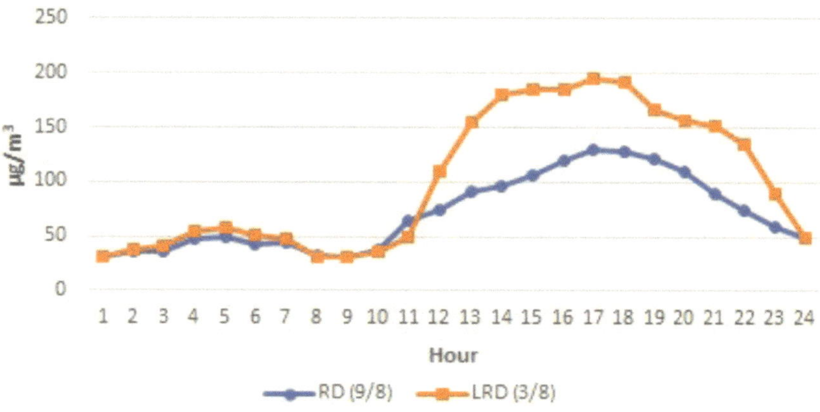

Figure 5.
Representative day (RD), least representative day (LRD) of O_3.

of air offshore, where the deposition rate is so slow that the ozone accumulates and is transported back to the coast when the daytime sea breezes resume, thus setting in motion mechanisms of photo-oxidant re-circulation [8, 9, 14].

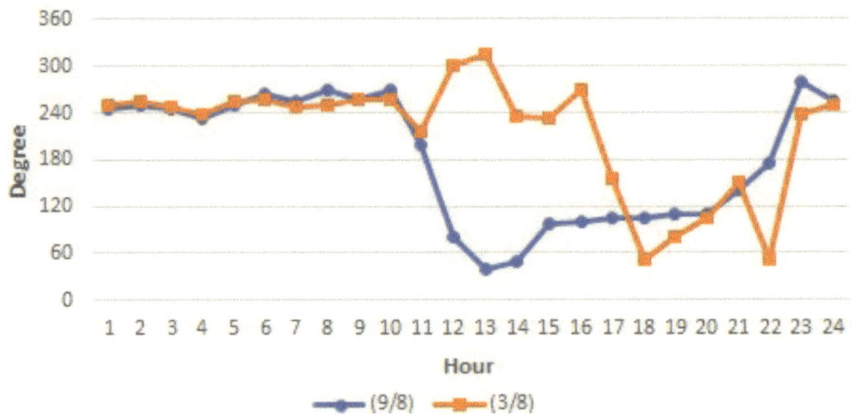

Figure 6.
Wind direction measured during RD and LRD of ***Figure*** *5.*

DI / gas	SO_2	O_3 - Figure 3	O_3 - Figure 5
RD	1.01	1.07	1.07
LRD	6.34	2.17	2.26

Table 1.
Representativity adimensional index DI of RD and LRD.

Table 1 provides the representative adimensional index DI of RD and LRD. It is possible to see the different behavior of SO_2 and O_3. In the case of O_3 the indices show that the RD are less representative and the LRD is less anomalous.

An explanation of this different behavior can be explained by the fact that SO_2 is emitted mainly from point industrial sources, so that at the beginning very concentrated plumes are composed. Therefore, the resulting concentrations on the ground are more subject to weather conditions and wind direction. On the contrary, ozone is a secondary gas (it is not directly emitted into the atmosphere) and therefore forms less concentrated plumes and offers greater inertia (compared to SO_2) to the production of different concentration scenarios on the ground.

4. Conclusions

Automatic monitoring networks are often used for study, control and management of local environmental problems. As a result, over time, a large mass of data is collected. While the individual data are very useful for real-time control and to report any alarms, it is necessary, for the study of the territory, for the understanding of the phenomena present in the area, to obtain a synthetic set of the measured data. Moreover, the processes that govern the transport and diffusion of pollutants are numerous and of such complexity that it is not possible to describe them without using mathematical models. Both the interpretation of the phenomena governing pollutant diffusion and the use of mathematical model requires a synthesis of the information given by temporal data series.

For this purpose, the most representative day constitutes a simple and immediate method through which to characterize the temporal structure of daily trends. We have called the "representative day" the day which, in a set of data composed of daily series, best represents the whole set of series. In mathematical terms, the one which minimizes the sum of squared differences with respect to all the daily trends in a temporal series. The attention is focused on the "day as a unit", without losing, however, the particular temporal structure, as partly occurs in the case of the "typical day".

Moreover, RD being an actual day, it allows the identification of the date on which it occurred and, thus, a knowledge of the meteorological and emission parameters which characterized it. Among other things, this allows it to be simulated with air pollution diffusion models.

The same approach also allows the identification of the "least representative day", that is, the day on which an anomalous, nearly always critical situation occurred, compared to the average trend recorded for that period. LRD is also an actual day. The study of the meteorological and emission parameters relating to this day, will allow a preliminary interpretation of the phenomena which brought about the situation.

By eliminating the data of that day from the original series and repeating the procedure on the remaining data, it is possible to highlight the second less representative day. Proceeding in the same way, the third, fourth, etc. LRD can be highlighted.

Both RD and LRD can be normalized so that the degree of representativity can be compared independently of the length of the measurement series, sampling period and characteristics of the area under study.

Author details

Tiziano Tirabassi*† and Daniela Buske[2]†

1 Federal University of Santa Maria, Santa Maria (RS), Brazil

2 Federal University of Pelotas, Pelotas (RS), Brazil

*Address all correspondence to: tiziano.tirabassi@gmail.com

† These authors contributed equally.

References

[1] Khan S A R, Zhang Y, Kumar A, Zavadskas E, Streimikiene D. Measuring the impact of renewable energy, public health expenditure, logistics, and environmental performance on sustainable economic growth. Sustainable Development. 2020; 28: 833–843.

[2] Khan S A R, Sharif A, Golpîra H, Kumar A. A green ideology in Asian emerging economies: From environmental policy and sustainable development. Sustainable Development. 2019; 27: 1063–1075.

[3] Tirabassi T. Operational advanced air pollution modeling. Pageoph. 2003; 160: 5–16.

[4] Seinfeld J H, Pandis S N. Atmospheric chemistry and physics. New York: Wiley; 2016. 1152 p.

[5] Tirabassi T, Rizza U. Flux parameterization for atmospheric-diffusion models in an industrialized area. Il Nuovo Cimento. 1993; 16C: 187–192.

[6] Tirabassi T, Nassetti S. The representative day. Atmospheric Environment. 1999; 33:2427–2434.

[7] Tirabassi T, Fortezza F, Vandini W. Wind circulation and air pollutant concentrations in the coastal city of Ravenna. Energy and Buildings. 1991; 1 (2): 699–704.

[8] Fortezza F, Strocchi V, Giovanelli G, Bonasoni P, Georgiadis T. Transport of photochemical oxidants along the northwestern Adriatic coast. Atmospheric Environment. 1993; 27A: 2393–2402.

[9] Latini G, Cocci Grifoni, R, Passerini G. Influence of meteorological parameters on urban and suburban air pollution. In: Brebbia C A, Latini G, editors. Air Pollution IX. Southampton (GB): WIT Press; 2002. p. 753–762.

[10] Latini G, Cocci Grifoni, R, Passerini G, Tirabassi T. Evaluation of Mixing Height over Complex Coastal Terrain. In: Brebbia C A, Jacobson M, Power H, editors. Air Pollution 99. Southampton and Boston: WIT Press; 1999. p. 1073–1081.

[11] Bower J, Broughton G, Stedman J, Williams M. A winter NO_2 episode in the UK. Atmospheric Environment. 1994; 28: 461–475.

[12] Ziomas I, Melas D, Zerefos C, Bais A, Paliatsos A. Forecasting peak pollutant concentrations from meteorological variables. Atmospheric Environment. 1995; 29 (24): 1352–2310.

[13] Colbeck, I, Mackenzie, A R. Air Pollution by Photochemical Oxidants. Amsterdan: Elsevier; 1994. 232 p.

[14] Latini G, Cocci Grifoni, R, Passerini G, Fava G. Applicability of a Photochemical Box Model over Complex Coastal Areas. In: Brebbia C A, Jacobson M, Power H, editors. Air Pollution 99. Southampton and Boston: WIT Press; 1999. p. 697–706.

Chapter 9

Why Airborne Pesticides Are So Dangerous

Tomaz Langenbach, Tácio M.P. de Campos and Luiz Querino Caldas

Abstract

More than four billion of tons of pesticides are used annually in agriculture worldwide. Part of it drifts down after pulverization, but a volatilized portion moves upwards. Pulverized pesticide applications are controlled by different parameters of fan and climate conditions. This can be mitigated with buffer zones, hedgerows and forest strips. Volatilization is determined by physicochemical parameters of the product and adsorption capacity to soils and leaves, and climate conditions. Prevention is the only efficient approach by banning high vapor pressure active ingredients. Volatilized pesticides are transported by air streams. Subsequently products are retained by mountains or eventually moved further by wind and descend in rain returning them to soil or vegetation. All regions of the planet are submitted to air pollution and nowadays pristine environments are very rare. These pollutants have hazardous effects on environment and toxic effects to skin and when they reach the blood stream directly via the lungs, are more intense to humans than from ingestion. The challenge of this overview highlights sustainability to avoid airborne pesticides by different strategies such as reduction of amounts sprayed through integrated pest management and mainly replacement of hazardous chemical pesticides by harmless ones or by biological control.

Keywords: pesticides airborne, air pollution, pulverization, volatilization, pesticide drift

1. Introduction

Before we trace the sinuous movement with hazardous effects of pesticides in the air, we need to take in account that humankind up to the twentieth century was threaten by hunger due to many factors including loss of agriculture production by pests [1]. Only by the control of essential factors of agricultural production as soil fertilization, dryness and pest management was it possible to get food security. Hunger today is addressed mainly to economic and political questions.

The first intensively used product was the insecticide DDT developed by Paul Muller at beginning of the 1940s, followed by new products applied as fungicides, acaricides, nematicides and bactericides [2]. Finally, the exclusion of herbs that compete for natural resources with crops, had changed from mechanical to chemical methods by herbicides. These processes dramatically increase productivity with less land used and resulted in intense migration of the rural population to the cities [3]. Herbicides in many countries now constitute about 50% of the amount of pesticide used.

IntechOpen

The mechanism of action of insecticides [4] and fungicides [5] are highly active on animal cells and therefore the toxic effects are generally high in humans and fauna. Many of the active mechanisms of herbicides inhibition act on plant and algae physiology like photosynthesis, hormones, and others [6], but other molecules that inhibit electron chains in plants are also very dangerous to humans and other animals. Poisoning side effects have been identified for many pesticides. The criteria of mutagenicity and carcinogenic were more required. The toxic effects are more permanent from persistent molecules, so biodegradation is one of the most important parameters in ecotoxicology [7].

The hazardous effects of pesticides in the environment were first denounced by Carlson with publication of "Silent spring" in 1962 [8]. Their deleterious effects have been sensed dramatically and increasing along with the chaotic shifts provoked in nature, fauna, and flora. The repercussion and significance of its effects in the natural history of our planet remains to be assessed.

After 1970, in the USA the "Environment Protection Agency" (EPA) organized the first registration process to control all pesticides some of which were banned or restricted, particularly, the most hazardous [9]. Later many countries created similar agencies. The legal requirements include physical–chemical parameters of the active ingredients of the pesticides that give information about volatility capacity, an important tool to restrict the use of the worst of them. In this process many other parameters were considered such as biodegradation/persistence, effects on non-target forms of life, and acute and chronic toxicology.

The increase and decrease of airborne particles in the environment follow the history of the pesticides not only in quality but also in quantity. The figures of worldwide consumption have increased (**Figure 1**). This process was started in developed countries and followed in developing countries. Control by banning of many products has occurred differently among countries, but in the overall the amount increased about 80% between 1990 and 2018 [10]. We are far from reaching sustainability in the pest control even in the countries with the best management.

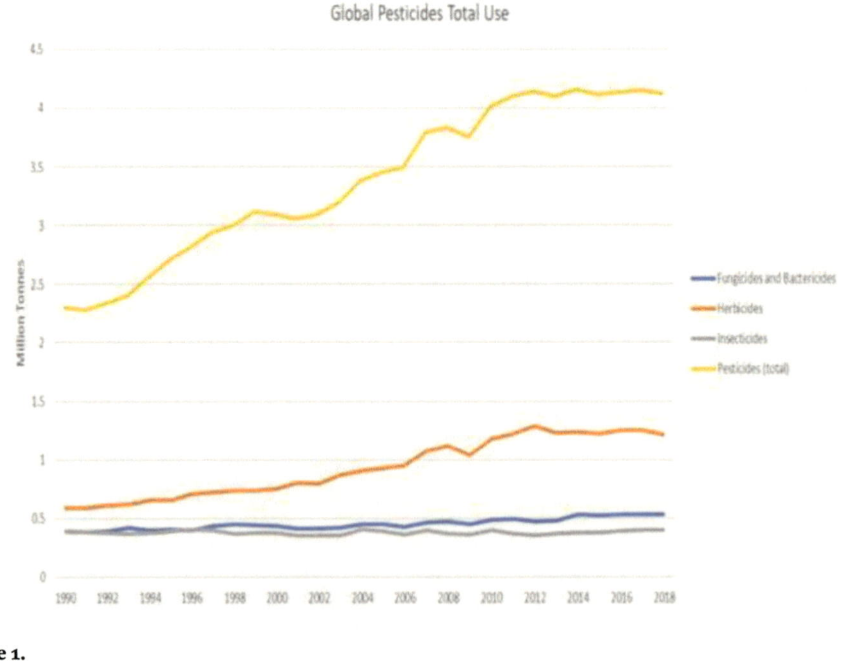

Figure 1.
Global pesticides use over 1990–2018.

The fate of airborne pesticides has some peculiarities: more disperse movement than those in polluting the whole world's water bodies [11, 12]; air movements are not under human control and the uptake by animals put them in direct contact with skin and blood. The purpose in this chapter is not only to describe airborne movement and the hazardous effects of pesticides, but also to identify new approaches in technical and management science that can overcome the bottlenecks toward sustainability.

2. How pesticides pollute air

Pulverization is a human decision of what, when, were and for what pests the pesticides will be applied. However, the step of volatilization in which molecules change to gas phase and are released into the air following application. Products with high volatilization are out of human control with scarce possibilities of mitigation.

2.1 Pulverization

Pesticides can be applied as dry powders that were mainly used in the past, as granules to control ants and other insects but the most frequent pesticide application is in water. In this form spraying can pollute air and is likely to drift to non-target areas. The modalities are wettable powder, emulcifiable concentrate and dilution of the commercial product in water for application.

Pulverization can be performed in different forms directly down to the soil, on the canopy of orchards or by floating for sanitary functions to control mosquitos. The efficiency of the methods used generally involves high amounts of loss to non-target forms of life in the environment. This process involves many parameters as fan position depending where the pests are in the soil, on upper leaf surface or on the underside of the leaves. The fan pressure is important for pesticide to arrive at the pest area and with high pressure directed toward the soil to reduce drift. Drop size higher than 5 u diameter weigh more than air and move down by gravity [13]. Smaller drops can float in the air. Climate conditions of wind speed and direction have important influence on the spraying through drift. All these parameters need to be considered for pesticide spraying.

The worst situation is of poor farmers using knapsack or pump with hose application in which the operator moves into the pesticide cloud with an enormous level of pesticide exposure [14]. Individual protection equipment in many instances are not used due to hot climate conditions and the protection itself has limits when using for longer periods. The use of tractors is much better because it moves away from the cloud and in tractors with a cabin the protection is quite good. The worst exposure to human populations that live in the countryside is spray from airplanes, forbidden in many countries or regions.

2.2 Volatilization

This process is when the active ingredient of the pesticide change to the gaseous phase and move upward into the atmosphere. Up to 0.5 cm above soil surface the molecules move only by dispersion while in the upper air layers gradual increase the wind speed and, in turbulence, mix this process these gases into the air [15]. Volatilization of each active ingredient depends on the physical–chemical conditions, mainly vapor pressure and Henri's Constant Law (partition of a substance between water and gas phase), strongly integrated with climate conditions. The highest volatilization rate occurs after rain followed by high temperatures due to Henri's Law. These conditions occur frequently in tropical areas.

Pesticide spraying is directed to soil or to vegetation for pest control. First, we begin with soil. In a report of different substances exposed under the same conditions shows that at low vapor pressure of less than 10^{-3} Pascals (Pa) no volatilization occurs, but after this value the increase is strong with amounts that can reach 90% of the pesticide sprayed. Volatilization on leaves with lower vapor pressure begins at 10^{-5} Pa and can arrive at high values from using higher Pa [15]. This parameter can be very useful to restrict dispersal of high volatile pesticides.

3. Factors influencing volatilization

Volatilization depends on the physical chemical properties of the active ingredient, climate, adsorption by surface, agriculture management and the interrelation between these factors. The humidity of the climate influences this residue adsorption, under dry conditions adsorption is more intensive and volatilization reduced [16, 17]. In **Table 1** climate and management factors influencing soil and leaf pesticide adsorption resulted in different amounts of volatilization [18].

Pesticide adsorbed on soil is a process in which aging is an important factor and this process can be reversible. Binding by entrapment and covalent bounding often occurs only after long contact time. Adsorption reduces the number of free molecules and therefore reduces the bioavailability that allows the molecule to eliminate pests. Likewise, adsorption can reduce the toxicity of the molecule and reduce free movement to leach or to volatilize. In practice, this is usually measured by adsorption of Freundlich isotherms or others [19]. The literature describes many different binding forms of pesticides depending strongly on the chemical characteristics of each product and soil composition as are mentioned below [20].

Covalent bonding – the pesticide reacts with some soil molecules resulting in a new substance that mischaracterize the pesticide. It is a strong chemical bounding.

Ionic charge - Compounds and their metabolites adsorbed by ionic bonding, or cation exchange, exist either in the cationic form in solution or can be protonated and become cationic.

Hidrogen bonding - Pesticide molecules compete with water for the binding sites on humic substances. H-bonding is suggested to play a vital role in the adsorption of several non-ionic polar pesticides,

Electron donor or acceptor complex - is a transfer of electron to acceptor pesticides with a part of humic substance that donated this electron. The resulting electrostatic attraction provides a stabilizing force for the molecular complex.

Climate conditions		**Pesticide on soil**	**Pesticide on leaves**
Temperature		10° C increase <3 to 4 times Pa > volat.	high temp. > volat.
Air humidity	Dry	> adsorption < volat.	> adsorption < volat.
	Dew	morning / late afternoon > volat.	
	Rain	leaching + enhanced > volat.	washout + penet. < volat.
Wind		> wind > volat.	> wind > volat.
		> turbulence > volat.	> turbulence > volat.
Pesticide management		**Pesticide in soil**	**Pesticide in leaves**
Small drop pulverization		less water < volat.	quick evaporation
Big drop pulverization		more water > volat.	faster adsorption

Table 1.
Parameters that influence pesticide volatilization from soil and leaves.

DOI: http://dx.doi.org/10.5772/intechopen.95581

Van der Walls forces - is a distance-dependent interaction between atoms or molecules. Unlike ionic or covalent bonds, these attractions do not result from a chemical electronic bond; they are comparatively weak and therefore more susceptible to disturbance.

Hidrophobic partition - Hydrophobic retention need not be an active adsorption mechanism but can also be regarded as a partitioning between a solvent and a non-specific surface.

Pesticides in soil are adsorbed in part by strong binding energy different than for plants in which the main adsorption occurs by hydrophobic partition, a much weaker binding.

Most pesticide applications are on the plants which have quite different chemical composition than soil. Covering the epidermal cells of leaves, a domain called the cuticular layer is rich in polysaccharides (**Figure 2**) enriched with waxes (which are hydrophobic, *i.e.* repel water). These wax deposits inside the cutin matrix are called intracuticular. On the surface, the cutin is covered with a film and epicuticular wax crystals that give the leaf a glossy appearance [21].

Figure 3a shows the dynamics of the fate of parathion volatilization, leaf penetration, and photo-degradation after spraying. In **Figure 3b** the fate of chlorotanoyl during an experiment where rain washed it from leaves to the ground [22].

In experiments with parathion and chlorotanoyl pulverization the fate of these molecules on the leaves is traced (**Figure 3**) showing the residue amounts on the leaves, volatilized, penetration of the cuticle and the photo-transformed quantities. More than twenty percent of the applied amount of Parathion was volatilized with the vapor pressure of 8.9×10^{-4} Pa and chlorotanoyl with lower vapor pressure of 7.6×10^{-5} Pa volatilized about 5%.

Volatilization of parathion is fast with high volatilization in the first day and strong reduction in the following period (**Figure 4a**). For chlorotanoyl the amounts per day are much smaller and the volatilization process gradually reduces (**Figure 4b**). In both cases no volatilization occurs during the night and residue volatilization is higher than for commercial products showing that formulation can reduce volatilization [23].

In all the figures obtained by SURFATM-Pesticides volatilization did not occur at night. High sensitivity of the model was shown for vapor pressure to promote volatilization and the constant of the relation octanol/water K_{ow} for the penetration in the wax layer of the leave. More recently models were developed that integrate pesticide volatilization on leaves and soil [24].

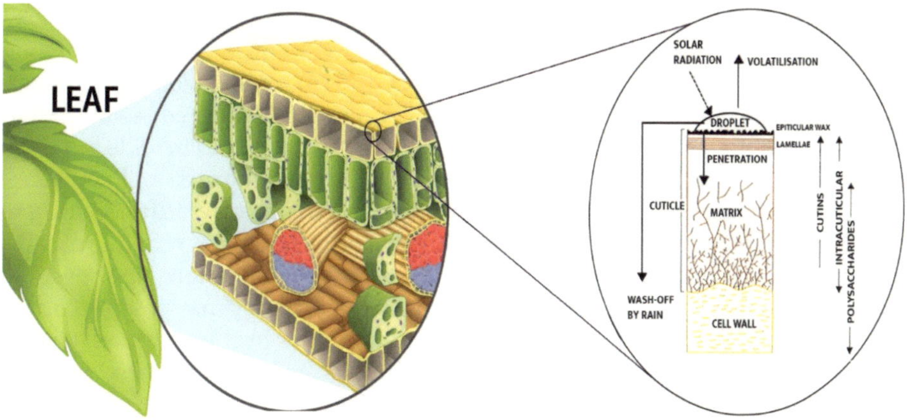

Figure 2.
The different process on fate of pesticide drops above leaves that are exposed to volatilization; photo-transformations; wash-off by rain and penetration on waxes. In focus is the cuticle with the wall (yellow) of upper epidermis.

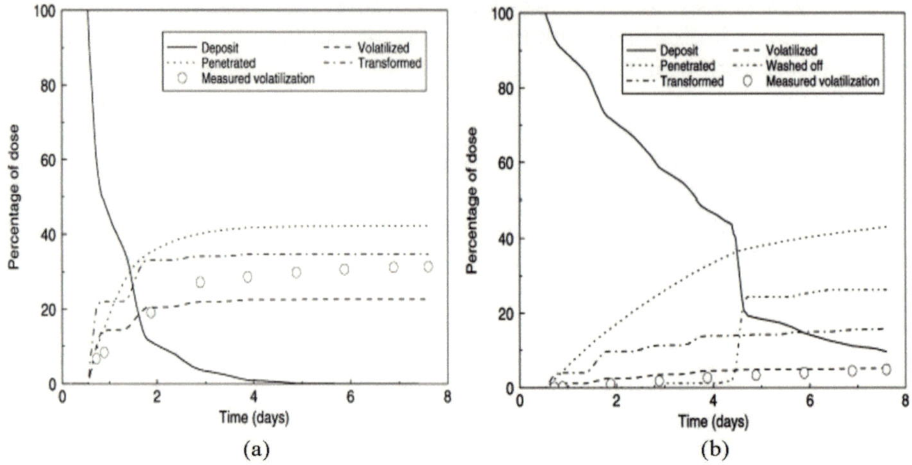

Figure 3.
(a) Decline of parathion and chlorotanoyl deposit on leaves, accumulation of volatilization, penetration into the plant and phototranformation in a computer program PEARL. The cumulative volatilization derived from measurements is plotted against (O). (b) Application was shortly after noon on the first day. For the chlorotanoyl experiment, the rain reduced residues by wash-off of 20% from the leaves.

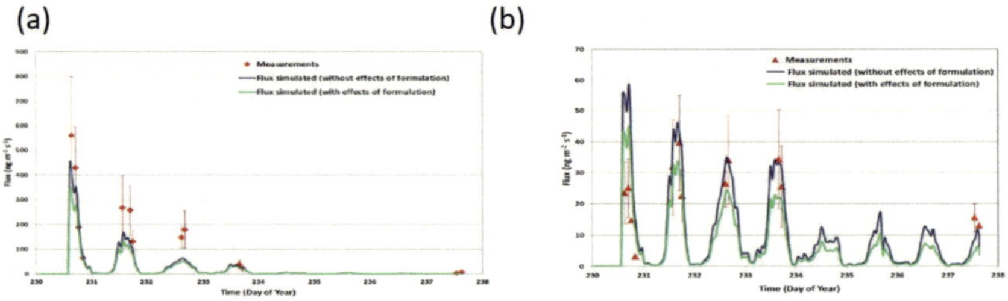

Figure 4.
Comparison of measured and modeled data by PEARL flux volatilization. Simulated without and with formulation: (a) parathion, and (b) chlorothalonil. The error bars represent the standard deviations based on the two measurement methods (AG and BR methods).

4. Prevention and mitigation of airborne pesticides

Many possibilities exist to mitigate pesticide pollution by spray drift. Directing the fan on the intended target to exclude as much general dispersion to non-target areas as possible. Some countries and in many cases, counties, cities and villages enforce buffer strips around the border in which spray is prohibited resulting in strong reduction of drift [25, 26]. This strip generally is about 10 m wide but can also be more. For small farms, these strips reduce the spraying area and land use. In such cases hedgerows of appropriate size can be used to mitigate drift pollution to neighbors or residential areas [27]. Every binding of residues can restrict volatilization. All wind break constructions as outdoor and indoor walls can retain pesticides [28]. Depending on the binding strength of the residues, they may escape to the air over time [29]. Indoor pollution with pesticides is a major problem by continuous exposure rural living habitants. In this case children are severely exposed by the hand to mouth movements [30]. Another possibility is to retain in hedgerows or forest strip to protect humans and environment [31].

Prevention of volatilization could be done by excluding all products with high vapor pressure. Mitigation can have some effects to preserve buildings and small villages when protected by a forest strip. Experiments have shown that riparian forest are able to retain volatilized pesticide residues on the canopy that by rain washing or leaf senescence move down to the soil [31, 32]. Phytoremediation with genetically modified poplar trees with introduction of Citochrome P450, shows that thrichloroethilene can be adsorbed by leaves and degraded efficiently [33]. In large scale pesticide dispersion up to remarkably high atmosphere altitudes are completely out of human control and no mitigation is possible. As a response to climate conditions, it is possible to change pesticide application management. Experiments in which sets of samples of the herbicide 2,4-D were applied in morning and then at 18 h in the evening showed samples from the afternoon volatilization were 9 to 30% less than when applied in the morning when measured 24 h after application [34]. These results suggest that dry soil in the afternoon adsorbed more residues yielding reduced volatilization as observed by others [35]. The best application management is to exclude loss factors as strong adsorption, photo degradation and volatilization mainly in the first period of application and this will occur at night [23, 36]. During this time residues need to have high bioavailability to act on the pest. Under these conditions it is possible to greatly reduce the amount of pesticide without efficiency loss.

5. Dynamics of pesticides in the air

After volatilization of pesticides from soil, leaves or water, these molecules move upward and through turbulence to arrive as a quite homogeneous mixture. Residues can move up and down powered by air streams following the directions of wind currents. The main deposition of pesticides are due to rain and snow. This occurs more frequently on mountains. When pesticides are airborne and arrive high altitudes, they can be transported for long-distances. Residues were found more than 50 years ago in the Antarctic [11]. This was due to wind and rain/snow deposits, but not from human activities. Pesticide residues coming from other regions were found recently on the top of the Itatiaia mountains in the State of Rio de Janeiro (Brazil) [12]. Much evidence shows that airborne pesticides arrive with impact in all environments around the world giving little chance of pristine environment without pesticides existing.

Photo-degradation occurs not only in the air but also on all kinds of surfaces exposed to solar irradiation. Many photo reactions are by direct action on molecules causing degradation while others have indirect effect in which some substances absorb photons and with this energy promote transformation reactions in other molecules. These processes have been shown to be a relevant pathway. The setup of laboratory conditions in which the experiments occur is very diversified and influence the reactions with different intensity [37].

Much of the airborne pesticide burden is removed from the atmosphere through deposit by rain in ocean and that could be considered as a sink. Nevertheless, the reverse movement in which pesticides volatilize to air occur also. The relative intensity of these two movements is expressed by the fugacity ratio $\left(\frac{fa}{fw}\right)$ of pesticide in air to the amount in the water. This define if the process is predominantly of deposition or volatilization [38]. Data show that some pesticides reduce fugacity rate with increase in surface water at higher latitudes (around 70° North) in the Arctic Ocean like a cold trap [39]. Persistent organochlorines compounds after long-range transport from the application site, pollute pristine environments as the artic in which the ecosystem is more vulnerable than those species on which risk assessment is based [40].

The fate of pesticides shows that all kinds of movements occur from air to soil/plants, to water and the reverse movement between all these environments. All flora and fauna are exposed to these residues in the air, but it is important to differentiate environments between high and low airborne pesticides as rural areas with intensive agriculture activity with cities and areas with little agriculture. In humans the toxicologic effects can be due to ingestion or to air pollution, but unprotected agricultural workers are exposed to high risk of acute toxic effects from pesticides.

6. Pesticides in environment and health

6.1 Air pollution in the environment

The potential degradation and accumulation of hazardous substances in the atmosphere of newly formed compounds within the environment are still quite unknown because of an enormous variety of possibilities generated by abnormal climatic conditions and anthropic actions around the globe [14]. When major, hard to predict, recurrent crucial environmental damages occur, they may irreversibly affect nature. Changes in its delicate balance that generally demand decades or centuries to recover, if they do at all. For instance, if the predator–prey relationship is severely disturbed due to the persistent exposure to airborne pesticides, it raises deep ecological changes in the local biota.

In humans, mammals, and reptiles we should pay special attention to the action of pesticides in two hormonal structures: gonadal and thyroid, because during the development, these glands (organs) are especially sensitive and, therefore, affected by exposure of low concentrations of sex steroids and thyroid hormones. It is recognized that there is a difference in the endocrine response of adults as compared to the embryonic/fetal/neonatal responses. Changes induced by exposure to these hormones during development are often irreversible, in contrast to reversible changes induced by exposure to transient hormones in adults [41]. Hayes and his coworkers [42–44], have studied isolated atrazine and a mixture of nine pesticides, on the impact in the environment, as an endocrine interfering factor. Examined larvae growth and development, sexual differentiation, and immune function in leopard frog found hormonal changes in sexual differentiation, body development and damage to the thymus resulting in immunosuppression. They concluded that the evaluation of each pesticide alone is inadequate to estimate the adverse impact on amphibian development or to link pesticides to the decline in the number of amphibians, as blends of pesticides provoked more effects than the isolated components.

In humans, apart from the bare skin exposure (explained further on), the lungs mainly of those occupationally exposed, usually remain the major cause of infirmities and the number of deaths among agricultural workers. Also, cancer and some effects in the endocrine system cannot be ignored [45–47].

6.2 Pesticides in the lungs

Inhalation of various volatile forms of aerosols, vapors, dust, or mist can be the source of respiratory diseases in agricultural workers, particularly for those without proper personal protective equipment or when they work in confined spaces. The main outcomes observed were bronchial asthma (BA), chronic obstructive pulmonary disease (e.g. emphysema) and lung cancer [48]. As explained below, the etiopathogenesis varies, some of them affecting the lung clearance by the mucocilliary tracheobronchial cells system, others affecting the bronchiolar smooth muscles physiology, or else destroying alveolar walls, burdening pneumocytes. Some pesticide molecules may deeply affect the oxygen supply and, therefore, will interfere with the exchange of carbon dioxide, the body's metabolic by-product. This, ultimately and decisively, will interfere in the metabolic turnover, and not rarely, leading to death.

In an average-sized person at rest, the O_2:CO_2 ratio is 1.25 (250 ml:200 ml/min) whereas when incrementing metabolic activity (e.g. exercise), the oxygen demand as well as concurrently its by-product CO_2 increases in order to preserve the body homeostasis. The bronchial ducts and alveolar sacs serve to conduct fresh air into the lungs (inhalation) where active gas exchange occurs between the environment and blood pressure gradients determine the gas exchange from the opposite direction, breathing out the body by-products. In the latter situation, the pulmonary ventilation rate may increase 20% with higher exposure to airborne pesticide.

Droplets larger than 10 μm are deposited in the upper respiratory tract (nose and mouth). The toxic effects of pesticides start as the air breathed in enters through the nose and contacts the sinus cavities where it irritates the fragile membranes and in response may lead to an inflammatory reaction with glands secreting significant mucous production which may ultimately become infected. When coarse particles (>5 μm) reach the upper respiratory tract, they are usually trapped into the conducting airways whereas fine (<0.2–5 μm) particles (such as pesticide drops) are deposited by impact and sedimentation (**Figure 5**). Other molecules may be adsorbed so that material dissolved in solvents or water vapor (hygroscopic) may get into contact with the airways ducts or alveolar septum with a 80–120-ml monolayer of blood.

That is, when these compounds provoke repeated cycle oxidation/reduction reactions generating toxic amounts of reactive oxygen species it can lead to diffuse pulmonary alveolitis culminating in a rampant acute or chronic fibrosis and, possibly, death [50].

This storm of clinical symptoms results in a profound change in the acid–base homeostatic mechanisms that lead to acidaemia, obnubilation culminating with convulsions and death. Supposedly, clean air, free from foreign substances, humidified and regulated to the body temperature must reach the alveoli where potentially injurious substances such as the airborne pesticides are in close contact with the alveolar system. In the tracheobronchial tree, smooth muscle spasms will resemble an asthmatic response.

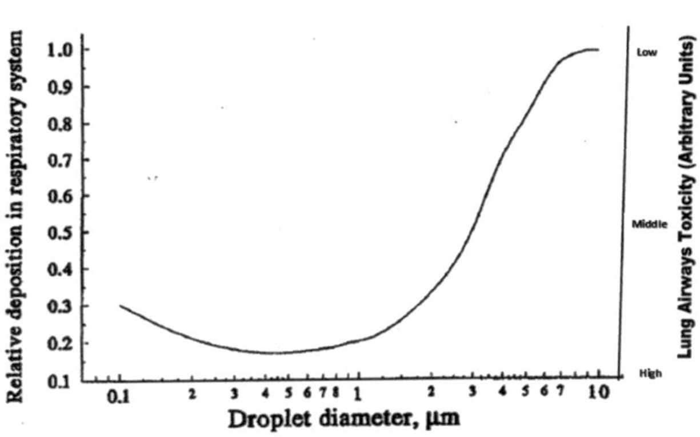

Figure 5.
Effect of aerodynamic diameter on deposition efficiency of inhaled droplets in human respiratory system and their reversal arbitrarily toxicity for the lungs adapted from Giles et al. [49].

6.3 Other systemic effects of airborne pesticides

The cutaneous hypersensitivity resulting from intimate contact of pesticides with the skin and/or the eyes since many spray adsorbent formulas can be seized and absorbed through the tissues.

Oxidative stress can be involved in many pathological conditions in the lungs and in the skin. It may alter irreversibly causing cell damage that evokes changes in proteins or DNA structures [51, 52] as well as mitochondrial disorders that lead to antioxidant enzyme suppression as dismutase and superoxidase [53]. Some compounds provoke repeated cyclic oxidation/reduction reactions generating toxic amounts of reactive oxygen species that can lead to diffuse pulmonary alveolitis culminating in a rampant acute or chronic fibrosis with possible, death [50].

7. Perspectives

Our focus here is on the future perspective of airborne pesticides. Airborne pesticide pollution can only be solved when sustainable products of pest control substitute harmful for harmless pesticides. The 1962 book Silent Spring of Carlson [8] was the first dramatic announcement of problems in flora and fauna due to pesticides use. After almost six decades no one, even the most pesticide-concerned countries, can commemorate agriculture with sustainable pest control [54]. In this period progress occurred with: the introduction of a registration process to ban the most hazardous pesticides worldwide; list of the worst products to be banned by the International conference of Stockholm 1972; introduction of the "International Code of Conduct on Pesticide Management" sponsored by FAO/WHO for a better practice of pesticide use; many scientific international conferences to improve technical advances and better policies of pesticide control and organic agriculture increased. It is difficult to know if the balance of these positive actions can overcome the hazardous exposure of environment and human health due to the enormous increase of pesticides used in the last decades. The conclusion is that we are far from the main target of sustainability in agriculture to avoid poisoning exposure but Improvement of sustainability would be a benefit in health, environment and social development [55].

The highly bureaucratic agencies of pesticide control together with the industrial interest shared with agricultural producers make the use of pesticides a steady state that does not promote progress toward sustainability due to the hegemonic economic interest. Approaching sustainability cannot achieved by banning of hazardous pesticides alone. Sustainability of pest control can be reached when applied pesticides are specific and not highly toxic to humans and to flora and fauna because of ecotoxic effects on non-target forms of life. An additional essential characteristic is to be completely biodegraded with no residue accumulation in the environment.

The new perspective for that is the intense use of recently developed biopesticides that can control insects, fungi, and nematodes efficiently. A lack of data makes it difficult to evaluate the intensity of use of these products up to now, but it seems that it was not very much. This demands a new specific policy to promote and push these changes to agriculture practice. To overcome the bottle neck of the few organic herbicides available, the most applied products, scientific advances with focus of new products with different approaches as allelopathic chemicals or development of reduction of herb populations by strategies of agriculture management or development of other solutions.

To arrive at sustainability all tools from restriction and banning of pesticides, integrated management supported by remote sensing [56] and the new approaches of nanotechnology [57], code of conduct of pesticide management and biopesticides, need to be used. This is a difficult task in which scientific development and information, creativity, flexibility and political will of the stakeholders, are essential.

Finally, a scenario in which hazardous chemical pesticides are substituted by non-hazardous organic control in a sustainable agriculture, would stop the introduction of more pesticides in the air. Nevertheless, to eliminate the huge amount of residual pesticides in the air, time would necessary to promote biodegradation, photo-transformation and covalent bounding to soil or other molecules, condition necessary to gradual elimination of these products from our environment.

Acknowledgements

The authors would like to thank the review of the manuscript to prof. Allen Hagler and the financial support of the workshop provided by CNPq and FAPERJ.

Author details

Tomaz Langenbach[1,2]*, Tácio M.P. de Campos[1] and Luiz Querino Caldas[3,4]

1 Civil and Environmental Engineering Department, Pontifical Catholic University of Rio de Janeiro, Brazil

2 Institute of Microbiology of the Federal University of Rio de Janeiro, Brazil

3 Clinic Toxicology Faculty of Medicine, Federal Fluminense University, Niteroi, Rio de Janeiro, Brazil

4 Harvard School of Public Health, Boston, USA

*Address all correspondence to: tomazlange@yahoo.com.br

References

[1] Handley J. Pesticides - A brief history and analysis. In Chemical & fertilizers and technical. Pitschcare Magazine. 2019;83

[2] Paul Müller – Biographical. NobelPrize.org. Nobel Media AB 2021. Tue. 26 Jan 2021. https://www.nobelprize.org/prizes/medicine/1948/muller/biographical/

[3] Migration Data Portal. https://migrationdataportal.org/themes/urbanisation-et-migration. 2020

[4] Ghosal A. Mode of Action of Insecticides. Chapter 1 in Applied Entomology and Zoology Volume – 1, Ed. B.S. Chandel. 2018

[5] Yang, C., Hamet, C., Vujanovic, W., and Gan Y. 2011. Review Article Fungicide: Modes of Action and Possible Impact on Non target. In Microorganisms International Scholarly Research ed. H. Sandersan Notices / 2011. Article ID 130289 https://doi.org/10.5402/2011/130289

[6] Sherwani I. S., Arif A. I. and Khan A. H. Modes of Action of Different Classes of Herbicides. In Herbicides Current Research and Case Studies in Use ed. Andrew Price, IntechOpen, 2015. DOI: 10.5772/61779.

[7] Poupin P. Biodegradability in Ecotoxicology. In: Férard JF., Blaise C. (eds) Encyclopedia of Aquatic Ecotoxicology. Springer, Dordrecht. 2013. ttps://doi.org/10.1007/978-94-007-5704-2_18

[8] Carlson R., Spring, 1962. 40th Anniversary Edition, Houghton Mifflin Harcourt, 2002. ISBN 0618249060, 9780618249060.

[9] EPA history FIFRA amendments – 1988. https://archive.epa.gov/epa/aboutepa/epa-history-fifra-amendments-1988.html

[10] FAO - UNITED NATIONS FAOSTAT. FAO. Ess website ess pesticide use

[11] Tatton, J.O.G., and J.H.A. Ruzicka. Organochlorine pesticides in Antarctica. Nature. 1967;215:346-348

[12] Meire, R.O., M. Khairy, A.C. Targino, P.M.A. Galva˜o, J.P.M. Torres, O. Malm, and R. Lohmann. Use of passive samplers to detect organochlorine pesticides in air and water at wetland mountain region sites (S-SE Brazil). Chemosphere. 2016;144:2175-2182

[13] Langenbach T., Mano D., Campos M.M., Cunha A.L. M.C., De Campos T.M. Pesticide dispersion by spraying under tropical conditions. J. of Env. Sci. and Health, Part B. 2017;52:843-849.. https://doi.org/10.1080/03601234.2017.1359040

[14] Langenbach T. and Caldas L.Q. PERSPECTIVE - Strategies for reducing airborne pesticides under tropical conditions. Ambio. 2018; 47:574-584; DOI 10.1007/s13280-017-0997-4

[15] FOCUS. 2008. Pesticides in air: Considerations for exposure assessment. Report prepared by the Working Group on Pesticides in Air (FOCUS Air Group)

[16] Vanclooster, M., J.D. Pineros-Garcet, J.J.T.I. Boesten, F. Van den Berg, M. Leistra, J. Smelt, N. Jarvis, S. Roulier, et al.: Effective approaches for adjustment of the pesticide risk index used in environmental policy in Flanders. APECOP 2003

[17] - Correa, F.V., A. Macrae, L.R.G. Guilherme, and T. Langenbach. Atrazine sorption and fate in an Ultisol from humid tropical Brazil. Chemosphere. 2007; 67:847-854

[18] Bedos C., Cellier P., Calvet R., Barriuso E., Gabrielle B. Review article Mass transfer of pesticides into the atmosphere by volatilization from

soils and plants: overview. Agronomie. 2002;22:21-33 21. DOI: 10.1051/agro: 2001003

[19] Farenhorst A., McQueen D.A.R.,. Saiyed I, Hilderbrand C., Li S., Lobb D.A., Messing P., Schumacher T.E., Papiernik S.K., Lindstrom M.J. Variations in soil properties and herbicide sorption coefficients with depth in relation to PRZM (pesticide root zone model) calculations.

[20] Gevao B., Semple K.T., Jones K.C. Bound pesticide residues in soils: a review. Environmental Pollution 2000;108:3-14

[21] Yeats T.H. & Rose J.K.C. The Formation and Function of Plant Cuticles. Plant Physiol. 2013;163:5-20

[22] Leistra M. and Van der Berg F. Volatilization of Parathion and Chlorothalonil from a potato Crop Simulated by the PEARL Model. Environ. Sci. Technol. 2007;4:2243-2248

[23] Lichiheb N., Personne E., Bedos C., Van den Berg F., Barriuso E. Im plementation of the effects of physicochemical properties on the foliar penetration of pesticides and its potential for estimating pesticide volatilization from plants. Science of the Total Environment 2016;550:1022-1031

[24] Taylor M., Lyons S.M., Davie-Martin C.L., Geoghegan T.S. and Hageman K.J. Understanding Trends in Pesticide Volatilization from Agricultural Fields Using the Pesticide Loss via Volatilization Model. Environ. Sci. Technol. 2020. https://dx.doi.org/10.1021/acs.est.9b04762

[25] Jong, F.; Snoo, G.; Zande, J. Estimated nationwide effects of pesticide spray drift on terrestrial habitats in the Netherlands. J. Env. Manage. 2008;4:721-730

[26] Schampheleire, M.; Baetans, K.; Nuyttens, D.; Spanoghe P. Spray drift measurements to evaluate the Belgian drift mitigation measures in field crops. Crop Prot. 2008;27:577-589

[27] Lazzaro, L., S. Otto, and G. Zanin. Role of hedgerows in intercepting spray drift: evaluation and modelling of the effects. Agriculture, Ecosystem, Environment. 2008;123:317-327

[28] Ucar, T.; Hall, F. Windbreaks as a pesticide drift mitigation strategy: a review. Pesticide Manag. Science. 2001;57:663-675

[29] Bennett, D.H.; Furtaw, E.J. Fugacity-Based Indoor Residential Pesticide Fate Model. Environ. Sci. Tech.. 2004;38:2142-2152

[30] Freeman, N.C.; Hore P.; Black , K. Contributions of children's activities to pesticide hand loadings following residential pesticide application. J. Expo Analytical Environ. Epidemiol. 2005;15(1):81-88

[31] Bicalho, S.T.T; Langenbach, T.; Rodrigues, R.R.; Correia, F.V.; Hagler, A.; Matallo, M.B.; Luchini, L.L. Herbicide distribution in soils of a riparian forest and neighboring sugar cane field. Geoderma, 2010;58:392-397

[32] Bicalho, S.T.T., and T. Langenbach. The fate of tebuthiuron in microcosm with riparian forest seedlings. Geoderma 2013;207-208: 66-70

[33] Doty S.L., James C.A., Moore A.L., Vajzovic A., Singleton G.L., Ma C., Khan Z., Xin G., Kang J.W., ParkJ.Y., Meilan R., Strauss S.H., Wilkerson J., Farin F. and Strand S.E. Enhanced Phytoremediation of Volatile Environmental Pollutants with Transgenic Trees. P.N.A.S. USA. 2007;104(43):16816-16821

[34] Costa, D., T. Campos, T. Langenbach, and A. Haddad-Nudi. Aplicacão e volatilizacão do 2,4-D na superfície de solos em diferentes horários. In

Conference: XVIII Congresso Brasileiro de Mecânica dos Solos e Engenharia Geotécnica. Minas Gerais: ABMS. 2016. DOI: 10.20906/CPS/CB-05-0036,

[35] Jansma J.W., Linders J.B.H.J., Volatilization of pesticides from soil and plants after spraying, National Institute of Public Health and Environmental Protection, Bilthoven (Netherlands), report no 679102030, 1995:48 p

[36] Pivato, A., A. Barausse, F. Zechinato, *L. palmeri*, R. Raga, M. Lavagnolo, and R. Cossu. An integrated model–based approach to the risk assessment of pesticide drift from wineyards. Atmospheric Environment 2015;111:136-150

[37] Hensen, B., Olsson, O., Kummerer, K. The role of irradiation source setups and indirect phototransformation: Kinetic aspects and the formation of transformation products of weakly sunlight-absorbing pesticides. Science of the Total Environment 2019;695:133808

[38] Li Y.F., Macdonald R.W. Sources and pathways of selected organochlorine pesticides to the Arctic and the effect of pathway divergence on HCH trends in biota: a review. Science of the Total Environment 2005;342:87-106

[39] Zhong G., Xie, Z., Cai M., M€oller A., Sturm R., Tang J., Zhang G., He J.,| and Ebinghaus R. Distribution and AirSea Exchange of Current-Use Pesticides (CUPs) from East Asia to the High Arctic Ocean. Environ. Sci. Technol. 2012;46:259-267. doi.org/10.1021/es202655k

[40] Bigsby R., Chapin R.E., Daston G.P., Davis B.J., Gorski J., Gray L.E., Howdeshell K.L., Zoeller R.T. and Van Saal F.S. Evaluating the effects of endocrine disruptors on endocrine function during development. Environmental Health Perspectives, 1999;107 n.4: 603-618

[41] Van Stralen, N.M. and Van Gestel, A.M. Ecological risk assessment of pesticides subject to long-range transport. Water air and soil poll.. 1999;115:71-81. DOI: 10.1023/A:1005245104606

[42] Hayes T.B., Collins A., Lee M., Mendoza M., Noriega N., Stuart A.A and Vonk A. Hermaphroditic, demasculinized frogs after exposure to the herbicide atrazine at low ecologically relevant doses. Proceedings of the National Academy of Sciences of the United States of America, 2002;99(8):5476-5480

[43] Hayes T.B,, Haston K., Tsui M., Hoang A., Haeffele C. and Vonk A. Atrazine-induced hermaphroditism at 0.1 ppb in American leopard frogs (*Rana pipiens*): laboratory and field evidence. Environmental Health Perspectives. 2003;11n.4:568-575

[44] Hayes T.B., Case P., Chui, S. Chung, D., Haeffele C., Haston K., Lee, M., Mai V.P., Marjuoa Y., Parker J. and Tsui M. Pesticide mixture, endocrine disruption and amphibian declines: are we underestimating the impact?. Environmental Health Perspectives. 2006;114(1):40-501

[45] Nerilo, S.B., Martins, F.A., Nerilo, L.B. Pesticide use and cholinesterase inhibition in small-scale agricultural workers in southern Brazil. Braz. J. Pharm. Sci. 2014;50:83-91

[46] Buralli, R.J., Ribeiro, H., Mauad, T. Respiratory condition of family farmers exposed to pesticides in the State of Rio de Janeiro, Brazil. Int, J. Environ. Res. Public Health. 2018;15(6):1203-12

[47] Ye, M., Beach, J., Martin, J.W., Senthilselvan, A. Pesticides exposures and respiratory health in general population. J. Environ. Sci. (China). 2017; 51:361-30

[48] Hernandez, A.F., Parron, T., Alaecon, R. Pesticides and asthma.

Curr. Opin. Allergy Clin, Immunol. 2011;11:90-96

[49] Giles, D.K.; Welsh, A.; Stenke, W.E.; Saiz, S.G. Pesticide inhalation exposure, air concentration, and droplet size spectra from greenhouse fogging. Transactions ASAE. 1995;38(5):1321-1326

[50] Huot, A.E., Hacker, M.P. Nitric Oxide in Craighead, J.E. (Pahology of Environmental and Occupational Disease, editor Mosby-Y.B. Inc. Ed., St. Louis. MI., chapter 22, 1995 p.357-372

[51] Atamas, S.P., Chapoval, S.P., Keegan, A.D. Cytokines in chronic respiratory diseases. F1000 Biol. Reports. 2013;5(3):1-13

[52] Günther, S., Hempel, D., Mathias-Dunkel, M., Rother, M., Preissner, R. SuperHapten, A. A comprehensive database for small immunogenic compounds. Nucleic Acids Research, Database issue. 2007;35: p.D906-D910

[53] Tarmure, S., Alexescu, T.G., Orasan, O.; Negrean, V., Sitar-Taut, A.V., Coste, S.C., Todea, D.A. Influence of pesticides on respiratory pathology – a literature review. Ann. Agric. Environ. Medicine. 2020;27(2):194-200.

[54] Report of a 2018 WHO/FAO survey. Global situation of pesticide management in agriculture and public health.

[55] Kahn, S.A.R., Sharif A. and Kumar, A. A Green Ideology in Asian Emerging Economies: From Environmental Policy and Sustainable Development. Sustainable Development, 2019

[56] Marei, Shahira. A review: application of remote sensing as a promising strategy for insect pests and diseases management. Envir. Sci. and Poll. Res. International, 2020;27(27:33503-33515

[57] Rodrigues, S. M., Demokritou P., Dokoozlian, N., Hendren, C.O., Karn, B., Mauter M.S., Sadik, O.A., Safarpour, M., Urine J., Viers, J., Welle, P. White, J. C., Wiesnerde M.R. and Lowry, G.V. Nanotechnology for sustainable food production: promising opportunities and scientific challenges. tutorial review: Environ. Sci. Nano. 2017;4, 767. DOI: 10.1039/c6en00573j rsc.li/es-nano.

Chapter 10

Air Pollution, Health and Perception

Banwari Dandotiya

Abstract

This chapter provides a general overview of the effects air pollution, pollutant sources in different regions and role of perception in assessment of pollution level and different health effects. The discussion in this chapter focuses basically on the different dimensions related to air pollution such as climatic and meteorological effects, anthropogenic and natural sources, ecosystem disturbances and their interrelationship in context of air quality. Basically, climate disturbances may be responsible for pollution episodes of certain regions. Discussions related to seasonal variations in air quality also included because seasonal meteorological elements exert different effects in different climatic regions. Air pollutants are emitted by a number of sources in the atmosphere such as urbanization, industrialization, transportation, and population growth, so these contributing factors and its effects of climatic events like temperature, wind speed and seasonal disturbances are relatively described in this chapter.

Keywords: air pollution, perception, health effects, climatic factors

1. Introduction

Air pollution in a great number of cities in India exceeds national and international standards. Air pollution in India is a growing threat to human health, especially in densely populated urban areas exposed to high concentrations of air pollutants. Meteorological parameters can contribute in high concentration episodes of air pollution [1]. Over the recent couple of years, air pollutant concentration and human exposure have received special attention due to increasing scientific evidences of causal relationship between exposure with air pollutants and negative health outcomes [2]. The detrimental effects of ambient air pollution on health of human being have been consistently observed in many epidemiologic studies globally, and it has been calculated that at least seven million deaths are annually attributable to the effects of air pollution.

In addition to the contributions of urban sources to air quality degradation, outdoor fires are a regional air pollution source dominated by fires in wheat and rice harvest [3–5]. For reducing health impacts from air pollution, it is important to know the sources contributing to human exposure. An important fraction of the exposure leading to those health impacts occurs in cities, due to the higher density of human activities and their emissions to the air. According to the region, however, high emissions can also occur from domestic fuel use in sub-urban or urban areas. Road traffic contributes significantly to the high levels of air pollution observed in urban areas around the world [6]. Ambient air pollution, especially in urban context, has demonstrated multiple health effects in human body [7]. High air pollutant concentrations increases risks for a wide range of diseases including respiratory [8], post-respiratory [9, 10], and is a leading environmental cause of cancer deaths [11]. Health effects of air pollution have been significantly associated with exposure time. Previous study of an urban area was reported that ambient air pollutants have comparatively stronger relationships with temperature than they do with any other meteorological parameter [12]. Degraded air quality, especially in industrial area, has demonstrated various negative human health impacts [13]. Higher concentration of air pollutants in ambient air increases risk for a wide range of health effects including respiratory [14] and cardiac [15, 16] and is a leading environmental cause of cancer related deaths [17]. For instance, a heart attack or stroke resulting from exposure during a day of high ambient air pollutant concentration may be a consequence of chronic disease progression associated with long-term exposure [18]. Some effects are long-term and causation can be difficult to prove because sometimes higher concentrations of air pollutants worked as catalyzing factor for some health effects. Respiratory particles have the largest contribution to the total particle number concentrations on in ambient air [19, 20]. Their major sources include direct emissions (vehicular exhaust, heating, burning) and atmospheric nucleation [21, 22]. Respiratory particles enter into the human body by respiratory track, and approx. 60% of them can be deposited in the respiratory system [23, 24]. They can cause adverse health effects mainly by inflammation in respiratory organs or oxidative stress [25, 26].

In order to take actions to reduce exposure to air pollution and hence the associated health impacts, it is essential to know the sources and activities contributing to local levels of pollution. For this reason, an increasing number of local studies on the contribution of sources to air pollution levels have been developed, most often at city level. Perceptional studies also important in this context because perception of people is the main indicator of pollution level and experienced health effects. On the basis of perception information researchers also identified that pollutant which is present in higher concentration in ambient air of studied area [27]. In the context of air pollution perception plays a major role in identification of pollutant responsible for degradation of air quality in that area in the sense of health effects of air pollution identified in perceptional studies.

2. Climate and air pollution

Multiple linkages connect air pollutants and meteorological parameters in different manners [28, 29]. Air quality degradation may leads to give pace to climatic changes in many polluted regions by changing air pollution meteorology, precipitation and by triggering some amplifying responses in atmospheric chemistry and in anthropogenic and natural sources. Climatic processes and air quality are inextricably connected. Many sources of conventional non-conventional air pollutants are also sources of CO2, other air pollutants and GHGs and/or particles that affect climate. These air pollutants interact with atmospheric solar and terrestrial radiation and perturb the planetary energy balance, leading to changes in climate [30]. Climate change can influences air pollution concentrations according to the changing pattern of climate by altering the frequency, severity, air stagnation events, precipitation, duration of heat waves, and other meteorology conducive to pollutant accumulation [31, 32]. A measure of the perturbation to the climatic system due to changes in various atmospheric constituents between the pre-industrial and present day atmosphere is radiative forcing. Positive radiative forcing induces a warming, whereas negative. Radiative forcing induces a cooling of the earth's surface and troposphere. The increases in the atmospheric burdens of GHGs, including CO2, and that of tropospheric O3 and its precursor methane (CH4) over the past few centuries have exerted a warming influence.

Climate simulations show that the population in tropical regions will be exposed to substantially more frequent daily temperature extremes due to global warming compared to the population at higher latitudes [33]. If anthropogenic air pollutants were a significant contributor to 20th-century climate, then it must have coincided with a large warming from increasing levels of CO2 and other GHGs to produce the observed increase in global mean surface temperature [34]. Tropical latitudes are highly populated regions and home to a large share of the global food and goods production. With high population densities and little incentives for improvement in the work environment, there are millions of people working in factories with poor or non-existent ventilation systems or in agriculture where they are fully exposed to heat and poor air quality.

3. Air quality monitoring in India

The Central Pollution Control Board had adopted first ambient air quality standards on November 11, 1982 as per section 16 (2) (h) of the Air (Prevention and Control of Pollution) Act, 1981. The air quality standards have been revised by the Central Pollution Control Board on April 11, 1994 and were notified in Gazette of India, Extraordinary Part-II Section 3, sub section (ii), dated May 20, 1994.

Central Pollution Control Board initiated National Ambient Air Quality Monitoring (NAAQM) programme in the year 1984 with 7 stations at Agra and Anpara. Subsequently the programme was renamed as National Air Monitoring Programme (N.A.M.P.). The number of monitoring stations under N.A.M.P. has increased, steadily, to 295 by 2000–2001 covering 98 cities/towns in 29 States and 3 Union Territories of the country. Steadily the air quality monitoring network got strengthened by increasing the number of monitoring stations from 28 to 365 during 1985–2009. During the financial year 2010–2011, 93 new stations were added and the number of stations under operation was raised to 456 covering 190 cities in 26 states and 5 Union Territories as on 31st March 2011. As on 31st October 2011 the number of stations under operation has been further raised to 503 distributed in 209 cities, 26 states and 5 UTs [35].

4. Health effects

In large urban areas around the world, the effects of higher concentrations of air pollutants on human health present a growing problem. In this context we can say that only government is not responsible for increasing concentrations of air pollutants in urban and rural areas, awareness and ethics of residents of concerned area also responsible for such situations. Urban residents comparatively more susceptible to effects of air pollution because ambient concentration is high in urban areas than indoor concentration and ambient pollutants also affects indoor concentrations [36], results increase in exposure time with higher concentrations. Urban populations experience heavier exposures to air pollutions. Time of exposures increases with the density of traffic and with traffic congestion, a trend common in many large urban centers and magnitude of exposure increases among higher density populations. In rural areas indoor air pollution in households is predominantly contributed from wood combustion, smoking, paints and varnishes, unpaved roads and agricultural residue burning, resulting in terrible health implications [37], but in rural areas ambient concentration of air pollutants not higher as much as in urban areas. The health effects of air pollution in rural and urban areas have been studied intensely in recent decades. Exposure to higher concentration of pollutants has been associated with increased mortality and hospital admissions (HAs) due to respiratory and cardiovascular diseases. These effects have been found in both short term and long-term studies rural, suburban and urban areas [38]. Exposure time have crucial role in health effect phenomenon in all type of residential areas. During covid-19 situation a number of studies were reported by researcher community in India and in the world, correlation has been drawn between number of covid patient and air quality indexes and associations between them reported in a great number of studies. Studies belongs to different areas of the world and depicted various observations but review of these kind of studies not indicated any kind of correlation of AQI and covid-19 patients because there were a number of places have number of patient with very good AQI. A perceptional study exhibits if the exposure time was long than not only old age residents were affected by air pollution but also residents of young age [39].

5. Role of perception

Perceptional studies have a crucial role in assessment of health effects and also a good indicator of responsible pollutant for AQI. We have done a perceptional study in Gwalior urban area for identification of impacts of air pollution and also awareness level of residents of different urban area such as residential, commercial, high traffic and greenery rich area. This perceptional study includes 22 types of major and minor health effects such as Asthma, lung problem, lung cancer, nose irritation, throat irritation, bronchitis, eye irritation, High blood pressure, nausea, sleeplessness, fatigue and cough etc. reported by respondents of study area. Study also includes questions regarding smoking habit, age group and exposure time. In the awareness section of this study questions included regarding awareness about air pollution, air pollutants, sources of air pollutants, gaseous and particulate pollutants and identification of air pollution as smoke and dust was one of the most interesting outcomes. This study identifies a fairly coherent knowledge regarding the health effects of air pollution and awareness in residents of Gwalior urban area. The results indicate that transportation and solid waste burning are important emitters of air pollutants in the urban area. In the perception of awareness respondents were experienced that the air pollutants harmful for health but they had very less aware about specific effect of air pollutants on health [40].

6. Seasonal effects on air pollutants

Local weather conditions must also be taken into account when assessing the air quality of a certain area, air quality depends on many meteorological and chemical variables [41]. Wind shear and turbulence exerts effects on the mixing and dilution of air pollutants [42]. Humidity is not always linked to the temperature and may be associated with the unequal distribution of moisture with height in the air masses [43]. Pollutants transported from Long distance are dominant during all seasons of the year [44]. Meteorological variables of temperature and wind speed are strongly related to area-wide air pollutant concentrations and a number of meteorological, topographic and settlement issues are significant factors that influence the accumulation, dispersion and chemical transformation processes of air pollutants [45]. In a correlation study of two years between air pollutants and meteorological parameters indicated that maximum concentrations of air pollutants occurred in winter season due to inversion and stagnant air masses, other seasons have intermediate trends according to areas. Minimum concentrations were found in rainy season in this study and there is no strong relationship between air pollutants concentrations and relative humidity. Concentrations of pollutants show comparatively stronger relationships with temperature than they do with any other meteorological parameter such as relative humidity and wind speed [46].

7. Impacts on ecosystems

Air pollution affects ecosystems in a number of ways, altering basic ecosystem functions such as biogeochemical cycling and primary production which in turn affect the ecosystem services and therefore the benefits that humans get from the ecosystem such as clean drinking water, timber wood, medicinal products and appreciation of nature. The complex causal chains by which air pollution subsequently affects a range of ecosystem services have been reviewed extensively by a great number of researchers [47–51]. Eutrophication (nitrogen), acidification (nitrogen and sulfur) and direct toxicity (ozone, ammonia and nitrogen oxides) are the main mechanisms of ecosystem affected by current pollution levels all of which affect a wide range of services due to their impacts on underlying ecosystem functions and processes (**Figure 1**). Pollutants such as sulfur can responsible for excess levels of acid in lakes and streams and damage forest soils and trees, increasing concentration of atmospheric nitrogen can reduce the biodiversity of plant communities and harm almost all species of fish and other aquatic life, higher ozone concentrations can damages tree leaves and negatively affects scenic vistas in protected natural areas, mercury and other heavy metal compounds emitted as

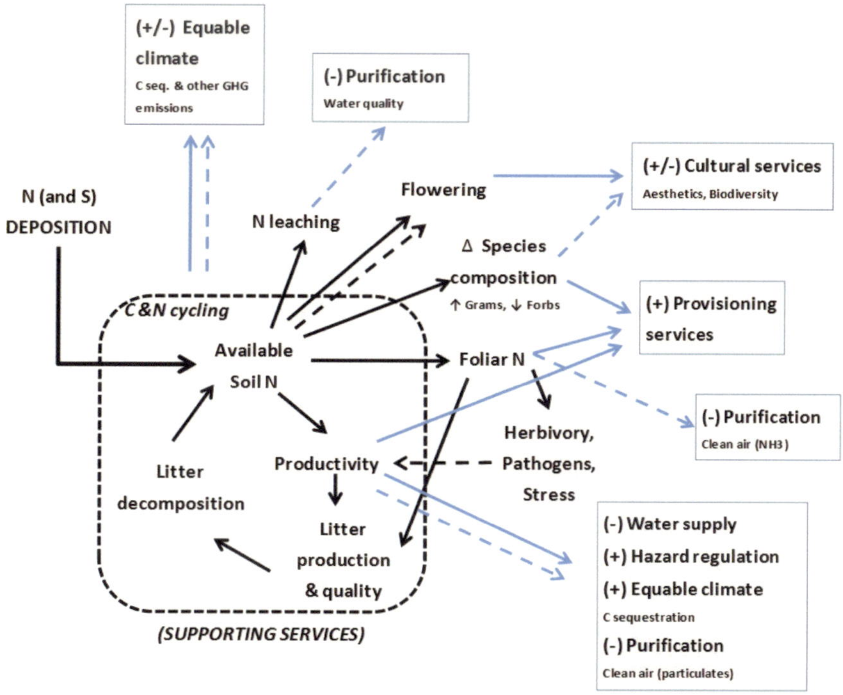

Figure 1.
Eutrophication [47, 48].

exhaust from fuel combustion can eventually accumulate in plants and animals, some of which are consumed by people [52]. Basically, increasing temperature and greenhouse gases derived from climate disturbances have the ability to induce this kind of phenomenons [53].

The northernmost terrestrial regions are experiencing warmer and wetter climate and the rate of warming is projected to increase in the future [54]. Climate change in concerned with many anthropogenic drivers like air pollution, of change is already rapidly affecting northern ecosystems with consequences for living beings, infrastructure and economic condition [55]. The impact that climate change has on natural terrestrial and other ecosystems, on human society and economies could be disastrous, potential effects range from sea level rise and melting ice at higher latitudes and altitudes, to changing regional weather patterns. The increase in the atmospheric abundance of greenhouse gases alters the energy balance of the climate system of terrestrial ecosystems and causes a variety of natural disturbances related to ecosystems [56].

8. Sources of air pollution

There are a number of sources of air pollutants such as fossil fuel, human-induced impacts on forestry, clearing for agriculture, degradation of soils, agricultural activities, waste management, energy use, fertilizer, industrial processes, refrigeration and use of a variety of consumer products [57]. Some other micro pollutants emitted in processes like energy production by combustion: entailing burning of wood, coal, and oil (As, Cd, Cu, Hg, Ni, Pb, Sb, Se, V, and Zn), metallurgical industry: emission of dust near the extraction and point of exploitation, high-temperature processing of ores emit aerosols rich in trace elements (Cd, Cu, Ni, Pb, V, and Zn). The proportion of individual elements emitted in the aerosols depends on the type of ore processed, other industrial processes: high-temperature processing and manufacturing (As, Cr, Cu, Ni, Pb, and Zn), Transport: road traffic (Cd, Cu, Fe, Ni, Pb, and Zn), erosion of brake pads (Cu, and Sb), erosion of train rails (Cu), waste treatment: incineration of household waste (As, Cr, Cu, Ni, Pb, Sb, Se, V, and Zn) (**Figure 2**) [58].

Air pollution in energy production emissions in Indian context increased by 1,563 MtCO2e (246%) from 1990 to 2014. International Energy Agency data show that total electricity generation quadrupled between 1991 and 2014, with an increasing share of coal and a decreasing share of hydropower [60]. Agriculture emissions increased 25% from 1990 to 2014, driven by emissions from synthetic fertilizers (47%) and enteric fermentation from livestock (30%) [61]. In India from 2002 to 2014, use of nitrogen fertilizers (total N) increased 62%, potash (K2O) 59%, and phosphates (P2O5) 51% [62]. India's GDP increased 357% from 1990 to 2014, while GHG emissions increased 180%. In 2014, India emitted more than twice the GHGs relative to GDP than the world average [63]. Globalization in tourism industry [64, 65], public health expenditure and Asian emerging economies can contribute in sustainable economic growth [66, 67]. India pledged to achieve electric power installed capacity of about 40% from non-fossil fuel-based energy resources by 2030 with the help of technology transfer and low-cost international finance, create an additional carbon sink of 2.5 to 3 billion tonnes of CO2 equivalent by 2030 through additional forest and tree cover, mobilize domestic and additional funds from developed countries to implement mitigation actions, and build capacity, create a domestic framework and international architecture for quick diffusion of cutting edge climate technology in India and for joint collaborative research and development for future technologies.

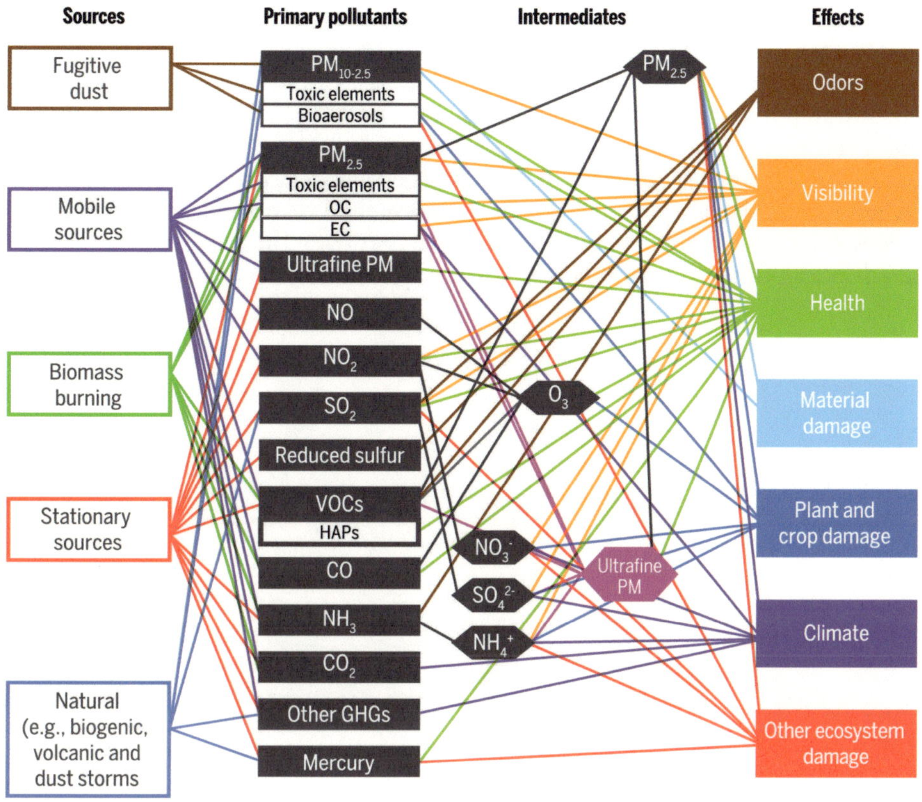

Figure 2.
Effects of air pollutants [59].

9. Conclusion

Air quality degradation is a growing threat to human health, especially in densely populated urban areas exposed to high concentrations of air pollutants. Meteorological parameters can contribute in high concentration episodes of air pollution. Over the last couple of years, air pollutant concentration and human exposure have received special attention due to increasing scientific evidences of causal relationship between exposure with air pollutants and negative health outcomes. Ambient air pollution, especially in urban context, has demonstrated multiple health effects in human health. Health effects of air pollution have been significantly associated with exposure time. Perceptional studies can play a crucial role in future; on the basis of perception information researchers also identified that pollutant which is present in higher concentration in ambient air of studied area. Interaction of air pollutants with atmospheric solar and terrestrial radiation and perturb the planetary energy balance, leading to changes in climate. Climatic changes can influences air pollution concentrations according to the changing pattern of climate by altering the frequency, severity, air stagnation events, precipitation, duration of heat waves and other meteorology conductive to pollutant accumulation. Exposure to higher concentration of pollutants has been associated with increased mortality and hospital admissions due to respiratory and cardiovascular diseases. These effects have been found in both short term and long-term studies in rural, suburban and urban areas. Exposure time have crucial role in health effect phenomenon in all type of residential areas.

Air quality depends on many meteorological and chemical variables. Wind shear and turbulence exerts effects on the mixing and dilution of air pollutants. Humidity is not always linked to the temperature and may be associated with the unequal distribution of moisture with height in the air masses. Pollutants transported from Long distance are dominant during all seasons of the year. Air pollution affects ecosystems in a number of ways, altering basic ecosystem functions such as bio-geochemical cycling and primary production which in turn affect the ecosystem services and therefore the benefits that humans get from the ecosystem such as clean drinking water, timber wood, medicinal products and appreciation of nature.

Author details

Banwari Dandotiya
Department of Environmental Science, Jiwaji University Gwalior, India

*Address all correspondence to: banwaridandotiya@gmail.com

References

[1] Banwari Dandotiya, Harendra K. Sharma & Nimisha Jadon (2020): Ambient Air Quality and Meteorological Monitoring of Gaseous Pollutants in Urban Areas of Gwalior City India, Environmental Claims Journal, DOI: 10.1080/10406026.2020.1744854

[2] Harendra K. Sharma, Banwari Dandotiya & Nimisha Jadon (2017). Exposure of Air Pollution and Its Health Effects in Traffic Police Persons of Gwalior City, India, Environmental Claims Journal, 29:4, 305-315, DOI: 10.1080/10406026.2017.1390357

[3] Vadrevu, K.P., Badarinath, K.V.S., Anuradha, E., 2008. Spatial patterns in vegetation fires in the Indian region. Environ. Monit. Assess. 147, 1-13. http://dx.doi.org/10.1007/s10661-007-0092-6.

[4] Vadrevu, K.P., Ellicott, E., Badarinath, K., 2011. MODIS derived fire characteristics and aerosol optical depth variations during the agricultural residue burning season, North India. Environ. Pollut. 159, 1560-1569. http://dx.doi.org/10.1016/j.envpol.2011. 03.001.

[5] Venkataraman, C., Habib, G., Kadamba, D., Shrivastava, M., Leon, J.F., Crouzille, B., Boucher, O., Streets, D.G., 2006. Emissions from open biomass burning in India: integrating the inventory approach with high-resolution Moderate Resolution Imaging Spectroradiometer (MODIS) active-fire and land cover data. Glob. Biogeochem. Cy. 20, GB2013. http://dx.doi.org/10.1029/2005GB002547.

[6] Harendra K. Sharma, Banwari Dandotiya & Nimisha Jadon (2017). Exposure of Air Pollution and Its Health Effects in Traffic Police Persons of Gwalior City, India, Environmental Claims Journal, 29:4, 305-315, DOI: 10.1080/10406026.2017.1390357

[7] Banwari Dandotiya, (2019). Health Effects of Air Pollution in Urban Environment. *Climate Change and Its Impact on Ecosystem Services and Biodiversity in Arid and Semi-Arid Zones.* IGI Global, 2019. 96-115. doi:10.4018/978-1-5225-7387-6.ch006,

[8] Guarnieri, M. and Balmes, J.R., 2014, *Outdoor air pollution and asthma* Lancet, **383**, 1581-1592. doi:10.1016/S0140-6736(14)60617-6.

[9] Brook, R.D., Rajagopalan, S., Pope, C.A., Brook, J.R., Bhatnagar, A., Diez-Roux, A.V., Holguin, F., Hong, Y., Luepker, R.V., Mittleman, M.A., Peters, A., Siscovick, D., Smith, S.C Jr., Whitsel, L., and Kaufman, J.D, 2010, Particulate matter air pollution and cardiovascular disease: an update to the scientific statement from the American Heart Association. Circulation, **121**(21), 2331-2378.

[10] EPA, 2009, *Integrated Science Assessment (ISA) for particulate matter (final report)*. Washington, DC: United States Environmental Protection Agency. Available online at: http://cfpub.epa.gov/ncea/cfm/recordisplay.cfm?deid=216546#Download (accessed 21 July 2017).

[11] IARC (International Agency for Research on Cancer), 2013, PRESS RELEASE No 221. Availableonline at: https://www.iarc.fr/en/media-centre/iarcnews/pdf/pr221_E.pdf (accessed 4 August 2017).

[12] Banwari Dandotiya, Nimisha Jadon & Harendra K. Sharma (2019). Effects of Meteorological Parameters on Gaseous Air Pollutant Concentrations in Urban Area of Gwalior City, India, Environmental Claims Journal, 31:1, 32-43, DOI: 10.1080/10406026.2018.1507508

[13] World Health Organization, 2013, *Review of evidence on health*

aspects of air pollution – REVIHAAP Project: final technical report. Available online at: http://www.euro.who.int/en/health-topics/environment-and-health/air-quality/publications/2013/review-of-evidence on health- aspects-of-air-pollution-revihaap-project-final-technical-report (accessed 24 June 2017).

[14] Guarnieri, M. and Balmes, J.R., 2014, *Outdoor air pollution and asthma* Lancet, **383**, 1581-1592. doi:10.1016/S0140-6736(14)60617-6.

[15] Brook, R.D., Rajagopalan, S., Pope, C.A., Brook, J.R., Bhatnagar, A., Diez-Roux, A.V., Holguin, F., Hong, Y., Luepker, R.V., Mittleman, M.A., Peters, A., Siscovick, D., Smith, S.C Jr., Whitsel, L., and Kaufman, J.D, 2010, Particulate matter air pollution and cardiovascular disease: an update to the scientific statement from the American Heart Association. Circulation, **121**(21), 2331-2378.

[16] EPA, 2009, *Integrated Science Assessment (ISA) for particulate matter (final report)*. Washington, DC: United States Environmental Protection Agency. Available online at: http://cfpub.epa.gov/ncea/cfm/recordisplay.cfm?deid=216546#Download (accessed 21 July 2017).

[17] IARC (International Agency for Research on Cancer), 2013, PRESS RELEASE No 221. Available online at: https://www.iarc.fr/en/media-centre/iarcnews/pdf/pr221_E.pdf (accessed 4 August 2017).

[18] Hime, N., Cowie, C., Marks, G., 2015, Review of the health impacts of emission sources, types and levels of particulate matter air pollution in ambient air in NSW Produced for the NSW Environment Protection Authority and NSW Ministry of Health, Environmental Health Branch. Available online at: http://www.carcre.org.au/images/content/Health%20impacts%20of%20 PM%20report_final%20for%20web.pdf.

[19] Kulmala, M., Vehkamäki, H., Petäjä, T., Dal Maso, M., Lauri, A., Kerminen, V.-M., Birmili, W., McMurry, P., 2004. Formation and growth rates of ultrafine atmospheric particles: review of observations. J. Aerosol Sci. 35, 143-176.

[20] Spracklen, D.V., Carslaw, K.S., Kulmala, M., Kerminen, V.-M., Mann, G.W., Sihto, S.-L., 2006. The contribution of boundary layer nucleation events to total particle concentrations on regional and global scales. Atmos. Chem. Phys. 6, 5631-5648.

[21] Wåhlin, P., Palmgren, F., Van Dingenen, R., 2001. Experimental studies of ultrafine particles in streets and the relationship to traffic. Atmos. Environ. 35, 63-69.

[22] Salma, I., Füri, P., Németh, Z., Balásházy, I., Hofmann, W., Farkas, Á., 2015. Lung burden and deposition distribution of inhaled atmospheric urban nanoparticles as the first step in their health risk assessment. Atmos. Environ. 104, 39-49.

[23] Kulmala, M., Kontkanen, J., Junninen, H., Lehtipalo, K., Manninen, H.E., Nieminen, T., Petäjä, T., Sipilä, M., Schobesberger, S., Rantala, P., Franchin, A., Jokinen, T., E., Äijälä, M., Kangasluoma, J., Hakala, J., Aalto, P.P., Paasonen, P., Mikkilä, J., Vanhanen, J., Aalto, J., Hakola, H., Makkonen, U., Ruuskanen, T., Mauldin III, R.L., Duplissy, J., Vehkamäki, H., Bäck, J., Kortelainen, A., Riipinen, I., Kurtén, T., Johnston, M.V., Smith, J.N., Ehn, M., Mentel, T.F., Lehtinen, K.E.J., Laaksonen, A., Kerminen, V.-M., Worsnop, D.R., 2013. Direct observations of atmospheric aerosol nucleation. Science 339, 943-946.

[24] Geiser, M., Rothen-Rutishauser, B., Kapp, N., Schürch, S., Kreyling,

W., Schulz, H., Semmler, M., Im Hof, V., Heyder, J., Gehr, P., 2005. Ultrafine particles cross cellular membranes by nonphagocytic mechanisms in lungs and in cultured cells. Environ. Health Perspect. 113, 1555-1560.

[25] Oberdörster, G., Oberdörster, E., Oberdörster, J., 2005. Nanotoxicology: an emerging discipline evolving from studies of ultrafine particles. Environ. Health Perspect. 113, 823-839.

[26] HEI Review Panel on Ultrafine Particles, 2013. Understanding the Health Effects of Ambient Ultrafine Particles. HEI Perspectives 3. Health Effects Institute, Boston.

[27] Banwari Dandotiya, (2019). Health Effects of Air Pollution in Urban Environment. *Climate Change and Its Impact on Ecosystem Services and Biodiversity in Arid and Semi-Arid Zones.* IGI Global, 2019. 96-115. doi:10.4018/978-1-5225-7387-6.ch006,

[28] Banwari Dandotiya, Nimisha Jadon & Harendra K. Sharma (2019) Effects of Meteorological Parameters on Gaseous Air Pollutant Concentrations in Urban Area of Gwalior City, India, Environmental Claims Journal, 31:1, 32-43, DOI: 10.1080/10406026.2018.1507508

[29] Banwari Dandotiya, Harendra K. Sharma & Nimisha Jadon (2020): Ambient Air Quality and Meteorological Monitoring of Gaseous Pollutants in Urban Areas of Gwalior City India, Environmental Claims Journal, DOI: 10.1080/10406026.2020.1744854

[30] Intergovernmental Panel on Climate Change. 2013. Climate Change 2013: The Physical Science Basis, Contribution of Working Group I to the Fifth Assessment Report of the Intergovernmental Panel on Climate Change. New York, NY: Cambridge University Press.

[31] Jacob, D. J., and D. A. Winner. 2009. Effect of climate change on air quality. Atmos. Environ. 43(1): 51-63. doi:10.1016/j.atmosenv.2008.09.051

[32] Vieno, M., A. J. Dore, D. S. Stevenson, R. Doherty, M. R. Heal, S. Reis, S. Hallsworth, L. Tarrason, P. Wind, D. Fowler, D. Simpson and M. A. Sutton. 2010. Modelling surface ozone during the 2003 heat-wave in the UK. Atmos. Chem. Phys. 10(16): 7963-7978. doi:10.5194/acp-10-7963-2010

[33] Harrington LJ, Frame DJ, Fischer EM, et al. Poorest countries experience earlier anthropogenic emergence of daily temperature extremes. Environ Res Lett. 2016;11:055.

[34] Arlene M. Fiore, Vaishali Naik & Eric M. Leibensperger (2015) Air Quality and Climate Connections, Journal of the Air & Waste Management Association, 65:6, 645-685, DOI: 10.1080/10962247.2015.1040526

[35] CPCB (2006) Central pollution control board, Ministry of environment&forests, Goverenment of India, Paarivesh Bhawan, East Arjun Nagar, Delhi. http://www.cpcb.nic.in/Annual%20Report/Previous%20Year/AnnualReport_2005-2006.pdf

[36] Banwari Dandotiya, (2019). Health Effects of Air Pollution in Urban Environment. *Climate Change and Its Impact on Ecosystem Services and Biodiversity in Arid and Semi-Arid Zones.* IGI Global, 2019. 96-115. doi:10.4018/978-1-5225-7387-6.ch006,

[37] World Health Organization. 2007. Indoor air pollution: national burden of disease estimates.

[38] Anderson JO, Thundiyil JG, Stolbach A. 2012. Clearing the air: a review of the effects of particulate matter air pollution on human health. J Med Toxicol. 8(2):166-175.

[39] Harendra K. Sharma, Banwari Dandotiya & Nimisha Jadon (2017). Exposure of Air Pollution and Its Health Effects in Traffic Police Persons of Gwalior City, India, Environmental Claims Journal, 29:4, 305-315, DOI: 10.1080/10406026.2017.1390357

[40] Banwari Dandotiya (2019). Ambient Air Quality Monitoring of Gwalior City and its Impact on Human Population, Corpus ID: 213382002

[41] Galina I. Y., R. Tsitouridou, S. L. Tsakovski & V. D. Simeonov. 2016. Urban air quality assessment using monitoring data of fractionized aerosol samples, chemometrics and meteorological conditions. Journal of Environmental Science and Health, Part A 51 (1): 544-552.

[42] Fisher B. 2002. Meteorological factors influencing the occurrence of air pollution episodes involving chimney plumes. Meteorological Applications 9, 199-210.

[43] Nygård T., Valkonen T. & Vihma T. 2014. Characteristics of Arctic low-tropospheric humidity inversions based on radio soundings. Atmospheric Chemistry and Physics 14, 1959-1971.

[44] Dekhtyareva A., Edvardsen K., Holmén K., Hermansen O. & Hansson H.-C. 2016. Influence of local and regional air pollution on atmospheric measurements in Ny-Ålesund. International Journal of Sustainable Development and Planning 11, 578-587.

[45] Banwari Dandotiya, Nimisha Jadon & Harendra K. Sharma (2018): Effects of Meteorological Parameters on Gaseous Air Pollutant Concentrations in Urban Area of Gwalior City, India, Environmental Claims Journal, DOI: 10.1080/10406026.2018.1507508

[46] Banwari Dandotiya, Harendra K. Sharma & Nimisha Jadon (2020): Ambient Air Quality and Meteorological Monitoring of Gaseous Pollutants in Urban Areas of Gwalior City India, Environmental Claims Journal, DOI: 10.1080/10406026.2020.1744854

[47] Jones, L. ; Provins, A. ; Harper-Simmonds, L. ; Holland, M. ; Mills, G.; Hayes, F.; Emmett, B.A.; Hall, J.; Sheppard, L.G. ; Smith, R.; Sutton, M. ; Hicks, K. ; Ashmore, M. ; Haines-Young, R. **2014** A review and application of the evidence for nitrogen impacts on ecosystem services. Ecosystem Services 7 76-88

[48] Jones, M.L.M.; Provins, A. ; Harper-Simmonds, L. ; Holland, M. ; Mills, G.; Hayes, F.; Emmett, B.A.; Hall, J.; Sheppard, L.J.; Smith, R.; Sutton, M. ; Hicks, K. ; Ashmore, M. ; Haines-Young, R. **2014** Using the Ecosystems Services Approach to value air quality.

[49] Mills, G.; Harmens, H. **2011** Ozone pollution: a hidden threat to food security.

[50] Harmens, H. ; Mills, G. **2012** Ozone pollution: Impacts on carbon sequestration in Europe. ICP Vegetation Programme Coordination Centre

[51] Mills, G.; Wagg, S. ; Harmens, H. **2013** Ozone pollution: Impacts on ecosystem services and biodiversity. ICP Vegetation Programme Coordination Centre

[52] 52.https://www.epa.gov/eco-research/ecosystems-and-air quality#:~:text=For%20example%3A%20pollutants%20such%20as,affects%20scenic%20vistas%20in%20protected (visited, 01/02/2021).

[53] Banwari Dandotiya and Harendra K. Sharma (2020). Climate Change and Its Impact on Terrestrial Ecosystems, Impacts of Climate Change on

Agriculture and Aquaculture. DOI: 10.4018/978-1-7998-3343-7.ch007

[54] IPCC. (2013). Summary for policy makers. In Climate Change 2013: The Physical Science Basis. Contribution of Working Group I to the Fifth Assessment Report of the Intergovernmental Panel on Climate Change (pp. 1-32). Cambridge University Press.

[55] Arctic Council. (2013). Arctic Resilience Interim Report 2013. Stockholm Environmental Institute and Stockhol Resilience Centre.

[56] IPCC, 2014. Intergovernmental Panel on Climate Change, *Contribution of Working Group III to the Fifth Assessment Report of the Intergovernmental Panel on Climate Change*

[57] IPCC. (2007). Climate Change 2007: The Physical Science Basis: Contribution of Working Group I to the Fourth Assessment Report of the Intergovernmental Panel on Climate Change. Cambridge University Press.

[58] J. M. Pacyna and E. G. Pacyna (2011). An assessment of global and regional emissions of trace metals to the atmosphere from anthropogenic sources worldwide. Environmental Reviews 9(4):269-298 DOI: 10.1139/er-9-4-269

[59] Saikawa E. et al. (2019) Air Pollution in the Hindu Kush Himalaya. In: Wester P., Mishra A., Mukherji A., Shrestha A. (eds) The Hindu Kush Himalaya Assessment. Springer, Cham. https://doi.org/10.1007/978-3-319-92288-1_10

[60] International Energy Agency (IEA). Statistics: India, Electricity and Heat, 1991 and 2014, viewed on September January 18, 2021. In 1991, 23% of electricity was generated by hydroelectric plants.

[61] FAOSTAT, 2018.

[62] FAOSTAT, India – Agriculture Use - Fertilizers by Nutrient (2002-2014), viewed on January 19, 2021.

[63] WRI CAIT 4.0, 2017.

[64] Arshian Sharif Danish, Iqbal Godil Bingjie Xu Avik Sinha Syed Abdul Rehman Khan Kittisak Jermsittiparsert (2020). Revisiting the role of tourism and globalization in environmental degradation in China: Fresh insights from the quantile ARDL approach, Journal of Cleaner Production, Volume 272, 1, 122906. https://doi.org/10.1016/j.jclepro.2020.122906

[65] Arshian Sharif, Sahar Afshan, Sesokunthideth Chrea, Aguir Amel & Syed Abdul Rehman Khan (2020). The role of tourism, transportation and globalization in testing environmental Kuznets curve in Malaysia: new insights from quantile ARDL approach, Environmental Science and Pollution Research, volume 27, pages25494-25509(2020).

[66] Syed Abdul Rehman Khan, Yu Zhang, Anil Kumar, Edmundas Zavadskas, Dalia Streimikiene (2020). Measuring the impact of renewable energy, public health expenditure, logistics, and environmental performance on sustainable economic growth, https://doi.org/10.1002/sd.2034

[67] Syed Abdul Rehman Khan, Arshian Sharif, Hêriş Golpîra, Anil Kumar (2019). A green ideology in Asian emerging economies: From environmental policy and sustainable development, https://doi.org/10.1002/sd.1958

Section 3

Environmental Sustainability Practices

Chapter 11

Aerotechnogenic Pollution of Boreal Forests in Northern Europe

Alexander Evdokimov

Abstract

This paper discusses the changes that boreal forest ecosystems undergo under the influence of gaseous waste from the processing of non-ferrous ores on the Kola Peninsula. These communities are represented primarily by pine forests growing on the northern border of their range. The main forest-forming species here is Scots pine main components for this local aeronautical emission are polymetallic dust and sulfur dioxide, which is the main by-product during the roasting of sulfide and polysulfide ores. The studies were carried out on the basis of materials obtained at 6 sample plots located at different distances from the pollution source. As a result, an exponential increase in the content of heavy metals in the soil, as well as in the assimilatory organs of the components of these communities, was shown when approaching the source of pollution (this pattern is different for each of the metals). Regularities of negative changes in the structure of some components of plant communities, such as phytomass, projective cover of the lower layers, and the vital structure of the tree layer were also identified.

Keywords: northern taiga, boreal forests, pine forests, aerotechnogenic emissions, heavy metals, sulfur dioxide, nonferrous metallurgy

1. Introduction

At present, atmospheric pollution is one of the most pressing environmental problems. An actively developing industry inevitably has a negative impact on the fragile structure of biocenoses. It is not only natural communities in the immediate vicinity of industrial centers that are under threat. The development of the transport system [1, 2], tourism [3] and, in general, the improvement of the quality of life of the population of the region has a negative impact. However, the most noticeable man-made effect of a local nature (including aero-man-made one) is produced by large enterprises. This problem is especially acute in the Russian Federation, where one of the main source of income for the state is the extraction and primary processing of natural resources (cleaning of raw materials, remelting ores). Basically, such pollution is of a local nature, and exposure to toxic substances occurs only in the area associated with the enterprise, as evidenced by various studies [4–6]. This study was carried out on the territory of the Kola Peninsula, the Murmansk region, where, in addition to the main object of pollution: Monchegorsk mining and metallurgical plant "Severonikel", there are a number of other industrial pollutants (Kola NPP, Kandalaksha aluminum plant, Apatity plant of nonmetallic materials, mining and metallurgical plant "Pechenganikel"). However, such enterprises have insignificant local atmospheric pollution. In the taiga zone of this region, the main plant communities are pine and spruce forests. Therefore, the species that make up such communities were selected by us as indicators of industrial atmospheric pollution.

At the moment, special attention is paid to assessing the level of pollution and assessing the state of plant and animal communities affected by this pollution. First of all, this is due to the general tendency to introduce such concepts as "ecological significance" and "environmental impact" into all aspects of our life. On the other hand, the very policy of the state develops in such a way that there is a shift in priorities from the predatory exploitation of natural resources to an attempt to maintain the ecological state of nature in a stationary primordial state and a model of sustainable development.

2. Aerotechnogenic impact on plant communities

To date, a significant amount of data has been accumulated on the impact of such emissions on the biocenoses of the Kola Peninsula. The effect of harmful effects on the layers in the community is especially pronounced in the areas where industrial enterprises are located. In these cases, toxic substances in the air form compounds that simultaneously affect different plant organs. The Severonikel Plant (Monchegorsk, Murmansk Region) is one of the largest non-ferrous metallurgy plants in the Russian Federation. The first permanent sample plots in the region were established in the 1970s. Since the second half of the 90s of the XX century, the plant has significantly reduced the amount of airborne industrial emissions (**Figures 1** and **2**), as evidenced by the data of various studies [7].

For several decades, atmospheric emissions have allowed toxic compounds to accumulate in biotic and abiotic components of the environment in the vicinity of the plant. The results of the analysis of various components of Scots pine (*Pinus sylvestris* L.) communities (soil, litter and various parts of plants) in the vicinity of the plant in the 1980s differed by almost two orders of magnitude from similar samples taken in the background areas, and were threshold for plant growth [8, 9].

Currently, special attention is paid to assessing the level of pollution and assessing the state of plant and animal communities subjected to this and similar pollution. First of all, this is due to the general tendency to introduce such concepts as "ecological significance" and "environmental impact" into all aspects of our life.

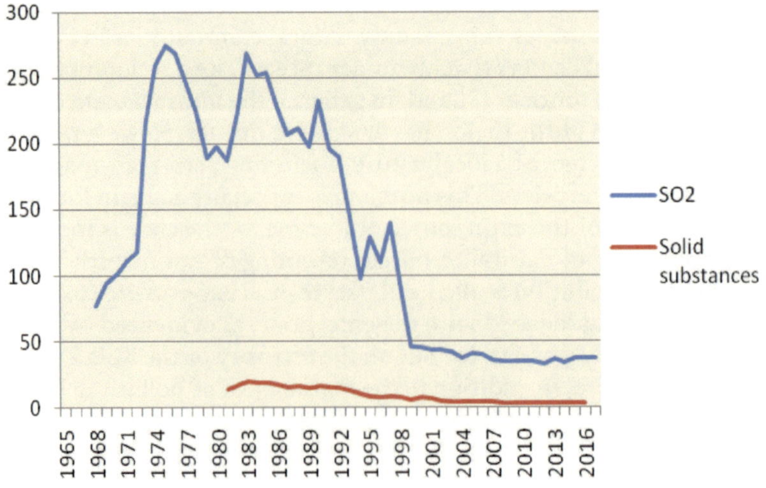

Figure 1.
Dynamics of the total volume of airborne industrial emissions from the Severonikel plant, thousand tons.

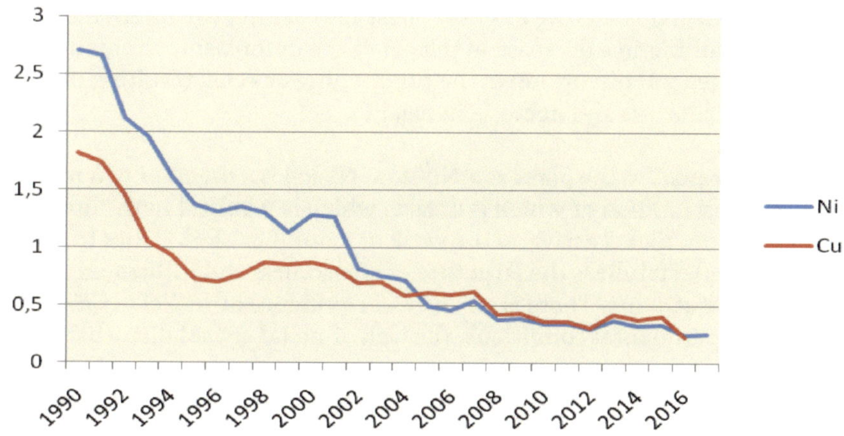

Figure 2.
Dynamics of heavy metals emissions from the Severonikel plant, thousand tons.

On the other hand, the very policy of the state develops in such a way that there is a shift in priorities from the active "predatory" exploitation of natural resources to an attempt to maintain the ecological state of nature in a stationary primordial state (according to the concept of sustainable development).

The progressive deterioration of the vital state and drying out of pine forests under the influence of industrial atmospheric pollution in almost all European countries, including Russia, is currently attracting special attention of researchers. Significant damage to pine forests is caused by uncontrolled accidental emissions even from relatively small industrial enterprises, in which the concentration of pollutants entering the atmosphere increases sharply. The influence of emissions is more distributed in the direction of the prevailing winds. On the basis of comprehensive and in-depth studies of the influence of industrial pollution on the growth, productivity and condition of Scots pine stands, methods and methodological approaches to assessing changes in forest phytocenoses under anthropogenic disturbances are proposed.

Studies have established that a decrease in the height gain in pine under the influence of air and soil pollution occurs much later than the manifestation of visual signs of damage to the assimilation apparatus of trees [10]. It was also found that the decrease in height gain and the process of top drying did not decrease with limited emissions [11]. This can be explained by the fact that the intensity of the drying process is due not only to the amount, but also to the concentration of harmful chemicals in the soil and tree tissues.

It was also found that the increase in diameter of 50-100-year-old individuals can serve as a reliable indicator of environmental pollution. These trees reduce the growth rate before serious damage or death of individuals occurs.

3. The main products aerotechnogenic emissions

Sulfur compounds.

These substances have an effect primarily on the assimilation apparatus of plants. The main atmospheric pollutant in this region is sulfur dioxide (SO_2), which, entering the leaves and needles through the stomata, turns into a highly toxic sulfite ion (SO_3^{2-}), which in turn is slowly oxidized to a less toxic sulfate ion (SO_4^{2-}). In addition to the selective effect on some enzymes and catalytic chains (disruption of the light and dark phases of photosynthesis, the effect on the state of chlorophyll, changes in the lamellar structure of gran. These ions acidify the internal environment of cells, resulting in a decrease in the pH of the cytoplasm, stroma and matrix, which leads to a general disruption in the functioning of cells, resulting in the development of chlorosis and necrosis in needles.

Heavy metals.

Nickel. It is present in the plant as a Ni^{2+} ion. Nickel is a member of a number of enzymes, the most studied of which is urease, which is involved in the breakdown of urea. In addition, nickel activates the work of a number of enzymes (peptidases, nitrate reductases), stabilizes the structure of ribosomes, and influences the supply and transport of nutrients. There is still no clear evidence of nickel deficiency in plants. Under experimental conditions, the lack of metal causes disturbances in the metabolism of urea.

Copper. The question of the form in which copper enters underground and aboveground organs (Cu^{+}; Cu^{2+}) is currently under discussion. In plants, up to 98% of the metal is in an insoluble bound state. Seeds and growing parts of the shoot are relatively rich in this element. In the leaves, most of the copper is concentrated in chloroplasts and almost half in the composition of plastocyanin, one of the electron carriers between PS I and PS II. Most of the functions of copper are associated with its participation in enzymatic redox reactions. In addition, copper promotes the formation of chlorophyll and slows down its destruction in the dark. It affects nitrogen metabolism, being a part of nitrite reductase and nitric oxide reductases, and enhances the process of binding nitrogen molecules. Copper Functions in the cytochrome oxidase complex of the mitochondrial respiratory chain. It also contributes to the intake of manganese, zinc and boron into the body, increases drought, frost, and heat resistance, takes an active part in protecting against pathogens. Lack of copper causes a delay in growth and flowering, leads to a loss of turgor in the leaves. A high content of copper in plants can lead to a change in pigmentation (blackening of the petals of angiosperms, the copper content can reach 0.021%).

Cobalt. For a long time, cobalt was considered as an element necessary only for animals and microorganisms. Currently, it is referred to as the metals necessary for higher plants. In plants, cobalt is found in free ionic (Co^{2+}; Co^{3+}) and bound forms. It concentrates in the generative organs, accumulates in the pollen, and accelerates its germination. Cobalt enhances protein biosynthesis, regulates growth processes, removes the inhibitory effect of auxin on cell division and inhibits ethylene biosynthesis. In addition, cobalt is a part of vitamin B12 and increases immunity to certain diseases, takes part in redox processes, increases the content of pigments in leaves, which is associated with an increase in the plastid apparatus due to the replication and growth of organelles. Along with magnesium and manganese, cobalt activates the glycolysis enzyme phosphoglucomutase.

The main source of heavy metals entering the environment is technogenic. The mechanism of entry of heavy metals from the soil into plants by the root route includes passive (non-metabolic) transfer of ions into the cell in accordance with their concentration gradient and active (metabolic) absorption by the cell against the concentration gradient. Also, heavy metals in the composition of aerosols and dust fall on the sheet, are retained on it in the form of surface deposits, some can be washed out by rainwater, and some enter the plant [12].

The influence of heavy metals on physiological processes.

Growth. Growth inhibition is one of the most common manifestations of the toxicity of heavy metals to plants. In the presence of high concentrations of heavy metals, the intensity of cell division slows down, the number of cells decreases at all phases of mitosis, and the duration of individual phases and the entire mitotic cycle increases. In addition, heavy metals can slow down the presynthetic and postsynthetic stages of cell division.

A reliable relationship was found between the radial growth of tree trunks and the volumes of emissions from the plant entering the atmosphere, while the relationship is nonlinear and depends on the level of anthropogenic load on trees at different distances from the plant, on the species of trees and location on the slopes of hills. In particular, the response of pine to technogenic stress was revealed. Despite a significant decrease in emissions in recent years, the radial growth of trees in the impact zone continues to remain significantly less than the background values. The main reason is the high content of heavy metals in the organogenic horizon of the soil and the persisting atmospheric precipitation.

Photosynthesis. A decrease in the intensity of photosynthesis in plants in the presence of heavy metals is primarily associated with their negative effect on photosynthetic pigments. In the presence of heavy metals, a decrease in the content of chlorophyll a and b was found. At the same time, pronounced chlorosis is observed on the leaves. The main reason for this effect is the suppression of chlorophyll biosynthesis, which is associated with the direct action of heavy metals on the active centers of enzymes. It is also possible to expel magnesium ions from the chlorophyll molecule.

Disruption of the chloroplast ultrastructure in the presence of heavy metals is also one of the most important reasons for a decrease in the content of pigments in plants and, in general, for a decrease in the intensity of photosynthesis. So, under the action of nickel and copper in high concentrations, the number of plastoglobules in plants increases, which indicates an increase in the degradation of organelles. In addition, the number of grains decreases and their structure is disturbed. Grana become irregular and contain fewer thylakoids. Heavy metals directly affect the transfer of electrons in photochemical reactions. In addition, in the presence of heavy metals in plants, cyclic and non-cyclic phosphorylation slows down, and ATP synthesis is suppressed.

4. Material and methods

The studies were carried out on the basis of material obtained on the territory of the Kola Peninsula in Monchegorsk (buffer and impact zones) and Kovdorsk

№	Sample plot name	Distance from pollution source, km	Community age, years	Height above sea level, m	Coordinates
Impact zone					
1	10 O	10	80	158	N 68 00.384, E 032 55.541
2	29	15	80	162	N 67 44.216, E 032 46.447
Buffer zone					
3	3	25	80	169	N 68 06.817, E 033 19.455
4	27 O	35	80	165	N 67 38.168, E 032 42.234
Non-contaminated zone					
5	20	65	80	145	N 67 30.055, E 031 46.886

Table 1.
Location of permanent sample plots in the study area.

Figure 3.
Sample plots location map.

(non-contaminated zone) districts on 5 previously established permanent sample plots oriented along the pollution gradient from the emission source to the background in the northern and southern directions (**Table 1**, **Figure 3**).

On each trial plot, samples of the upper organic horizon of Al-Fe-podzol were taken in the amount of 5 samples for each trial plot, from which the undecomposed upper elements of forest litter were excluded. These samples were used to prepare a soil extract (extract with 1.0 N HCl solution), for which the content of acid-soluble forms of Ni, Cu, Co was further determined by atomic absorption spectrometry. Similarly, the content of heavy metals was determined in the assimilation organs of plants growing on these test plots (the tree layer and the undergrowth canopy were mainly formed by the species *Pinus sylvestris* L.; the herb-dwarf layer was formed by the species: *Vaccinium vitis-idaea* L., *Vaccinium myrtillus* L., Empetrum hermaphroditum Hagerup., *Arctostaphylos uva-ursi* (L.) Spreng.; moss-lichen layer is formed by species: lichens of the genus *Cladonia* (*C. stellaris* (Opiz.) Brodo, *C. rangiferina* (L.) Nyl., *C. mitis* (Sandst.) Hustich), *C. coccifera* (L.) Willd. and *Pleurosium schreberi* (Brid.) Mitt.). The work presents the average values of the available sample for these sample plots.

For the ground layers, the total projective cover of the species was determined, as well as the ground phytomass of plants. The characteristics of the tree layer were also obtained on these permanent sample plots. We assigned all individuals of the forest-forming species with a trunk diameter of 4 cm or more at a height of 1.3 m to the tier of the stand. For all individuals of the stand, trunk diameter, tree height and vitality class were determined. We have identified 5 categories of vitality class:

I - healthy individuals;

II - defoliated individuals;

III - severely defoliated individuals;

IV - withered individuals;

V - dead individuals.

The vitality structure of the tree layer was analyzed. The index of the vital state of the stand was also calculated using the formula (Eq. (1)):

$$I_n = \frac{n_1 + 0{,}7n_2 + 0{,}4n_3 + 0{,}1n_4}{n} \tag{1}$$

where I_n – life condition index;
n_1 – the number of individuals classified as "healthy";
n_2 – the number of individuals classified as "defoliated";
n_3 – the number of individuals classified as "severely defoliated";
n_4 – the number of individuals classified as "withered";
n – total number of individuals.

5. Results

Content of heavy metals in community components.

As can be assumed, with approaching the source of pollution, the amount of heavy metals in the components of the community is steadily increasing. However, the nature of this increase may vary depending on the specific component of the community (including its species) and the pollutant.

Organic soil horizons.

The content of nickel in the non-contaminated zone averages 7.5 mg/kg, copper - 16.75 mg/kg, cobalt - 1.5 mg/kg. In studies of past years, the value of 10 mg/kg for the first two metals is taken as background values. Consequently, a slight increase in the copper content is observed, which is most likely associated with the peculiarity of the location of the constant sample plot relative to the prevailing winds, as well as with the change in the activity of the plant (the use of various ores, the change in the nature of production and purification, etc.). We take the cobalt content as the background value.

In the buffer zone, the nickel content is about 50 mg/kg, which is 5 times higher than the background values. The content of copper ranges from 185.5 to 76.9 mg/kg (different sample plots), cobalt - from 2.72 to 5.63 mg/kg, which also significantly exceeds the background values.

In the impact zone, the concentration in all three cases sharply increases the concentration. For nickel, the content can reach 598.5 mg/kg (60 times), for copper - 3582 mg/kg (almost 360 times), for cobalt - 28.7 mg/kg (almost 20 times) (**Figure 4**).

The data obtained are consistent with the earlier stages of the study of this region. It should also be noted that this picture is observed with a decrease in the intensity of emissions into the atmosphere by the Severonikel plant. This indicates a strong fixation of heavy metals in the organic horizon and a very low rate of their leaching into the underlying horizons.

Community Component Characteristics.

Tree layer.

The main component of pine forests is a tree stand, the analysis of the vitality structure of which can indicate not only the state of the entire cenopopulation at the moment, but can also help to reveal the natural processes of formation and development, as well as reflect the effect of stress factors (in our case, this is aerotechnogenic pollution) To assess the vitality class, as a rule, a number of quantitative and qualitative signs are used. On the basis of these features, from 3 to 6 classes are distinguished and the vital state index is calculated (in our work we will limit ourselves to 5 classes).

In the non-contaminated zone (65 km and more from the source of pollution), healthy individuals of *Pinus sylvestris* L. dominate (from 61–68%). The ratio of defoliated individuals varies from 20–9%, those of severely defoliated ones - from 15–17%. The number of withered trees is minimal and ranges from 2–4%. Such a significant number of defoliated and severely defoliated trees can be explained by the competition between trees for water and nutrients in the soil, and further

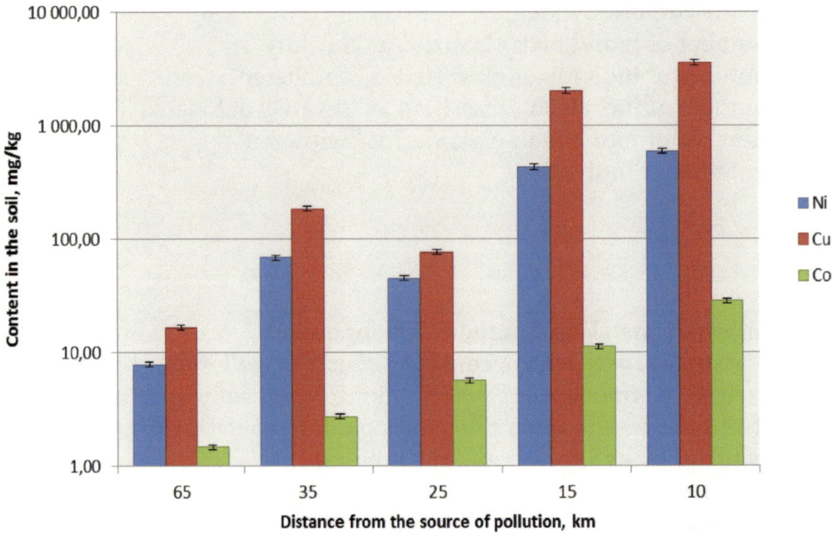

Figure 4.
The content of heavy metals in the organic soil horizon at different distances from the source of pollution (logarithmic scale of ordinates).

differentiation of populations of the same age by categories of vitality class. The vitality class index was 0.81.

In the buffer zone, the proportion of healthy plants does not exceed 64%. The share of defoliated and severely defoliated trees is 10% and 12%, respectively. These data are quite similar to what we observed in the non-contaminated regions. But at the same time, the share of withered trees is about 12%, which is significantly higher than the indicators of the non-contaminated zone. These data most likely indicate a decrease in the airborne anthropogenic load on forest communities (a large number of withered and dead trees indicates a higher level of pollution in the past). This is also confirmed by data from past years, which speaks of a significant suppression of the vital state. Thus, part of the trees with a decrease in the intensity of the aerial anthropogenic load passed into another group of vitality. The vitality class index is similar to that of the background areas and is 0.79.

In the vitality spectrum of the impact zone (15 km or less from the source of pollution), the situation changes dramatically. The proportion of healthy trees ranges from 10 to 25%. The proportion of defoliated individuals reaches 30%, and the proportion of severely defoliated ones - 45%. Moreover, the share of withered trees ranges from 8–32%. Here we see the predominance of weakened and strongly weakened individuals, which indicates a strong oppression of the entire community as a whole. And first of all, this is determined not only by the proximity to the source of pollution, but also by a change in the nature of the aerotechnological load (in the impact area, plants are affected not only by sulfur dioxide, but also by aerosol forms of heavy metals, which almost do not penetrate into the buffer zone). The vitality class index is 0.53 (**Figure** 5).

Ground tiers.

Assessment of the condition of the lower layers of boreal forests is no less important than assessing the condition of the tree layer. Moreover, here one of the most important criteria is the total projective cover.

In the non-contaminated zone, the total projective cover of the herb-dwarf shrub layer is about 20%. The dominant species are *V. vitis-idaea* L. and *V. myrtillus* L.,

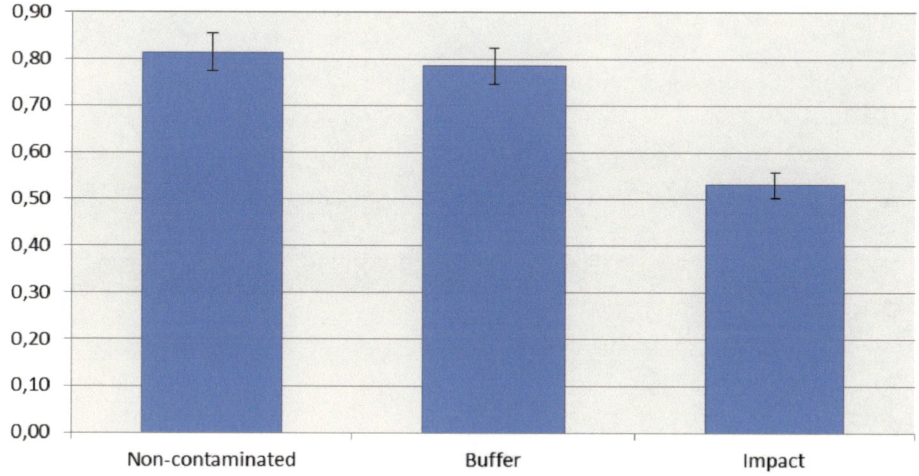

Figure 5.
Changes in the vital status index depending on the level of pollution.

occupying 9.5% and 12.5%, respectively. According to the data of previous studies, it can be done that the total projective cover of the herb-dwarf shrub layer gradually increases (on average, for the species of the genus *Vaccinium*, the total projective cover increased from 6% to fifteen%). First of all, this can be explained by the post-fire restoration of pine forests. And also a decrease in the intensity of emissions from the Severonikel plant (to a lesser extent).

The total projective cover of the moss-lichen layer is about 60% and is represented by lichens of the genus *Cladonia* (*C. stellaris* (Opiz.) Brodo, *C. rangiferina* (L.) Nyl., *C. mitis* (Sandst.) Hustich) and moss *Pleurosium shreberi* (Brid.) Mitt. It should be noted that the total projective cover of mosses is insignificant and amounts to no more than 1%. Such an insignificant amount of moss indicates a disturbance of the moss-lichen layer.

In the buffer zone, the total projective cover of the lower tiers begins to decrease, which indicates an increase in the airborne industrial load on the communities. The cover of the grass-dwarf shrub layer remains almost unchanged and amounts to 15-18%. In general, the herb-dwarf shrub layer of pine forests in the buffer and background regions does not differ significantly in terms of the total coverage and composition of dominant species.

The situation is different with the moss-lichen layer of the buffer zone. The total projective cover of the genus *Cladonia* decreases significantly and amounts to about 15%. At the same time, *Cladonia coccifera* (L.) Willd appears. - a species typical for successional communities. The total coverage of this type can be up to 30%. In this area, mosses are almost completely absent (hepatic mosses are occasionally found, but their effect on the total projective cover is insignificant).

Despite the reduction in the volume of emissions from the Severonikel plant, there is a noticeable deterioration in the lower tiers. Projective cover by *C. coccifera* (L.) Willd. increased almost 2 times. This is also evidenced by the almost complete absence of the species *P. shreberi* (Brit.) Mitt. In this case, it can be concluded that the moss-lichen layer is in a depressed state and is significantly disturbed.

In the impact zone, the total projective cover of the grass-dwarf shrub layer is on average about 5%. At the same time, the species composition changes sharply. *V. vitis-idaea* L. and *V. myrtillus* L. have a very depressed state. The number of *E. hermaphroditum* Hagerup increases, which becomes the dominant species and *Arctostaphylos uva-ursi* (L.) Spreng appears. This fact indicates a higher level of tolerance to

Figure 6.
Ground layer of pine forests in the non-contaminated (top left), buffer (top right) and impact (bottom) zones.

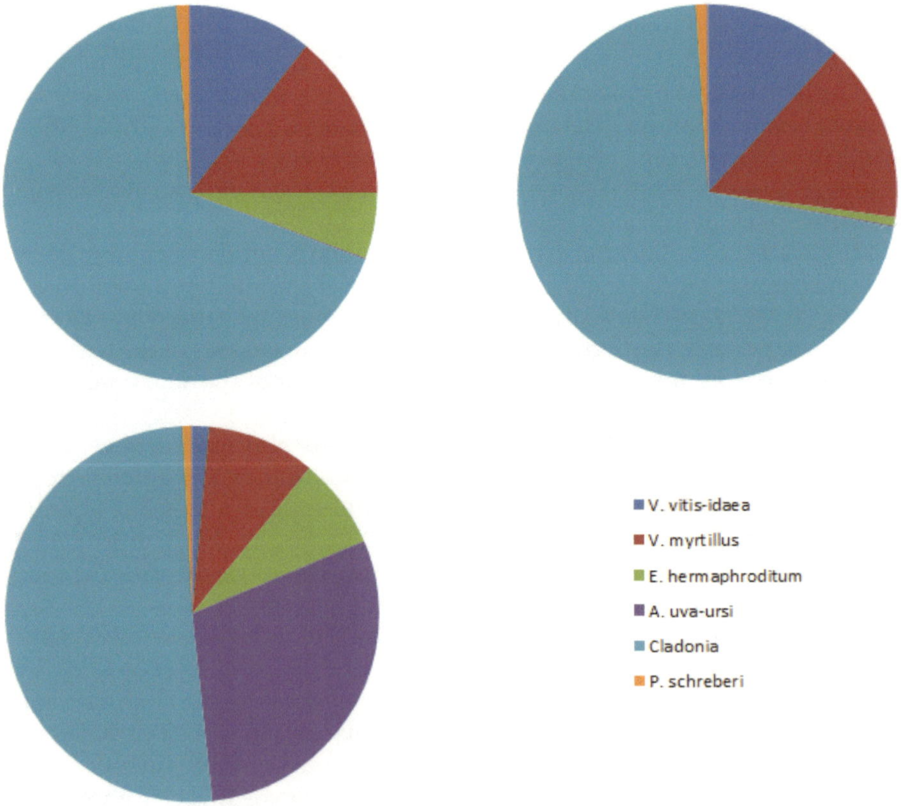

Figure 7.
Percentage ratio of species of the ground layer of pine forests in the non-contaminated (top left), buffer (top right) and impact (bottom) zones.

aerotechnogenic impact in these species compared to representatives of the genus *Vaccinium* (the projective cover is no more than 1.5%). This is also due to the destruction of the moss-lichen layer and litter, which play an essential role in maintaining the moisture capacity of organic soil horizons.

The total projective cover of the moss-lichen layer reaches 5%, which is significantly lower than in the buffer zone (in some cases, this indicator may be even lower). *C. coccifera* (L.) Willd becomes the dominant species. The lichens of the genus *Cladonia* disappear almost completely. Mosses are completely absent. Despite the decrease in the intensity of the airborne technogenic load, there was no significant change in the state of the lower tiers in the impact zone.

The study of the projective cover of ground layers can vary greatly depending on the distance to the source of airborne industrial pollution (**Figures 6** and 7).

6. Conclusions

The study made it possible to reveal some regularities in the effect of airborne pollution on the vitality structure of pine forests. First of all, our data are consistent with the results of other studies. General tendencies of the formation of forest communities under the conditions of aerotechnogenic load are revealed. The study showed that the vital state of the communities directly depends on the intensity and nature of the aerial anthropogenic impact.

Some features have also been identified. First of all, this concerns changes in the structure of communities in comparison with the data of previous years (there is a general trend towards an improvement in the living condition due to a decrease in emissions from the Severonikel plant). Based on the data obtained, it can be concluded that the pollution with sulfur dioxide and heavy metals is of a local nature.

The vitality structure of pine forests varies depending on the distance from the pollution source (and, accordingly, on the intensity of the aerial anthropogenic load). In the non-contaminated zone, the vitality index is 0.81; in the buffer zone - 0.79; in the impact zone - 0.53. Under the conditions of airborne industrial pollution, noticeable disturbances in the growth and development of the tree layer occur. There is a deviation from the monopodial type of branching (due to the death of the apical meristem of the leading shoot), a noticeable decrease in the increase in phytomass and a violation of the assimilation apparatus (the appearance of pronounced chlorosis and necrosis on the needles). The total projective cover and ground phytomass of the lower tiers noticeably decrease when approaching the pollution source. Moreover, in addition to the general oppression, there is a change in the dominant species in the synusia of the herb-dwarf shrub and moss-lichen layers.

Author details

Alexander Evdokimov
Herzen State Pedagogical University of Russia, St. Petersburg, Russia

*Address all correspondence to: evdokimov89@gmail.com

References

[1] Khan S A R, Zhang Y, Kumar A, Zavadskas E, Streimikiene D. Measuring the impact of renewable energy, public health expenditure, logistics, and environmental performance on sustainable economic growth. Sustainable Development 28; 2020; 833– 843. DOI: 10.1002/sd.2034

[2] Khan S A R. The Effect of Green Logistics on Economic growth, Social and Environmental Sustainability: An Empirical Study of Developing Countries in Asia. Preprints 2019; 2019010104. DOI: 10.20944/preprints201901.0104.v1

[3] Zhang Y, Khan S., Kumar A, Golpira H, Sharif A. Is tourism really affected by logistical operations and environmental degradation? An empirical study from the perspective of Thailand. Journal of Cleaner Production 227; 2019; 158-166. DOI: 10.1016/J.JCLEPRO.2019.04.164

[4] Kozlov M V, Haukioja E, Yarmishko V T, Aerial pollution in Kola peninsula Proc. Int. Workshop St. Petersburg, Apatity, 1993. 417 p.

[5] Nöjd, P, Kauppi, P, Growth of Scots pine in a changing environment. In: Tikkanen, E. & Niemelä, I. (eds.). Kola Peninsula pollutants and forest ecosystems in Lapland. Final report of The Lapland Forest Damage project. Finland's Ministry of Agriculture and Forestry. The Finnish Forest Research Institute, 1995. p. 61-64.

[6] Yarmishko V T Scots pine and aerial pollution in the European north. Research Institute of Chemistry, St. Petersburg State University, 1997. 210 p.

[7] Lyanguzova I V, Dynamic trends of heavy metal contents in plants and soil under different industrial air pollution regimes. Russ J Ecol 48; 2017; 311-320. DOI: 10.1134/S1067413617040117

[8] Lyanguzova I V, Maznaya E A, Dynamic trends in Vaccinium myrtillus L. Cenopopulations in the zone affected by a copper-nickel smelter complex: Results of 20-year monitoring. Russ J Ecol 43; 2012; 281-288. DOI: 10.1134/S106741361204008X

[9] Lukina N V, Chernen'kova T V, Pollution-induced successions in forests of the Kola Peninsula. Russ J Ecol 39; 2008; 310-317. DOI: 10.1134/S1067413608050020

[10] Madejski A, Rucinski P, Changes of height growth in the Scots pine under the impact of the emission of nitrogen compounds. Sylwan. 1976; 120, 07; p.79-82

[11] Havas P,Huttunen S. The effect of air pollution on the radial growth of Scots pine (Pinus sylvestris L.), Biological Conservation; 4, 5; 1972, p. 361-368, DOI: 10.1016/0006-3207(72)90052-3

[12] Titov A F, Talanova V V, Kazina N M, Physiological foundations of plant resistance to heavy metals Petrozavodsk: Karelian Scientific Center of the Russian Academy of Sciences; 2011. 77 p.

Chapter 12

Influence of Microenvironments and Personal Activities on Personal $PM_{2.5}$ Exposures among Children and Adults

Mahima Habil, David D. Massey and Ajay Taneja

Abstract

Environmental issues are a major worldwide problem of significant concern. Due to the growing human population and advancement in every sector, the environmental related issues are multiplying in recent years. Scalable exposures assessments approach that captures personal exposure to particles for purposes of epidemiology are currently limited, but very valuable especially for a country like India. The high levels of indoor particulate matter and the apparent scale of its impact on the global burden of disease underline the importance of particulate as an environmental health risk and the need for monitoring them. Human exposure especially to fine particles can have significant harmful effects on the respiratory and cardiovascular system. To investigate daily exposure characteristics to $PM_{2.5}$ with ambient concentrations in an urban environment, personal exposure measurements were conducted for different age groups of people residing in different indoor environments. To account for $PM_{2.5}$ exposure and measurements personal environment monitors (PEM) and medium volume sampler APM 550 was used to measure $PM_{2.5}$ concentration. On comparing the annual average $PM_{2.5}$ concentration with National Ambient Air Quality and WHO standards the concentrations were found to be many folds higher for personal and ambient monitoring at homes, schools, and offices. Moreover, the questionnaire data study explains the fact that the health hazards experienced by occupants linked to various activity patterns pose a greater risk in different indoor environments as compared to outdoor environments. The presented research method and analysis can help develop environmental awareness in identifying these pollutants and can also help in elucidating these contaminants. A real understanding of these possible causes of airborne contaminant is crucial for selecting and developing suitable and effective control methods.

Keywords: personal exposure monitoring, $PM_{2.5}$, indoor air quality

1. Introduction

Indoor air quality in sensitive areas like homes and public places has caught the interest not only of scientists but of the general public as well. Increasing public awareness is focusing on this issue as more and more individuals spent time inside than outside, particularly those who are most susceptible to the effects of poor air quality, such as the elderly, the young, and those with poor health [1, 2]. A series of epidemiological studies reported that there are robust associations between short-term and long-term exposure to fine particles (particulate matter with a diameter less than $PM_{2.5}$ μm), they are responsible for harmful effects on human health, including cardiac and respiratory diseases. The American Cancer Society has noticed that for every 10 $\mu g/m^3$ increase in $PM_{2.5}$, a 6% increased risk of mortality and morbidity, which have recently increased to 10%, resulting in premature deaths, according to California Air Resource Board [3].

Mostly all individuals are potentially affected, but as a subpopulation child are more prone to suffer health effects due to surrounding emission sources [4]. Since their lung structure and immune system are not fully developed and have a higher metabolic rate than adults, which means that they breathe in more air per unit of body weight and are generally more susceptible to the effects of indoor air pollutants. Studies have reported an association of pupil's health with both school and domestic exposure [5, 6]. These pollutants may emanate from a variety of sources (building and construction materials and furnishings, building occupants and activities, inadequate building design, lack of maintenance), including the infiltration of outdoor pollutants, such as dust, soil, and fuel consuming products and internally from smoking, cooking, incense burning, building, and furniture materials, consumer products, shed skin cells and organic fibers. Therefore, children's personal exposures to air pollutants differ from that of adults. Nowadays the highest importance is attributed to aerosols, especially fine particles because they represent a complex mixture of organic and inorganic substances with potentially toxic, carcinogenic, inflammatory, allergens, and other adverse properties. Furthermore, varied sources of PM results in an extensive range of particle sizes, i.e., the lesser the diameter size, the further deeply it will deposit in the respiratory tract. In the nasal-breathing mode, the mucus and cilia act as a very particle deposition inside the respiratory area depend not only on particle characteristics, but also human physiology, i.e., is the individual's behavior throughout the day like breathing deeply. Personal monitoring is the most accurate approach for determining direct exposure to airborne environmental contaminants because it incorporates complex human activity patterns into the exposure assessment. Only a few data are available on personal measurements of fine particles and its characterization of chemical species is lacking in comparative studies for both adults and children's activity patterns in different microenvironments which is important for developing a regulatory outline.

Thus the present study aimed to pinpoint the integrated actions essential to reduce the particulate pollutant and eliminate the toxicological environmental impacts of the urban environment. Identification of sources and controls of this location may allow for better protection for adults and children's health and understanding to control them.

2. Site description

Agra (27°10'N, 78°05'E,) is situated in the state of Uttar Pradesh in India. It is surrounded by the Thar Desert in its south-east, west, and north-west peripheries and is therefore, it's a semiarid area. Agra experience three different seasons; summer (March–June), monsoon (July–October), and winter (November–February) in a year. In the summer period Agra experiences hot weather with dry westerly winds. During this time temperature ranges between 30–46°C. The relative humidity is quite low between 18% and 48% during this time duration. In the monsoon it is hot and humid, temperature varieties from 24 to 36° C, and the humidity varieties

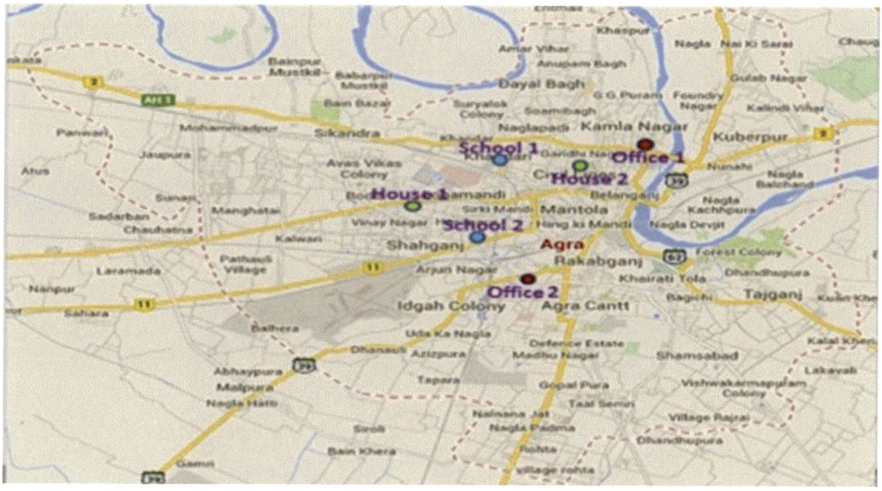

Figure 1.
Site map of Agra showing major regions, roads, and highways and sampling sites. Two homes, two schools, two offices.

from 70–90%. The pre-monsoon and monsoon periods experience strong northeast and southeast winds and in the winter, the temperature varies from 5–25°C. Thunderstorms and dust storms are frequently observed during the months from March to June. The wind speeds diverge from 2.6 to 6.9 Km/h with a maximum during summer and monsoons and a minimum in winters [7]. Agra is the one of the most populated cities in Uttar Pradesh, India. Agra has a population of about 1.5 million (Females = 47% and Males = 53%), with a birth rate of 28/1000 people and a death rate of 7/1000 people (Agra Nagar [8]). Tones of solid wastes are generated every day that can be seen in the form of piles along roadsides and streets due to poor municipal services. It is also a major cause for adding contamination to soil and groundwater. In the present study, personal and ambient monitoring was carried out for $PM_{2.5}$ in three different locations (two homes, two schools, and two offices) in the city of Agra (**Figure 1**).

3. Sampling sites

Two homes were selected in different colonies of the urban area of the city Agra, personal and ambient monitoring was done throughout the study. In the urban colonies, there is usually the old type of houses, having parking garages inside their houses or beneath the house due to lack of spaces. The houses are very near to each other making the environment very congested which is believed to be due to lack of proper planning. Adults were selected especially women, for personal sampling at homes as they spend most of their time indoors. Similarly, two schools were selected adjacent to roads in the city of Agra. Near schools usually, their environment is of high traffic during the morning and late afternoon hours when the schools get over. Personal and ambient monitoring was also carried out here during the study period. Different children were selected for personal sampling at schools. Similarly, two offices were selected in the commercial areas of the city of Agra. The offices are situated adjacent to a busy road in the city which experiences heavy traffic throughout the day. Personal and ambient monitoring was also carried out here during the study period. Different adults were selected for personal exposure monitoring carrying out routine activities in the offices and during their outdoor visits (**Figure 2**).

Figure 2.
Sampling done at homes, schools and offices.

4. Sampling and analytical method

The study was conducted from December 2013 to February 2015 to determine the mass concentration of $PM_{2.5}$ in personal and ambient samples from two homes, two schools, and two offices located in the urban area of the city Agra. A total of 180 samples were collected from personal and ambient monitoring over 15 months from December 2013 to February 2015, i.e. 90 samples from personal monitoring and 90 samples from ambient monitoring. The monitoring was done for 24 hours for personal exposure and ambient sampling. For personal monitoring, the time period contains the occupant's daily indoor and outdoor activities. The number of samples gathered in a month duration was 12; six for the ambient environment (i.e. two from schools same way from homes and offices) and six samples of personal exposure (i.e. two from school and in similar fashion from homes and offices) respectively. Personal exposure measurements were carried out, using PEM (Personal Environmental Monitor, SKC Inc., USA) (**Figure 3**) with the Leland legacy sampling pump at 10 L/min, fitted within a waist pack to each individual participating. Rechargeable lithium-ion (Li-Ion) battery pack will provide 24-hour run time with impactors and other sampling devices with low back pressures. PEM is a lightweight, personal sampling device consisting of a single-stage impactor and an after-filter. Aerosol particles are sampled through the single-stage impactor to remove particles above the 50% cut-point of 2.5 μm in aerodynamic diameter. These large size particles are gathered on a greased ring and are rejected after sampling. Particles lesser than the 50% cut-point is authorized to pass through the impactor and get collected on a 37-mm after-filter paper. To investigate personal exposure, the filter is investigated gravimetrically for particle mass and chemical analysis. The sampling pump provides the required airflow through the PEM. The PEM operates on the principle of inertial separation of airborne particles using an impactor. Particle-laden air is accelerated into the sampler through the round nozzles located in a circle around the outer edge of the cover [9]. For $PM_{2.5}$ ambient samples were collected with a fine particulate dust sampler (APM 550, Envirotech) shown in **Figure 3**. The air inlet has a circular symmetrical hood designed to keep out of the rain, insects, and very large particles. The Inlet section leads to an Impactor stage, which traps particles larger than 10 microns aerodynamic diameter. The airstream down in the tube carries particles. For the monitoring of $PM_{2.5}$, WINS Impactor is attached. It is designed to trap medium size particles between 2.5 and 10 microns. To avoid sampling fault due to bouncing of small size particulates from the impaction surface, a 37 mm diameter GF/A paper engrossed in silicon oil is used. The airstream now leaving the WINS Impactor consists of only fine particles with an aerodynamic diameter smaller than 2.5 microns. APM 550 uses an oil-less rotary pump to produce the suction pressure and critical flow control orifice (as recommended by [10]) for maintaining a constant airflow rate of 16.7 L/ min.

Filter papers were weighed thrice before and after sampling using five digits microbalance. Before weighing the samples were homogenized in desiccators at 20-30°C with a standard relative humidity range of 30–40% 24 hours. Filter cassettes were utilized to transport weighed filter papers to the sampling site, their filters were shifted to filter holders and positioned on the sampling plate. The exposed filter and the filter holder was then enfolded with aluminum foil, and were later placed back in the desiccators (**Figure 4**). Field blank filters were collected to lessen gravimetric prejudice due to filter handling after and during sampling duration. After evaluating, samples were refrigerated at 4°C to prevent the loss of volatile components as per the USEPA, Compendium Method, 1999. Exposed filter paper after conditioning and weighing was cut into two equal pieces. One part was used for acid extraction and the other half was preserved for later analysis. The other half of the exposed filter paper was put in pre-washed Borosil beakers such that the exposed paper faces toward the bottom of the beaker, then added 6–8 ml analytical grade (Merck) HNO_3 and HCl in the ratio 1:3 and kept on a hot plate at the temperature of 40-60°C for 90 minutes [11]. The beakers were covered with a watch glass filled with water to avoid the loss of volatile compounds due to evaporation. The solution was then filtered with prewashed Whatman filter paper. The

Figure 3.
Personal environmental monitor (SKC Inc., USA) and fine particulate dust sampler (APM 550).

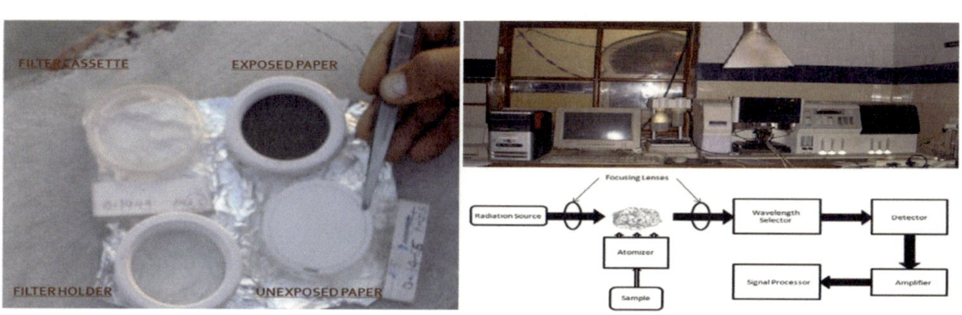

Figure 4.
Filter paper before and after sampling and atomic absorption spectrophotometer with schematic diagram (make: Perkin Elmer, AAnalyst 100).

wall of the beaker was washed 2–3 times with distilled water and allowed to stand for 30 min and then filtered without shaking. Finally, the extract was diluted up to 50 ml with distilled de-ionized water and stored in pre-washed polypropylene till analysis. Zinc (Zn), lead (Pb), nickel (Ni), iron (Fe), chromium (Cr), Cadmium (Cd) manganese (Mn), copper (Cu), Barium (Ba), and Mercury (Hg) were analyzed with this extract. Analysis of these trace metals was done on AAS regularly within two months of extraction.

5. Questionnaire survey

A Questionnaire survey is a common way to relate illness prevalence with air quality for small groups of individuals targeted for a specific purpose. It also gives us an idea about the potential sources of indoor air pollution, lifestyle, and different human activities (**Table 1**). A questionnaire (respondent schedule) was prepared from the personal interview of occupants from homes, schools, and offices, in all 475 were filled (i.e. 300 children and 275 adults) in different sampling locations, i.e. (homes, schools, and offices) of the Agra region. The questionnaire contains questions, which inquired about the personal exposure parameters and the occupant's perceptions of the air quality around them. The questionnaire includes daily time/ activity diary, age, sex, socio-demography, house/school/office characteristics, air exchange conditions, timing of different activities such as cooking in homes, cleaning, heating, number of occupants present near the source, other considerable characteristics. The information on the health-related issues of the occupants was also investigated like allergies, skin problems, cold and flu, vomiting or giddiness, eyes or nose annoyance, headache, and mental exhaustion. In the last part of the questionnaire survey, it was also investigated that the participants, if they were conscious of the harmful effects of polluted air? The results are discussed based upon the percentage of the data obtained below.

6. Health effects

Indoor air pollution is associated with a variety of health effects of both an immediate and long-term nature [12]. Short-term health effects may show up after single or repeated exposures (**Table 1**). They may include annoyance in the eyes, throat, and nose, headaches, cold, wooziness, and exhaustion. On the other hand, long-term health effects may appear or prevail after exposure has happened. These may comprise respiratory and heart diseases, and may also lead to cancer. As this study is of short term and limited to fifteen months, therefore only the short-term health-related issues are discussed as long-term exposure needs more years of study. According to the survey (**Table 1**), the most frequently occurring symptoms were headache (53% in the homes, 65% in the schools, and 52% in the offices), eyes, and nose irritation (53% in the homes, 60% in the schools and 48% in the offices), cold and flu (50% in the homes, 62% in the schools and 38% in the offices) allergies (35% in the homes, 45% in the schools and 20% in the offices) sinus, mental fatigue, nausea, and dizziness were at a somewhat lower percentage in the different microenvironments. The results show that these problems were more dominant in schools in comparison to homes and offices. The schools are also located near busy roads and streets of the city, vehicular emission causes the formation of fine particulate that penetrates indoors (higher values of r) and in turn is supposed to increase health-related problems [13, 14]. **Table 1** also shows a higher percentage of smoking with 42% was reported from homes than in comparison to offices whereas,

Parameters	Homes	Schools	Offices
Short-term health effects			
1. Headaches	53	65	52
2. Eyes and nose irritation	53	60	48
3. Cold and flu	50	62	38
4. Allergies	35	45	20
5. Sinus	20	22	14
6. Mental fatigue	28	35	22
7. Nausea and dizziness	18	20	13
Types of fuel used in homes			
1. L.P.G	95	NA	NA
2. Wood	3	NA	NA
3. Kerosene	2	NA	NA
4. Cow dung cakes	1	NA	NA
Cooking activities			
1. Refined oil	25	NA	NA
2. Mustard oil	70	NA	NA
3. Ghee	5	NA	NA
Other activities			
1. Candles or diyas	50	0	0
2. Incense sticks	75	0	60
3. Mosquito Coils	40	0	0
4. Smoking	42	0	35
5. Cleaning and Sweeping	Twice a day morning and evening	In evening after school hours	In evening after office hours
Ventilation inside the houses			
1. Ventilation acceptable	40	20	60
2. Ventilation not accepted	60	80	30
1. Aware	55	80	64
2. Not aware	45	20	36

Table 1.
Questionnaire survey analysis (in percentage).

the percentage reported from offices was also quite high with 35%. It was consistent with the results that health symptoms occurred more in indoor places where smoking was more common than in non-smoking [13, 15].

As the questionnaire data were collected from a different kind of socioeconomic status, family members living in different urban areas of the city, working in different places like offices, and their children studying in different schools. Their living stats are also different like the use of different kinds of fuels like wood, dung, and kerosene with the use of clean fuel like LPG (liquefied petroleum gas). As in the urban houses the per capita income increases, families generally shift to cleaner energy-efficient systems for their home energy needs. While particulate levels seem to be closely associated with health effects, they normally seem in association with other contaminants. Thus, there is a need to distinguish different types of particles and how they can be best controlled [16].

7. PM2.5 results

The result obtained from the analysis, observation tables are drawn and presented below. $PM_{2.5}$ and its metal content in personal and ambient samples from homes, schools, and offices respectively during the sampling period from December 2013 to February 2015. During the study period, annual average concentration and standard deviations (SD) for PM2.5 (personal and ambient) at homes were 132.88 ± 43.28 μgm^{-3} and 122.78 ± 36.27 μgm^{-3}, at schools were 137.77 ± 42.18 μgm^{-3} and 127.19 ± 36.18 μgm^{-3} and at offices site 123.60 ± 41.44 μgm^{-3} and 109.20 ± 36.88 μgm^{-3} respectively. The concentration trend varied from 103.21 to 177.26 μgm^{-3} and 79.18 to 185.57 μgm^{-3} for personal and ambient concentrations at homes, at schools trend varied from 119.18 to 208.16 μgm^{-3} and 83.88 to 190.18 μgm^{-3} respectively. At offices $PM_{2.5}$ concentration varied from 93.05 to 161.78 μgm^{-3} and 78.16 to 181.15 μgm^{-3} respectively. This indicates that air quality in schools and homes is more affected by $PM_{2.5}$ and at offices have relatively better air quality. On comparing the annual average $PM_{2.5}$ concentration with National Ambient Air Quality Standard (40 μgm^{-3}) [17], It was found to be 3–4 times higher from personal and ambient concentration data collected from all the sampling places. We also compared our data with World Health Organization [18] standards (10 μgm^{-3} annual means respectively), our results exceeded 12.3–13.3 times at homes for personal and ambient monitoring and 12.8–13.8 times at schools and 10.9–12.4 times at offices.

High concentrations at the sampling sites may be attributed to heavy vehicular traffic flow, especially at schools, emissions from nearby industries and re-suspension of road and soil dust, emission from solid waste incineration, and construction activities. In homes and offices, cooking and smoking are the dominant activities, which should have contributed to the formation of fine particulate emissions. In the case of ambient, concentration levels were found to be elevated depending upon the outdoor activities.

On applying the one-way Anova to the mean values of particulate concentrations at all the sites for each location significance values $PM_{2.5}$ were adjacent to 1 or were around 1. For the homes, it ranged from 0.951 to 0.999, for the offices they ranged from 0.911 to 0.997 and for the schools it ranged from 0.917 to 0.997 signifying that there is no substantial variance between the concentrations of two offices, homes and schools and thus have similar kind of sources which lead to the generation of particulate pollutant in their environment. Due to the above reason, the discussion made in this report is explained based on the average concentration of two homes, two schools, and two offices located in each microenvironment rather than two houses, two schools, and two offices at different sampling sites.

8. Chemical constituent of PM2.5

Characterization of PM components, including inorganic elements, is of central importance in proposing mechanisms for health effects and in source apportionment studies [19, 20]. Data obtained by chemical analysis for ten metals in, $PM_{2.5}$ collected from personal and ambient environments at homes, schools, and offices.

The sum of the average concentration of ten parameters determined in $PM_{2.5}$ personal and ambient was found to be 19.09 μgm^{-3} and 12.12 μgm^{-3} at homes and it ranged from 0.03–10.98 μgm^{-3} and 0.03–5.68 μgm^{-3} respectively. At schools, the sum of the average concentration was 21.48 μgm^{-3} and 12.93 μgm^{-3} and it ranged from 0.04–12.40 μgm^{-3}, 0.03–5.73 μgm^{-3} respectively; whereas at offices the sum of the average concentration was 16.93 μgm^{-3} and 11.05 μgm^{-3} and it ranged from

Parameters	$PM_{2.5}$ Personal (ugm^{-3})				$PM_{2.5}$ Ambient Environment(ugm^{-3})			
Homes	Mean	Sd	Max	Min	Mean	Sd	Max	Min
PM Conc.	132.88	43.28	177.26	103.21	122.78	36.27	185.57	79.28
Ba	10.98	4.33	16.78	4.67	5.68	1.93	7.89	4.05
Fe	2.63	1.99	7.21	0.63	1.33	1.56	4.11	0.12
Pb	1.12	1.12	3.55	0.06	1.12	0.67	2.17	0.17
Mn	0.06	0.07	0.25	0.02	0.22	0.28	0.49	0.13
Cu	0.40	0.35	1.07	0.04	0.36	0.55	1.10	0.03
Ni	0.60	0.58	1.85	0.03	0.68	0.85	2.95	0.05
Cr	2.39	1.35	6.09	0.30	0.60	1.03	3.06	0.03
Zn	0.53	0.51	1.89	0.03	1.75	2.26	3.33	0.26
Cd	0.35	0.09	0.38	0.18	0.35	0.12	0.37	0.23
Hg	0.03	0.01	0.04	0.02	0.03	0.01	0.03	0.01
Schools								
PM Conc.	137.77	42.18	208.16	119.18	127.19	36.18	190.18	83.88
Ba	12.40	4.85	17.73	5.20	5.73	2.00	7.95	4.66
Fe	2.70	2.09	7.38	0.80	1.51	1.74	4.22	0.14
Pb	1.15	1.19	3.62	0.12	1.37	0.86	2.43	0.20
Mn	0.09	0.08	0.32	0.03	0.23	0.28	0.50	0.16
Cu	0.45	0.32	1.08	0.04	0.39	0.53	1.16	0.05
Ni	0.72	0.60	2.02	0.06	0.70	0.84	3.29	0.08
Cr	2.99	1.70	6.27	0.36	0.71	1.20	3.12	0.06
Zn	0.58	0.57	1.96	0.04	1.84	2.21	3.48	0.24
Cd	0.36	0.10	0.42	0.20	0.42	0.13	0.38	0.23
Hg	0.04	0.02	0.04	0.02	0.03	0.01	0.04	0.01
Offices								
PM Conc.	123.6	41.44	161.78	93.05	109.20	36.88	181.15	78.16
Ba	9.45	4.85	15.12	4.35	4.86	1.92	7.65	4.06
Fe	2.39	1.73	5.74	0.43	1.36	1.55	4.07	0.09
Pb	1.02	1.09	3.51	0.05	1.12	0.53	2.09	0.13
Mn	0.04	0.06	0.23	0.01	0.20	0.29	0.44	0.08
Cu	0.35	0.33	1.03	0.02	0.34	0.43	1.05	0.04
Ni	0.57	0.64	1.62	0.02	0.63	0.68	2.83	0.03
Cr	2.37	1.45	5.77	0.31	0.55	0.78	2.88	0.03
Zn	0.42	0.45	1.67	0.01	1.68	2.07	3.28	0.20
Cd	0.30	0.10	0.31	0.26	0.30	0.06	0.30	0.21
Hg	0.02	0.01	0.03	0.01	0.01	0.01	0.02	0.01

Table 2.
Average metal concentrations (ugm^{-3}) in $PM_{2.5}$ by personal and ambient monitoring at homes, schools and offices.

0.02–15.76 μgm^{-3}, 0.01–4.86 μgm^{-3} in personal and ambient environment respectively. The total analyzed parameters contributed 13.69% to 14.36% and 9.87% to 10.16% for personal and ambient monitoring at homes, schools, and office sites. The unanalyzed part of the total $PM_{2.5}$ mass collected from personal and ambient monitoring may include carbon, organics, silicates, and phosphates [21, 22] (**Table 2**).

9. Multivariate analysis

Global emissions reported by many authors [23–28] have revealed as anthropogenic and natural sources can contribute to the principal aerosol classes, but values variate rendering to the local scenario (coarse, fine and ultrafine) of atmospheric particulate matter. Approximately 10–20% of the aerosols can be categorized as anthropogenic on an international scale [20, 29], but these values may drastically change due to local scenarios, human activities, and the prevailing particle cut-off. Correlation analysis is the statistical procedure applied for this assessment. The levels of various elements in $PM_{2.5}$ vary by diverse orders of scale and hence the correlation analysis was utilized. Correlation analysis was done to study the connection between ambient and personal sources among metallic species. **Table 3** shows the interrelation between metallic species at the schools, homes, and offices, during the sampling period. The samples used for the analysis were (n = 90) and the eight variables in each sample of $PM_{2.5}$. Fe presented the maximum concentration in indoor homes, schools, and offices followed by Cr, Pd, Zn, Ni, Cu, Cd, Mn, Ba, and Hg. The coefficient of variance was in the order of Fe > Ni > Zn > Cr > Mn > Cu > Ba >Cd > Pd > Hg. Comparable kinds of trends are found in trace metal concentrations in all three types of sites, indicating that there could be one or additional similar kinds of sources contributing to a somewhat similar kind of urban environment of the city area. Correlation analysis was done to achieve the association between individual trace metals and to show that two or more components may associate either due to common cause (**Table 3**) Zn and Ni showed good to strong correlation with Cr (R^2 = 0.712 and 0.824) and Mn (R^2 = 0.758 and 0.792) at homes sites, (R^2 = 0.876 and 0.996) and (R^2 = 0.502 and 0.687) at school sites and (R^2 = 0.798 and 0.975) and (R^2 = 0.821 and 0.946) at office sites. These correlations indicate smoking done by the occupants in the indoor working environment and incense burning could be the probable source of these trace metals [30]. Fe too showed strong correlation with Cu at these three sites (R^2 = 0.913, 0.921 and 0.902), followed by Cr with Mn (R^2 = 0.736, 0.826 and 0.872) and Ni with Cr (R^2 = 0.824, 0.501 and 0.821). Anthropogenic activities and the use of different mechanical and electrical apparatus like computers, printers, and photocopiers, etc. The work environment can give rise to emissions of such metals [29, 31]. Ni also showed a correlation with Cd (R^2 = 0.512 and 0.501) at homes and school sites which may be due to ambient sources. Ba showed a strong correlation with Fe and Cu (R^2 = 0.645 and 0.503) at home sites, (R^2 = 0.678 and 0.512) at school sites and, (R^2 = 0.719 and 0.567) at office sites, indicating outdoor sources like vehicles brake lining and other anthropogenic can contribute to it [32]. Whereas; Hg showed no significant correlation with any of the other metal species at any of the sampling sites.

10. Conclusion

Through this research work, we have been able to identify that personal exposure monitoring is a practical technique for improving knowledge about

Home										
	Zn	Pd	Ni	Fe	Cr	Cd	Mn	Cu	Ba	Hg
Zn	1.000									
Pd	0.455	1.000								
Ni	0.687*	0.264	1.000							
Fe	0.211	0.421	0.475	1.000						
Cr	0.712*	0.487	0.824*	0.664*	1.000					
Cd	0.216	0.124	0.512*	0.227	0.427	1.000				
Mn	0.758*	0.348	0.792*	0.216	0.736*	0.178	1.000			
Cu	0.372	0.367	0.218	0.913*	0.689*	0.246	0.215	1.000		
Ba	0.116	0.123	0.216	0.645*	0.056	0.187	0.114	0.503*	1.000	
Hg	0.009	0.050	0.079	0.010	0.098	0.102	0.008	0.009	0.004	1.000
School										
Zn	1.000									
Pd	0.368	1.000								
Ni	0.497	0.168	1.000							
Fe	0.318	0.336	0.587*	1.000						
Cr	0.876*	0.421	0.501*	0.478	1.000					
Cd	0.325	0.248	0.501*	0.246	0.416	1.000				
Mn	0.996**	0.485	0.226	0.364	0.826*	0.325	1.000			
Cu	0.116	0.167	0.697*	0.921**	0.348	0.348	0.139	1.000		
Ba	0.120	0.131	0.203	0.678*	0.045	0.201	0.128	0.512*	1.000	
Hg	0.010	0.061	0.062	0.012	0.078	0.117	0.009	0.006	0.005	1.000

Home										
	Zn	Pd	Ni	Fe	Cr	Cd	Mn	Cu	Ba	Hg
Office										
Zn	1.000									
Pd	0.358	1.000								
Ni	0.826*	0.276	1.000							
Fe	0.296	0.331	0.379	1.000						
Cr	0.798*	0.167	0.821*	0.334	1.000					
Cd	0.058	0.256	0.428	0.387	0.342	1.000				
Mn	0.975**	0.428	0.946**	0.421	0.872*	0.315	1.000			
Cu	0.018	0.348	0.167	0.902*	0.216	0.246	0.216	1.000		
Ba	0.118	0.136	0.198	0.719*	0.044	0.201	0.103	0.567*	1.000	
Hg	0.007	0.072	0.045	0.0013	0.045	0.110	0.010	0.005	0.008	1.000

**Correlation is significant at the 0.05 level (2-tailed).*
*** Correlation is significant at the 0.01 level (2-tailed).*

Table 3.
Correlation matrix at home, school and office sites.

individual-level exposure to environmental stressors. This exploration work summaries the elemental characterization, correlation analysis, and health risks concomitant with fine particle $PM_{2.5}$ that is generated during different activities at homes, schools, and offices.

Acknowledgements

This work was supported by the DST/SERB Project no.SR/FTP/ES-58/2012. The authors like to thank Principal Prof. S. P. Singh and Dr. Susan Jaison. Head Department of Chemistry, St. John's College, Agra for providing the necessary facility.

Author details

Mahima Habil[1*], David D. Massey[1*] and Ajay Taneja[2]

1 Department of Chemistry, St. John's College, Agra, India

2 Department of Chemistry, Dr. B.R. Ambedkar University, Agra, India

*Address all correspondence to: mahi.habil@gmail.com
and davidmassey22@gmail.com

References

[1] Khan SAR, Sharif A, Golpira H, Kumar A. A green ideology in Asian emerging economies: From environmental policy and sustainable development. Sustainable Development. 2019;**27**(6):1063-1075

[2] Khan SAR, Zang Y, Kumar A, Zavadskas E, Streimikiene D. Measuring the impact of renewable energy, public health expenditure, logistics, and environmental performance on sustainable economic growth. Sustainable Development. 2020;**28**(4):833-843

[3] California Environmental Protection Agency. Methodology for Estimating Premature Deaths Associated with Long-Term Exposures to Fine Airborne Particulate Matter in California. Draft: Report; 2009

[4] Zhang Y, Khan SAR, Kumar A, Golpira H, Sharif A. Is tourism really affected by logistical operations and environmental degradation? An empirical study from the perspective of Thailand. Journal of Cleaner Production. 2019;**227**:158-166

[5] Habil M, Massey D, Taneja A. Personal and ambient PM2.5 exposure assessment in the city of Agra. Data in Brief. 2016;**6**:495-502

[6] Lueker J, Bardha R, Sarkar A, Norfor L. Indoor air quality among Mumbai's resettled populations: Comparing Dharavi slum to nearby rehabilitation sites. Building and Environment. 2020;**167**:106419

[7] Indian Meteorological Department (IMD). Climate of Uttar Pradesh. New Delhi: Government of India Press; 1989

[8] Nigam AN. Department of Registration for Birth and Death. Agra: M.G.road; 2009

[9] USEPA, Compendium Method, (1999).

[10] Guidelines on Speciated Particulate Monitoring. USEPA. Research Triangle Park: Office of Air Quality Planning and Standards; 1998

[11] Technical Report No. 3, (2002). Review of data on Heavy Metals in Ambient air in Australia. Department of the Environment and Heritage (DEH), Australian Government.

[12] Colome S, Kado N, Jaques P, Kleinman M. Indoor-outdoor air pollution relations; particulate matter less than 10 μm in aerodynamic diameter (PM10) in homes of asthmatics. Atmospheric Environment. 1992;**26**:2173-2217

[13] Oosterlee A, Drijver M, Lebret E, Brunekreef B. Chronic respiratory symptoms in children and adults living along streets with high traffic density. Occupational and Environmental Medicines. 1996;**53**:241-247

[14] Lu KF, He HD, Wang HW, Li XB, Peng ZR. Characterizing temporal and vertical distribution patterns of traffic-emitted pollutants near an elevated expressway in urban residential areas. Building and Environment. 2020;**172**:106678

[15] Burge PS. Sick building syndromes. Occupational and Environmental Medicines. 2004;**6**:185-190

[16] Smith KR, Jantunen M. Why particles? Introduction to special issue, methodologies of assessing exposure to combustion products: Particles and their semi-volatile constituents. In: Smith, K.R., Jantunen, M, Goldstein BD. Chemosphere. 2002;**49**(9):865-871

[17] Hazadrous Air Pollutants Monitoring & Control. National Ambient air Quality Standards (NAAQS). Central Pollution Control Board Delhi: Indian Association for Air Pollution Control; 2008

[18] WHO, Air Quality Guidelines, (2006). www.who.int/phe/health/_topics/outdoor_apg/en/index.html; 2005. Global Update.

[19] Butler, O. T., Cook, J. M., Davidson, C.M., Harrington, Miles, D.L. (2009). Atomic spectrometry update. Environmental analysis. Journal of Analytical Atomic Spectrometry, 24, 133-177.

[20] Massey D, Kulshrestha A, Taneja A. Particulate matter concentrations and their related metal toxicity in rural residential environment of semi-arid region of India. Atmospheric Environment. 2013;**67**:278-286

[21] Pitts BJ, Pitts JN Jr. Atmospheric Chemistry, Fundamentals and Experimental Techniques. New York: Wiley; 1986

[22] Nair PR, George SK, Sunilkumar SV, Parameswaran K, Jocab S, Abraham A. Chemical composition of aerosols over peninsular India during winter. Atmospheric Environment. 2006;**40**:6477-6493

[23] Chung SH, Seinfeld JH. Global distribution and climate forcing of carbonaceous aerosols. Journal of Geophysical Research. 2002;**107**:509-512

[24] Stortini AM, Freda A, Cesari D, Cairns WRL, Contini D, Barbante C, et al. An evaluation of the PM2.5 trace elemental composition in the Venice lagoon area and an analysis of the possible sources. Atmospheric Environment. 2009;**43**:6296-6304

[25] Kulshrestha A, Satsangi PG, Masih J, Taneja A. Metal concentration of PM2.5 and PM10 particles and seasonal variations in urban and rural environment of Agra, India. Science of the Total Environment. 2009;**407**:6196-6204

[26] Nihalani SA, Khambete AK, Jhariwala ND. Source apportionment of particulate matter- a critical review for Indian scenario. Environmental Processes and Management. 2020:249-283

[27] Sharif A, Afshan S, Chera S, Amel A, Khan SAR. The role of tourism, transportation and globalization in testing environmental Kuznets curve in Malaysia: New insights from quantile ARDL approach. Environmental Science Pollution Research International27. 2020a;**20**:25494-25509

[28] Sharif A., Godil, I., Danish Xu, Bingjie Sinha, Avik Abdul, Rehman Khan S. A.R., Jermsittiparsert Kittisak, (2020b). evisiting the Role of Tourism and Globalization in Environmental Degradation in China: Fresh Insights from the Quantile ARDL Approach. Journal of Cleaner Production, 101156

[29] Massey D, Habil M, Taneja A. Particles in different microenvironmets-its implications on occupants. Building and Environment. 2016;**106**:237-244

[30] Horemans B, Grieken RV. Speciation and diurnal variation of thoracic, fine thoracic and sub-micrometer airborne particulate matter at naturally ventilated office environments. Atmospheric Environment. 2010;**44**:1497-1505

[31] Massey. D. & Taneja, A. (2011). Emission and formation of fine particles from hard copy devices: the cause of indoor air pollution, Monitoring, Control and Effects of Air Pollution, In-Tech Publication, ISBN: 978-953-307-204-3.

[32] Dundar, M. S., Altundag, H., Deryaoglu, N., Ozdemir, F., Karacoban, Z. (2011). Determination of Germanium, Silver, Barium, Potassium and Iron in indoor and outdoor airdust of Sakarya using ICP-OES Techniques, SAU,Fen Bilimleri, 15, Cilt, 3.Say1, 105-112.

Chapter 13

Fossil Fuel Fires: A Forgotten Factor of Air Quality

Łukasz Kruszewski

Abstract

Spontaneous fossil fuel fires, especially coal fires, are known worldwide. They occur in numerous sites, both completely natural (coal seam outcrops) and anthropogenic (burning mining waste heaps, or BMWHs). Coal and waste/barren rock fires produce gaseous emanations, acting within exhalative processes. This factor is rarely being considered as influencing quality of the atmospheric air. The paper shortly discusses most important available methods for field gas analysis, with an emphasis on a portable FTIR spectrometer. It summarizes results of gas analyses from Polish BMWHs, using a multi-tool approach. It also lists a number of additional analyses from 53 vents of these environmentally important objects, with the main purpose of enlarging the knowledge of the span of concentrations of the particular compounds. This is especially true for formaldehyde, pyridine, CO, 1,1,1-trichloroethene, 1,1-dichloroethene, cumene, SO_2, and, to a lesser extent, NO_2, CCl_4, ethane, propane, ethene, and thiophene. The latter, and DMS, are confirmed as gaseous S source more frequent and rich than SO_2.

Keywords: natural spontaneous coal fires, combustion gas emissions, in situ FTIR gas analysis

1. Introduction – fossil fuel fires

Spontaneous fires of fossil fuels – mainly coals but also bituminous shales and oil shales – are known worldwide. They both concern natural environments and their anthropogenic analogues – burning mining waste heaps (BCWH). The CWHs are, more or less, permanent elements of the environment of coal basins. Although sometimes under reclamation, their recultivation procedures may also negatively influence the surroundings. The phenomena taking place in the BCWH are described, e.g., in Nasdala & Pekov [1], Cebulak et al. [2], Sokol et al., [3], Stracher [4], and papers of Ł.K. The later largely characterize complex products of gaseous emissions related to both coal and barren rock – mutually known as mining waste – burning. This chapter characterizes the composition of these emissions, by juxtaposing published concentrations and their related mean values with new data obtained for new BCWH-type object. As such, the chapter extends knowledge about the geochemical charge of the BCWH gaseous emissions and, as such, their potential atmospheric input.

2. Environmental gas emission measurement methods

Numerous methods of gas analysis in the environment exist. One of the most simple one, based on colorimetric chemical reactions, uses indicatory tubes (IT).

This method is based on colorimetric interaction of measured gaseous species with a chemical filler. In particular, Dräger tubes allow to detect and measure amounts of gases like O_2, CO_2, CO, NO_2, SO_2, NH_3, PH_3 (phosphine), acetic acid, acetone, propane, benzene, toluene, styrene, *o*-xylene, butadiene, total mercaptans (thiols), methanol, *i*-propanol, trichloroethene (TCE), vinyl chloride, methyl *tert*-butyl eter (MTBE), and others. However, the IT method brings large errors due to cross-sensitivity and numerous coincident reactions of the emanation-contained gaseous species, and humidity. Positive determinations for the BCWHs gases were thus single, and the following substances were observed (with semi-quantitative due to the above factors): H_2S (up to 1140 ppm), HCN (single determination (s.d.), 16 ppm), acetaldehyde (possibly up to 1150 ppm), diethyl ether (up to 1100 ppm), trimethylamine (and/or other amines; ca. 57 ppm), ethyl formate (s.d., <23 ppm), and I_2 (s.d., 1.7 ppm). Gas Chromatography (GC) is a method of choice for the analysis of environmental organics. A sample is put into specialized columns, where retention time of a particular molecule, related to its mass and charge (*m*/*z* parameter), is measured. However, it is relatively rarely used for gas analysis due to a need of a more sophisticated sample loader. This is overcame by a method of Colman et al. [5], where a sample sucked into a steel can and sent to laboratory (here: overseas) is reheated (to the temperature measured *in situ*), divided into aliquots with various pre-treatments including (1) passing heated aliquots over a glass for low-volatile compounds exerting and (2) water-immersion-driven revolatization, and (3) chromatographic separation. Analyses of such portioned sample using 3 detection methods: Mass Spectrometry (MS), Flame Ionization (FI), and Electron Capture (EC), both shown in Kruszewski et al. [6] and this chapter, proven to be problematic, as explained below.

The GC method is, however, useful in the environmental gas analyses if coupled with tools like Nitrogen-Phosphorus-Detector and cryo-focusing. A good example is a work of Wickenheiser et al. [7], who analyzed gases emitted from Italian wetland bogs. The compounds included PH_3, ethane, ethene, and NO_x. GC coupled with Inductively Coupled Plasma – Mass Spectroscopy (ICP-MS) allowed them to address heavy organo(semi/non)metallic gases like trimethylarsine (TMA), $(CH_3)_3As$, and trimethylstibine (TMS), $(CH_3)_3Sb$, and also metallic Hg, emitted from algal mats. The same method allowed to Feldmann et al. to detect (via cryo-trapping) trimethylbismuthine, $(CH_3)_3Bi$, as a common gas in municipal solid waste and sewage gas. Traces of tetramethyltin and TMS were also detected this way (*vide* [8]). Another method mentioned by the latter author is hydride generation. The use of tedlar bags, a gas trapping solution (with HNO_3 and H_2O_2), charcoal sorbent tubes, preconcentrators, and analysis with GC–MS and GC-PID (GC with photo-ionization detection) is also widely exploited, e.g., to measure TMA and propanethiol [9]. A method to be exploited by the author (Ł.K.) is a GC in conjunction with Atomic Emission Spectroscopy (AES). This two-step method involved very-low-detection-limit analysis, both qualitative and quantitative, of mainly (semi)metals in a gas sample, followed by analysis of their immediate surroundings for proposing types of organic and inorganic (semi)metal forms (R. Stasiuk, *pers. comm.*).

3. Mining waste heaps and products of their fires

A large number of coal mining waste heaps bear numerous spontaneous fire foci. In these burning coal-mining waste heaps (BCWHs), the fire incidents are due to criss-crossing influence of coal petrography (i.e., maceral composition), sulfide mineral content (especially pyrite), coal rank, and microbial activity. The fires induce three types of mineral-forming phenomena: a high-temperature solid–solid and gas–solid transformation of the waste, known as pyrometamorphism (up to ~1200°C in the coal case; [3]); medium-temperature exhalative processes; and low-temperature supergene weathering processes ([6, 10–12], and references therein). Of the Air Quality interest is, of course, the second group of processes, involving both gas emission and gas-waste interface reactions. The latter include direct gas desublimation (condensation) and pneumatolysis-like gaseous extraction of various waste-contained metals followed by hydrothermal mineralization. The first process mainly produces minerals like native sulfur (S_8), salammoniac (NH_4Cl), and a number of less frequent species like kremersite, $(NH_4,K)_2[FeCl_5(H_2O)]$ and other chlorides. The second one is responsible for vast, thick sulfate crusts mainly comprising godovikovite-sabieite solid solution, $(NH_4)(Al,Fe)(SO_4)_2$, millosevichite-mikasaite solid solution, $(Al,Fe)_2(SO_4)_3$, steklite, $KAl(SO_4)_2$, tschermigite, $(NH_4)Al(SO_4)_2{\cdot}12H_2O$ (natural ammonium aluminum alum), alunite-supergroup minerals, and many others. Pyrometamorphic processes and their product in Polish BCWH – within both the Upper and Lower Silesian Coal Basins (USCB and LSCB, respectively) was extensively studied, e.g., by Kruszewski [13, 14] and Kruszewski et al. [15, 16], with process imitation experiments described, e.g. by Kruszewski [10]. Mineralogy of the exhalative processes and gas phase composition of the local fumaroles was largely addressed by Kruszewski [6, 12, 17–19]. Fabiańska et al. [20] and Lewińska-Preis et al. [21] addressed some environmental aspects of the gas emissions in question. Supergene mineralogy was described in Kruszewski [11]. Presentation of the BCWH as models of various natural environments, including extraterrestrial ones, was shown by Kruszewski et al. [22, 23]. Biological aspects of the BCWH environment were brought up by Kruszewski & Matlakowska [24].

The fumaroles bear numerous minerals rich in trace and toxic elements, like zinc, copper, nickel, arsenic, thallium, lead, bismuth, selenium, bromine, iodine, indium, silver, and others. The mineral segregations are, obviously, related to the gas phase composition. Analyzing the latter was somewhat pioneering, as we could not find any literature sources showing the use of a portable FTIR (Fourier-Transformed InfraRed) spectroscopy for *in situ* analyzing of gaseous emissions, at least in the BCWH or the coal-fire environment in general. The IR method is a type of spectroscopy where vibrations of chemical bonds in molecules are being addressed, and depicted by their interaction with IR laser (a similar method is Raman spectroscopy). Various types of vibrations (i.e., stretching, bending, rocking, and other types) are responsible for various peaks in the spectra observed. Most compounds show response to the IR light (i.e., IR laser), by a pattern more or less characteristic for the particular molecule. Some exceptions include H_2S (hydrogen sulfide), which – in the variation of the IR method described here – gives only weak signals, thus making the aforementioned IT method somewhat more useful. The main components were shown (in [6]) to be H_2O and CO_2, with minor but variable add of CH_4 and CO. However, the composition was shown to be much more complex. The portable FTIR GASMET DX-4000 (OMC ENVAG) system was thoroughly characterized in Kruszewski et al. [6, 12]. It system a tool of choice for analysis of complex, hot, chemically aggressive and char- and ash-rich emanations, including combustion/exhaust gases. It comprises a probe with stainless-steel tip, connected with special wires with gas conditioning system (with a pressure control, pump, and system of 2 μm filters for catching any solid and liquid contaminants) and then the FTIR spectrometer. The interferometer has a ZnSe beam splitter; the sample cell has its path length of 5.0 m, volume is 0.4 L; Viton gaskets, MgF_2 protective coating, and BaF_2 window are present, too. The whole sampling system is internally coated by protective layers of rhodium and, gold and nickel.

FTIR results obtained for total 52 fumaroles in four BCWHs located in Pszów, Rybnik-Rymer, Radlin, and Rydułtowy (USCB), respectively, showed up to [in ppm, unless noticed; whole-range maximums underlined]: H_2O 57.5, 19.3, 12.5, 36.2 vol.%; CO_2 67.2, 7.63, 6.82, 30.6 vol.%; CO 2690, 694, 21, 347; NO 434, 38, 123, 151; N_2O not observed (n.o.), 0.42, 1.2, 8.7; NO_2 16430, 116, 24, 191; NH_3 1715, 646, 14, 98; SO_2 582, 74, 64, 226; HCl 58, 23, 2.4, 8.9; CCl_4 22, 1.5, 6.0, 14; HF 4.0, 2.2, n. o., 5.1; SiF_4 1890, 228, 504, 1980; AsH_3 8.2, 0.49, 0.18, 0.64; CH_4 82970, 1050, 838, 888; ethane 511, 306, 42, 316; propane 1446, 100, 16, 284; hexane 921, 123, n.o., 262; ethene 92, 28, 21, 21; dichloromethane (DCM) 5472, 1730, 241, 1980; 1,1-dichloroethane (1,1-DCE) 2110, 580, 175, and 742; 1,2-DCE 573, 28, 7.4, n.o.; 1,1,1-trichloroethane (1,1,1-TCE) 7.7, n.o., 40, 23; 1,2-dichloropropane (1,2-DCP) 4900, 12, n.o., 44; 1,1-dichloroethene (1,1-DCEe) 51, 3.3, 34, 140; vinyl chloride 1700, 809, n.o., 1980; chlorobenzene 416, 71, 92, 100; cumene (*i*-propylbenzene) 194, 84, 35, 75; phenol 43, 348, 37, and 103; *o*-cresol (2-methylphenol) 1620, 99, n.o., 99; furan 27, 29, 130, 12; tetrahydrofuran (THF) 598, 372, n.o., 2830; thiophene 781, 578, 773, 550; acetic acid 7000, 12, 12, 650; dimethyl sulfide (DMS) 6650, 2230, n.o., 6780; dimethyl disulfide (DMDS) 518, 36, n.o., 97; formaldehyde 5.7, n.o., n.o., and 3.1. Pyridine was observed only in Radlin, in very constant amounts, 10–11 ppm. Although certified (as in the case of other compounds in the calibration library), the maximum contents of germanium tetrachloride, $GeCl_4$, i.e., 3130, 209, 333, and 2098 should be treated with care due to possible coincidence as yet unresolvable by the Calcmet software. Geometric means of the concentration values (Pszów, Rybnik-Rymer, Radlin, Rydułtowy, whole series) are: H_2O 31, 12, 3.0, 21, and 19 (n_{total} = 46); CO_2 31, 4.0, 0.22, 11, and 7.0 (n_{total} = 50) [vol.%]; CO 84, 186, 9.6, 81, and 81 (n_{total} = 41); NO 87, 15, 14, 66, and 42 (n_{total} = 24); NO_2 334, 38, 14, 42, and 41 (n_{total} = 26); N_2O -, 0.10, 0.66, 4.3, and 0.83 (n_{total} = 17); NH_3 287, 22, 3.4, 59, and 88 (n_{total} = 18); SO_2 110, 18, 17, 48, and 56 (n_{total} = 31); HCl 7.4, 4.2, 0.56, 3.0, and 3.8 (n_{total} = 46); CCl_4 3.2, 0.18, 0.91, 2.5, and 1.6 (n_{total} = 51); HF 4.0, 2.2, –, 3.3, and 3.4 (n_{total} = 9); SiF_4 16, 114, 94, 182, and 65 (n_{total} = 29); AsH_3 1.1, 0.19, 0.17, 1.0, and 0.58 (n_{total} = 26); CH_4 1945, 500, 23, 537, and 457 (n_{total} = 47); ethane 46, 114, 15, 75, and 59 (n_{total} = 37); propane 148, 70, 16, 27, and 46 (n_{total} = 27); hexane 160, 25, –, 15, and 38 (n_{total} = 26); ethene 7.9, 7.3, 11, 8.3, and 8.2 (n_{total} = 28); DCM 230, 119, 160, 295, and 214 (n_{total} = 45); 1,1-DCE 235, 190, 98, 99, and 139 (n_{total} = 32); 1,2-DCE 153, 28, 7.4, –, and 91 (n_{total} = 9); 1,1,1-TCE 5.4, –, 40, 9.5, and 7.7 (n_{total} = 19); 1,2-DCP 1038, 5.7, –, 20, and 166 (n_{total} = 9); 1,1-DCEe 19, 2.6, 31, 25, and 20 (n_{total} = 35); vinyl chloride 329, 38, –, 394, and 289 (n_{total} = 32); chlorobenzene 24, 36, 92, 32, and 32 (n_{total} = 12); cumene 28, 30, 35, 15, and 22 (n_{total} = 38); phenol 14, 36, 3.8, 32, and 19 (n_{total} = 32); *o*-cresol 115, 21, –, 99, and 73 (n_{total} = 15); furan 11, 9.8, 72, 12, and 31 (n_{total} = 18); THF 126, 372, –, 643, and 195 (n_{total} = 10); thiophene 251, 200, 90, 156, and 186 (n_{total} = 40); formaldehyde 3.5, –, –, 0.54, and 0.82 (n_{total} = 8); acetic acid 189, 6.6, 8.1, 83, and 54 (n_{total} = 22); DMS 517, 433, –, 921, and 533 (n_{total} = 16); DMDS 68, 11, –, 31, and 41 (n_{total} = 35); and pyridine -, –, 11, –, and 11 (total 8 records).

GC results were also published in the paper, with confirmed occurrence of carbonyl sulfide, COS, carbon disulfide, CS_2, freons (CCl_3F, CCl_2F_2, $CHClF_2$), *i*-butane, *n*-butane, *i*-pentane, *n*-pentane, *n*-hexane, *n*-heptane, *n*-octane, *n*-nonane, *n*-decane, propene, 1-butene, *i*-butene, *trans*- and *cis*-2-butene, *trans*- and *cis*-2-pentene, ethyne, 1,3-butadiene, isoprene (2-methyl-1,3,-butadiene), 2,3-dimethylbutane; 2- and 3-methylpentanes; benzene, toluene, *m*/*p*- and *o*-xylenes, styrene, ethylbenzene, *n*- and *i*-propylbenzene; 2-, 3,- and 4- (or *m*-, *p*- and o-) ethyltoluene; 1,2,3-, 1,2,4-, and 1,3,5-trimethylbenzenes; and α- and β-pinene. As shown in the paper, the GC results may be quite unreliable due to their non-*in situ* character and possible intra-gas and gas-steel interactions, and are thus not resumed here. In turn, we have later used a second and third mode of the FTIR spectra reading. The first one is an external library search, where the spectra are read and calculated using libraries containing other compound sets, thus reporting semi-quantitative results with fit factor (r^2, in %), as described in Kruszewski et al. [12]. Any misfits are due to recording the standards in different conditions than in the DX-4000 calibration library case. Applying this method allowed to detect additional compounds for the previously listed 4 BCWH sites [in ppm, with results for fit ≥90%, 75–90%, 50–75%, and < 50%, and whole-data maximums underlined]: acetylene, C_2H_2 (up to 0.81; up to 27; up to 38; up to 288), *n*-butane (–; –; 7.1; 1.5), *i*-butane (–; –; 9.7; 0.25), propene (–; –; up to 101; up to 30), *n*-pentane (–; –; 4.0; 1.9), *i*-pentane (–; –; 11; 0.91), heptane (–; –; up to 2.1; –), octane (–; –; up to 2.3; –), nonane (–; –; up to 2.1; –), decane (–; –; up to 2.0; –), undecane (–; –; up to 2.0; –), 1,3-butadiene (3.2; –; up to 144; up to 169), cyclohexane (–; –; up to 2.7; –), α-pinene (–; –; up to 4.0; up to 1.1), limonene ($C_{10}H_{16}$; –; –; up to 4.9; 2.7), 3-carene ($C_{10}H_{16}$; 512; up to 2.2), benzene (8.8; up to 5.1; up to 52; up to 5700), toluene (–; –; up to 74; up to 18), styrene (–; 88; 0.76; up to 154), *m*-xylene (–; –; 19; up to 51), *p*-xylene (–; –; 16; up to 23), ethylbenzene (–; –; –; up to 8.4), 1,3,5-TMB (–; –; up to 729; up to 32), 1,2,4-TMB (–; –; up to 1610; up to 27), 1,2,3-TMB (–; –; up to 1360; up to 23), tetrachloroethene (–; up to 4.3; up to 28; up to 27), methanol (11; 5.4; up to 18; up to 75), ethanol (16; 5.4; up to 38; up to 126), *i*-propanol (isopropanol; –; –; –; up to 16), *i*-butanol (isobutanol; –; –; –; 5.4), *n*-propanol (–; –; 982; –), methanethiol (methylmercaptan), CH_3SH (–; –; –; up to 55), ethanethiol (ethylmercaptan), C_2H_5SH (–; –; 2500; up to 14), HCN (up to 8.4; up to 16; up to 88; up to 65), acrylonitrile (prop-2-enenitrile, $CH_2 = CHCN$; –; 6.0; up to 63; up to 82), isocyanic acid (–; –; –; up to 717), formic acid, HCOOH (3.0; 8.7; up to 29; up to 48), trimethylamine, $(C_2H_5)_3N$ (–; –; –; up to 1.5), acetaldehyde (up to 45; up to 97; up to 1810; up to 6270), propionaldehyde (propanal), $(C_2H_5)CHO$ (–; –; –; up to 24), 2-ethylhexylaldehyde ($C_4H_9CH(C_2H_5)CHO$; –; –; up to 342; –), acrolein (propenal, $CH_2 = CHCHO$; –; 1.6; up to 57; up to 25), acetone (propan-2-one) (–; –; –; up to 98), methyl ethyl ketone (MEK, or butan-2-one), $CH_3C(O)C_2H_5$ (–; –; –; up to 28), methyl isobutyl ketone (MIBK, or 4-methylpentan-2-on), $(CH_3)_2C_2H_3C(O)CH_3$ (–; –; –; up to 2.6), diethylether (ethoxyethane, $(C_2H_5)_2O$; –; –; 1.7; up to 24), MTBE (–; –; –; up to 9.4), 2-ethoxyethanol, $(C_2H_5)O(CH_2)O(C_2H_5)$ (–; –; up to 47; up to 32), 2-ethoxyethyl acetate (–; –; –; up to 19), butyl acetate (–; –; –; up to 15), 2-(2-butoxyethoxy) ethyl acetate (–; –; –; up to 13), methyl metacrylate (methyl 2-methylprop-2-enoate; –; –; –; up to 10), PH_3 (phosphine; –; up to 43; up to 144; up to 152), COS (up to 0.88; up to 6.1; up to 0.40; –), and last but not least SF_6 (–; –; up to 1.6; up to 1.5). The last compound is environmentally very important, as it is said – by the Intergovernmental Panel on Climate Change – to be the most potent greenhouse gas [25]. The measured BCWH emanation concentrations are also much higher (over 170000 times) than the highest ones measured at Mauna Loa fumaroles [26]. Calculated geometric means (whole series; with values for fit ≥50% in the parentheses): 13 (2.3) for acetylene (n = 14 (31)), 25 (51) for propene (n = 9 (3)), 17 (29) for 1,3-butadiene (n = 15 (5)), 0.76 for α-pinene (n = 6), 3.5 for limonene (n = 3), 6.2 for 3-carene (n = 4), 55 (9.7) for benzene (n = 34 (14)), 7.4 (21) for toluene (n = 11 (3)), 9.6 for styrene (n = 9), 9.9 (10) for *m*-xylene (n = 11 (8)), 13 for *p*-xylene (n = 7), 13 (13) for 1,3,5-TMB (n = 11 (8)), 11 for 1,2,4-TMB (n = 10), 4.5 for 1,2,3-TMB (n = 6), 15 (5.5) for methanol (n = 24 (4)), 32 (8.6) for ethanol (n = 26 (7)), 6.9 for *i*-propanol (n = 7), 23 for ethanethiol (n = 4), 4.2 (1.4) for tetrachloroethene (n = 31 (9)), 7.2 (5.9) for HCN (n = 47 (33)), 293 for isocyanic acid (n = 18), 1.2 for trimethylamine (n = 3), 47 (47) for acrylonitrile (n = 12 (9)), 15 (12) for formic acid (n = 35 (7)), 62 (28) for acetaldehyde (n = 50 (45)), 9.4 for propionaldehyde

(n = 4), 37 for 2-ethylhexylaldehyde (n = 3), 23 (28) for acrolein (n = 13 (9)), 21 for acetone (n = 22), 5.2 for diethylether (n = 10), 6.8 for 2-ethoxyethanol (n = 10), 7.7 for 2-ethoxyethyl acetate (n = 20), 4.8 for butyl acetate (n = 8), 5.1 for methyl metacrylate (n = 9), 12 for MEK (n = 8), 2.0 for MIBK (n = 5), 2.7 for MTBE (n = 4), 0.37 (0.41) for COS (n = 16 (13)), 72 (40) for PH_3 (n = 28 (10)), and 1.1 for SF_6 (n = 10). As such, acetaldehyde, HCN, PH_3, tetrachloroethene, ethanol, benzene, COS, methanol, acetylene, and 1,3-butadiene, isocyanic acid, acrolein, and likely acetone and 2-ethoxyethyl acetate seem to be the most frequent admixing gases in the BCWH exhausts studied.

The third operation mode is qualitative analysis of residual spectra, as thoroughly described in both my previous papers. This method allowed to list proposals of additional, very interesting, admixing gases, many of which were likely first documented in the nature. They include neutral hydroxides of Ca, Mg, Al, Fe(II), Fe (III), Zn, Cu; nitrosyls and carbonyls of Ti, V, Mn, Fe, Ag, Mo, Fe, Cu; hydrides of Al, Cu, Zn, Ge, Mo, Sb, and Hg; nitriles, azo and related compounds (azacyclopropenylidene, dicyanoacetylene, cyanogen isocyanate, cyanogen *N*-oxide, diazomethyl radical, hydrogen isocyanide, isocyanic acid, *m*-hydroxybenzonitrile, phenylnitrene radical; 2,4,6-trinitrene-1,3,5-triazine); amines (methyl(nitrosomethyl)amine); hypobromous and hydroiodic acids; hydrocarbons and halocarbons (cyclohexene, dibenz[*a*,*h*]anthracene, difluorovinylidene, hexachlorobenzene, hexachloroethane, 5-methyl-1,3-didehydrobenzene, pentacene, phenanthrene, triphenylene); nitrosyl chloride and iodide, phosgene; organoboron compounds (fluoroisocyanatoborane) and compounds like CBrO and B_2O_2; organosulfurs (thiirene, thioacetaldehyde, thioxoethenylidyne radical), organophosphorus compounds (methylphosphine), and organosilicons (difluorosilane, disinale, silanenitrile, tribromosilane), organoiodine compounds (iodosomethane – an I^{3+}-bearing compounds; iodocyanoacetylene), $HAlCl_2$, ClO_2^*, and dimeric NO, to mention some. Due to multiple coincidence possible these results should, however, be treated with care.

4. New *in situ* FTIR gas analysis results of the USCB heaps

Results presentation within this chapter has its main goal in enlarging the span of the knowledge on the concentration range of various (major and minor) components of the BCWH combustion gases, both by pFTIR and GC methods. **Table 1** shows data from Czerwionka-Leszczyny (18, that is, 10 vents / vent zones from zone CLD and 8 from the CL one). **Table 2** juxtaposes data for 10 additional, differently mineralized vents from the Radlin heap (RD), with that from a BCWH in Bytom (BTM, 7 vents / vent zones). **Table 3**, in turn, juxtaposed data for vents in a BCWH in Świętochłowice (SWC, 11 vents / vent zones), "Starzykowiec" heap in the Chwałowice part of Rybnik (RCH, 1 vent, surface and deep part), and "Ruda" heap in Zabrze-Biskupice (ZBB, 5 vents / vent zones). In total, data for additional 53 vents is reported. As in the case of the data presented in Kruszewski et al. [6, 12], gases were probed at the surface and from deeper parts of the vents, whenever possible. Temperatures were measured using an IR pyrometer.

Following are values describing maximum and geometric-mean concentrations of gaseous species as detected within fumarolic vents of the CLD, CL, RD, BTM, SWC, and ZBB sites (whole-series-maximums are underlined): H_2O, 18.12, 14.74; 7.30, 2.83; 27.14, 23.04; 11.15, 9.83; 6.42, 4.68; 25.63, 23.19; CO_2, 2.85, 2.29; 27.00, 0.20; 29.89, 20.85; 8.12, 6.05; 38.41, 33.89 [vol.%]; CO, 135, 110; 163, 9.4; 2430, 1002; 3590, 2675; 1090, 303; 26700, 3257; NO, 112, 96; 10, 6.4; 7.3, 7.3; –, –; 19, 15; –, –; NO_2, 44, 22; 368, 155; 2.0, 2.0; 1430, 1430; –, –; 66, 45; N_2O, 3.5, 2.3; 0.06, 0.02; 4.4, 4.4; 2.8, 2.8; 1.3, 1.3; –, –; NH_3, 21, 7.7; 30, 2.5; 19, 7.1; 65, 55; 4.1, 2.4; 8.3,

vent [1]	**CLD1**	**CLD1o**	**CLD2**	**CLD3**	**CLD5**	**CLD5S**	**CLD5o**	**CLD6o**	**CLD6o2**	**CLD7**	**CLdA**	**CLdAr**	**CLdU**	**CLd**	**CL1**	**CLdo**	**CL2a**	**CL2aA**
T [°C]	40	45	45	45	25	35	35	35	35	60	90	90	82	82	50	82	30	45
pFTIR																		
inorganics, vol.%																		
H_2O[2]	9.77	12.21	12.22	12.27	16.43	17.36	16.88	17.62	17.65	18.12	2.68	2.65	2.57	2.58	0.75	2.62	6.10	7.30
CO_2	1.96	2.32	2.34	2.40	1.90	2.26	2.08	2.39	2.54	2.85	0.03	0.03	0.03	0.03	bdl	0.03	18.00	27.00
inorganics, ppm																		
CO	bdl	bdl	bdl	bdl	**135**[3]	**101**	**101**	**132**	94	**103**	6.4	3.1	bdl	0.98	bdl	1.5	145	163
N_2O	2.8	3.4	3.5	3.5	bdl	bdl	bdl	bdl	bdl	0.57	0.02	0.04	0.06	0.01	bdl	0.02	bdl	bdl
NO	64	**107**	**112**	**112**	bdl	bdl	bdl	bdl	bdl	bdl	6.9	9.9	bdl	9.7	1.6	10	bdl	bdl
NO_2	bdl	bdl	bdl	bdl	bdl	bdl	bdl	bdl	11	44	bdl	bdl	bdl	bdl	bdl	bdl	**368**	65
NH_3	4.4	3.9	3.6	3.7	10	15	21	10	11	9.4	bdl	1.0	0.23	1.4	bdl	1.4	18	30
SO_2	4.2	bdl	bdl	bdl	bdl	87	bdl	bdl	20	120	bdl	bdl	bdl	1.9	bdl	2.4	**119**	**671**
HCl	0.04	bdl	0.24	0.57	11	10	10	8.7	7.6	6.6	1.5	0.76	0.08	0.58	0.01	0.63	6.5	5.8
CCl_4	bdl	bdl	bdl	bdl	bdl	bdl	bdl	bdl	bdl	bdl	bdl	bdl	bdl	bdl	bdl	bdl	bdl	6.6
HF	bdl	bdl	bdl	bdl	bdl	0.62	bdl	bdl	bdl	bdl	bdl	bdl	bdl	bdl	bdl	bdl	bdl	0.03
SiF_4	bdl	0.06	0.21	0.16	3.7	5.3	3.8	3.4	6.3	4.6	2.2	1.9	1.9	2.1	1.1	2.1	20	31
AsH_3	bdl	0.08	0.17	0.15	0.03	bdl	bdl	bdl	bdl	bdl	bdl	bdl	0.16	bdl	bdl	bdl	0.20	**1.7**
aliphatic and aromatic hydrocarbons and their derivatives, ppm																		
CH_4	26	31	31	31	**244**	**259**	**262**	**251**	**253**	**248**	4.5	4.8	6.8	4.6	0.51	4.8	**811**	**2950**
ethane	bdl	bdl	bdl	bdl	bdl	bdl	bdl	bdl	bdl	bdl	bdl	bdl	bdl	bdl	bdl	bdl	30	bdl
propane	6.9	3.8	4.4	4.4	34	42	36	bdl	40	37	bdl	4.2	3.1	13	11	8.2	bdl	**729**
hexane	1.8	0.20	0.31	0.35	2.6	4.6	2.3	7.0	4.6	11	bdl	bdl	bdl	bdl	bdl	bdl	**152**	17

vent [1]	CLD1	CLD1o	CLD2	CLD3	CLD5	CLD5S	CLD5o	CLD6o	CLD6o2	CLD7	CLdA	CLdAr	CLdU	CLd	CL1	CLdo	CL2a	CL2aA
ethene	bdl	bdl	bdl	bdl	bdl	1.3	bdl	bdl	3.7	6.6	2.2	bdl	1.3	bdl	1.9	bdl	37	79
DCM	17	10	16	15	73	**104**	71	141	57	20	52	65	56	50	18	51	**368**	**159**
1,1-DCE	bdl	bdl	bdl	bdl	bdl	bdl	bdl	7.2	bdl	bdl	bdl	5.1	bdl	7.4	bdl	7.2	17	12
1,2-DCE	bdl	bdl	bdl	bdl	bdl	bdl	bdl	bdl	bdl	bdl	bdl	bdl	bdl	bdl	bdl	bdl	77	bdl
vent	**CLD1**	**CLD1o**	**CLD2**	**CLD3**	**CLD5**	**CLD5S**	**CLD5o**	**CLD6o**	**CLD6o2**	**CLD7**	**CLdA**	**CLdAr**	**CLdU**	**CLd**	**CL1**	**CLdo**	**CL2a**	**CL2aA**
1,1,1-TCE	bdl	bdl	bdl	bdl	**99**	bdl	bdl	bdl	bdl	21	bdl	bdl	bdl	6.4	bdl	6.5	**417**	bdl
1,2-DCP	bdl	bdl	bdl	bdl	bdl	bdl	bdl	31	bdl	bdl	bdl	bdl	bdl	bdl	bdl	bdl	56	bdl
1,1-DCEe	4.3	5.3	4.4	4.8	57	77	54	63	43	68	35	45	49	48	25	57	**226**	**347**
ClB	bdl	bdl	bdl	bdl	20	bdl	18	39	bdl	bdl	15	14	22	6.0	6.0	6.3	bdl	**186**
cumene	4.4	6.3	6.4	5.5	29	30	31	29	34	31	6.1	1.3	7.6	1.9	2.4	1.7	**399**	bdl
phenol	2.4	4.4	4.2	4.0	29	bdl	29	28	17	17	4.6	3.2	7.0	bdl	2.9	bdl	bdl	bdl
o-cresole	0.29	bdl	0.41	0.37	19	46	17	20	23	24	1.9	2.2	0.28	7.1	0.81	6.9	66	10
heterocyclic organic compounds, ppm																		
furan	bdl	0.14	bdl	bdl	bdl	bdl	bdl	bdl	bdl	bdl	bdl	bdl	bdl	bdl	bdl	bdl	bdl	bdl
THF	0.29	bdl	1.3	0.18	bdl	bdl	bdl	bdl	bdl	bdl	bdl	bdl	bdl	bdl	bdl	bdl	bdl	**177**
py	bdl	bdl	bdl	bdl	bdl	7.1	bdl	bdl	bdl	bdl	0.27	2.3	6.3	12	23	12	86	bdl
tph	bdl	bdl	bdl	bdl	**192**	**137**	**191**	bdl	**127**	**103**	bdl	bdl	bdl	bdl	8.3	bdl	bdl	**173**
other organic compounds, ppm																		
fm	0.11	bdl	bdl	bdl	12	7.0	6.0	5.3	5.3	5.7	0.58	1.2	0.47	0.42	0.45	0.42	13	13
DMS	bdl	13	7.7	11	62	6.9	27	33	31	51	bdl	0.15	bdl	bdl	bdl	bdl	**893**	**165**
DMDS)	12	23	32	28	**99**	**104**	40	7.0	85	5.6	3.5	37	31	36	17	36	bdl	bdl

GC – additional compounds, *ppm*																		
vent	**CLD1**	**CLD1o**	**CLD2**	**CLD3**[4]	**CLD5**	**CLD5S**	**CLD5o**	**CLD6o**	**CLD6o2**	**CLD7**	**CLdA**	**CLdAr**	**CLdU**	**CLd**	**CL1**	**CLdo**	**CL2a**	**CL2aA**
CH_3Cl	0.002			0.001	0.003			0.002		0.005					0.002	0.08	0.03	0.01
ethyne	0.001			0.001	0.001			0.001		0.01					0.003	0.01	0.01	0.0002
propene	0.0002			bdl	1.7			5.2		1.4					0.09	0.51	0.36	0.09
i-butane	0.0003			0.002	1.6			3.2		3.5					0.08	0.69	8.3	2.3
n-butane	0.001			0.002	2.4			5.4		3.6					0.15	1.1	12	3.9
1-butene	0.0001			0.001	0.14			0.40		0.13					0.02	0.03	0.08	0.03
i-butene	0.0001			0.002	0.41			0.32		0.52					0.03	0.04	1.3	0.42
t^{-2}-bu	0.00004			0.001	0.52			1.4		0.50					0.02	0.04	0.59	0.24
c-2-bu	0.00003			0.0005	0.29			0.80		0.23					0.01	0.02	0.17	0.06
vent	**CLD1**	**CLD1o**	**CLD2**	**CLD3**	**CLD5**	**CLD5S**	**CLD5o**	**CLD6o**	**CLD6o2**	**CLD7**	**CLdA**	**CLdAr**	**CLdU**	**CLd**	**CL1**	**CLdo**	**CL2a**	**CL2aA**
i-pentane	0.001			0.01	0.92			1.8		1.8					0.06	0.35	6.5	1.7
n-pentane	0.0003			0.01	1.1			2.3		1.5					0.06	0.35	6.0	1.9
t-2-pte	bdl			0.002	0.20			0.57		0.19					0.01	0.01	0.36	0.18
c-2-pte	bdl			0.001	0.08			0.23		0.07					0.004	0.003	0.09	0.04
n-heptane	bdl			0.003	0.44			0.84		0.34					0.02	0.07	1.7	0.68
n-octane	bdl			0.002	0.36			0.65		0.10					0.01	0.04	0.64	0.32
n-nonane	bdl			0.0003	0.25			0.40		0.03					0.01	0.01	0.22	0.04
n-decane	bdl			0.0001	0.16			0.20		0.003					0.001	0.002	0.03	0.04
2,3-DMBu	bdl			0.01	0.07			0.13		0.16					0.003	0.02	0.58	0.14
2-MPT	bdl			0.01	0.41			0.76		0.75					0.02	0.11	3.0	0.82

vent	CLD1	CLD1o	CLD2	CLD3	CLD5	CLD5S	CLD5o	CLD6o	CLD6o2	CLD7	CLdA	CLdAr	CLdU	CLd	CL1	CLdo	CL2a	CL2aA
3-MPT	0.0001			0.005	0.16			0.30		0.30					0.01	0.05	1.3	0.34
cpt	bdl			0.001	0.19			0.42		0.26					0.01	0.05	1.0	0.30
benzene	0.0004			0.002	2.1			3.3		0.06					0.05	0.27	1.1	0.41
toluene	0.001			0.02	2.0			3.4		0.01					0.02	0.11	0.05	0.02
EtB	0.0001			0.003	0.27			0.43		0.01					0.01	0.02	0.12	0.05
m/*p*-X	0.0003			0.01	1.3			1.9		0.19					0.02	0.09	0.06	0.22
o-X	0.0002			0.003	0.39			0.53		0.03					0.01	0.03	0.02	0.01
styrene	0.001			0.0001	bdl			bdl		bdl					bdl	bdl	bdl	bdl
i-PrB	bdl			0.0002	0.03			0.05		0.03					0.001	0.001	0.06	0.03
n-PrB	bdl			0.0005	0.06			0.08		0.02					0.001	0.002	0.02	0.01
m-EtT	0.0001			0.001	0.18			0.24		0.06					0.004	0.01	0.01	0.01
p-EtT	bdl			0.0005	0.07			0.10		0.02					0.001	0.003	0.01	0.01
o-EtT	bdl			0.0005	0.08			0.10		0.02					0.003	0.002	0.01	0.01
1,3,5-TMB	0.0001			0.001	0.14			0.19		0.07					0.003	0.005	0.01	0.005
1,2,4-TMB	0.0001			0.002	0.27			0.32		0.05					0.01	0.01	0.04	0.03
1,2,3-TMB	0.0001			0.001	0.10			0.10		0.02					0.004	0.003	0.06	0.05

Values in parentheses denote overrun of the upper measurement range.
[1]*"A" – samples taken from the depth of 0.8–1 m; "a" and "o" - nearby vents; "r" – repeated analysis; "S" – sulfur-mineralized vent.*
[2]*DCM – dichloromethane, DCE – dichloroethane, DCEe – dichloroethene, TCE – trichloroethane, DCP – dichloropropane, ClB – chlorobenzene, THF – tetrahydrofuran, py – pyridine, tph – thiophene, fm – formaldehyde, DMS – dimethyl sulfide, DMDS – dimethyl disulfide; t(c)-2-bu –* trans(cis)*-2-butene, c(t)-2-pte –* cis(trans)*-2-pentene, DMBu – dimethylbutane, MPT – methylpentane, cpt – cyclopentane, EtB – ethylbenzene, X – xylene, PrB – propylbenzene, EtT – ethyltoluene, TMB – trimethylbenzene; vinyl chloride, acetic acid, isoprene, and 1,3-butadiene were analyzed but were below their detection limits.*
[3]*Notable (>100 ppm) enrichment given in bold.*
[4]*GC data for a nearby vent.*

Table 1.
Results of the pFTIR and GC gas analyses of BCWH in Czerwionka-Leszczyny.

DOI: http://dx.doi.org/10.5772/intechopen.96294

vent[1]	RD07	RD07A	RD08N	RD08NA	RD08kr	RD08krA	RD08o	RD11L	RD11U	RD11o	BTM1	BTM1A	BTM1o	BTM1o2	BTM1o3	BTM1o4	BTM2
T [°C]	60	90	133	187	77	107	76	76	76	76	115	144	73	150	79	115	60
pFTIR																	
main components, vol. %																	
H_2O[2]	19.99	24.62	26.02	18.26	27.14	22.25	26.61	16.94	24.80	26.65	6.92	9.94	10.41	10.67	11.15	11.02	9.43
CO_2	25.00	29.89	20.21	13.94	26.16	21.32	21.24	11.96	21.14	24.64	4.08	5.43	5.52	5.25	7.58	8.12	7.49
inorganics, ppm																	
CO	**1580**	**2430**	**923**	**502**	**1110**	**813**	**832**	**673**	**1030**	**1100**	**1730**	**2830**	**3090**	**2640**	**3090**	**3590**	**2210**
N_2O	bdl	bdl	bdl	bdl	bdl	bdl	bdl	4.4	bdl	bdl	bdl	bdl	bdl	bdl	bdl	bdl	2.8
NO	bdl	bdl	bdl	bdl	bdl	bdl	7.3	bdl	bdl	bdl	bdl	bdl	bdl	bdl	bdl	bdl	bdl
NO_2	bdl	2.0	bdl	bdl	bdl	bdl	bdl	bdl	bdl	bdl	bdl	bdl	bdl	bdl	**1430**	bdl	bdl
NH_3	9.2	9.2	5.5	6.9	12	19	5.0	1.8	3.1	16	42	60	60	61	62	65	41
SO_2	**388**	**311**	bdl	57	bdl	**193**	bdl	79	bdl	bdl	86	**139**	**147**	**139**	**532**	**378**	**281**
HCl	4.5	3.1	6.1	2.9	2.5	2.6	6.2	bdl	5.0	2.3	2.7	3.3	3.6	5.4	19	15	6.3
CCl_4	6.3	5.1	10	4.4	6.8	6.9	3.2	7.4	11	7.8	bdl	bdl	bdl	bdl	1.1	bdl	bdl
HF	0.12	bdl	bdl	bdl	bdl	bdl	bdl	bdl	bdl	bdl	bdl	0.59	0.07	1.1	bdl	bdl	bdl
SiF_4	15	13	21	16	15	15	18	17	14	13	1.7	1.8	1.7	0.95	bdl	3.2	bdl
AsH_3	bdl	**1.3**	0.93	0.49	**1.2**	**1.4**	0.56	**1.7**	bdl	0.11	0.81	**2.9**	**1.3**	**2.0**	0.79	**2.8**	**3.4**
aliphatic and aromatic hydrocarbons and their derivatives, ppm																	
methane	**3190**	**3470**	**1120**	**713**	**1120**	**1160**	**1140**	**500**	**1120**	**1150**	**285**	**437**	**421**	**359**	**808**	**819**	**554**
ethane	bdl	**142**	bdl	bdl	bdl	bdl	bdl	bdl	bdl	bdl	bdl	bdl	bdl	bdl	**281**	bdl	bdl
propane	**387**	**601**	**202**	bdl	**291**	**230**	**213**	bdl	**195**	**249**	bdl	bdl	bdl	bdl	**694**	**242**	bdl
hexane	71	**139**	70	53	90	67	71	53	64	76	88	**148**	**160**	**309**	**608**	**543**	**138**
ethene	bdl	0.71	12	5.4	9.3	7.6	12	12	3.1	6.9	4.7	16	17	39	84	78	35

vent[1]	RD07	RD07A	RD08N	RD08NA	RD08kr	RD08krA	RD08o	RD11L	RD11U	RD11o	BTM1	BTM1A	BTM1o	BTM1o2	BTM1o3	BTM1o4	BTM2
DCM	8.3	18	**126**	29	**104**	25	**111**	**94**	**102**	34	11	68	74	84	bdl	bdl	bdl
1,1-DCE	bdl	bdl	bdl	bdl	bdl	bdl	bdl	bdl	bdl	bdl	bdl	bdl	bdl	16	bdl	bdl	bdl
1,2-DCE	39	bdl	bdl	bdl	bdl	bdl	bdl	bdl	bdl	bdl	bdl	bdl	bdl	38	bdl	bdl	bdl
vent	**RD07**	**RD07A**	**RD08N**	**RD08NA**	**RD08kr**	**RD08krA**	**RD08o**	**RD11L**	**RD11U**	**RD11o**	**BTM1**	**BTM1A**	**BTM1o**	**BTM1o2**	**BTM1o3**	**BTM1o4**	**BTM2**
1,1,1-TCE	29	15	21	33	29	36	bdl	21	bdl	bdl	bdl	bdl	1.0	bdl	59	bdl	20
1,2-DCP	**124**	**384**	75	42	**264**	**175**	**224**	**108**	**251**	**248**	bdl	bdl	39	bdl	**568**	bdl	**519**
1,1-DCEe	81	30	**195**	**101**	**164**	**123**	**176**	**155**	**189**	**162**	bdl	11	11	bdl	24	28	bdl
ViCl	**416**	**96**	**276**	**187**	**248**	**283**	**269**	bdl	**203**	**278**	38	29	47	63	**137**	**205**	**359**
ClB	bdl	bdl	bdl	bdl	11	bdl	74	bdl	77	62	62	46	61	**112**	73	**206**	bdl
cumene	**94**	**126**	**119**	36	**101**	**108**	**123**	**105**	**114**	90	38	46	50	**264**	**117**	bdl	**107**
phenol	bdl	15	49	bdl	21	1.6	51	bdl	11	17	18	39	42	47	60	67	35
o-cresole	37	65	25	40	34	50	26	39	30	31	27	46	48	53	81	73	76
heterocyclic organic compounds, ppm																	
THF	43	**96**	bdl	3.1	bdl	9.3	8.7	4.1	6.1	12	bdl	bdl	bdl	**232**	bdl	bdl	**293**
tph	**164**	**257**	**337**	**385**	**363**	**369**	**344**	**301**	**556**	**391**	63	bdl	bdl	bdl	**689**	**201**	**388**
other organic compounds, ppm																	
fm	bdl	bdl	2.6	1.4	2.0	2.0	5.5	1.4	1.3	1.6	0.74	0.77	0.91	1.2	**17**	**17**	2.1
acac	bdl	9.1	bdl	bdl	bdl	bdl	bdl	bdl	bdl	bdl	bdl	bdl	2.2	bdl	35	9.9	50
DMS	**131**	**126**	47	**401**	bdl	16	70	**255**	27	15	**172**	**219**	**243**	**1540**	**1370**	**1200**	bdl
GC – additional compounds, *ppm*																	
vent	**RD07**	**RD07A**	**RD08N**	**RD08NA**	**RD08kr**	**RD08krA**	**RD08o**	**RD11L**	**RD11U**	**RD11o**	**BTM1**	**BTM1A**	**BTM1o**	**BTM1o2**	**BTM1o3**	**BTM1o4**	**BTM2**
CH_3Cl	0.001		0.05	1.4	0.03	0.03	0.02	0.01			0.10						0.02
COS	7.4		0.78	2.6	0.88	0.16	2.1	0.43			1.3						0.98

vent	RD07	RD07A	RD08N	RD08NA	RD08kr	RD08krA	RD080	RD11L	RD11U	RD110	BTM1	BTM1A	BTM10	BTM102	BTM103	BTM104	BTM2
ethyne	0.01		0.01	0.001	0.003	0.003	0.002	0.04			0.002						0.004
propene	0.47		0.91	0.93	0.01	0.03	3.8	1.5			8.5						0.62
i-butane	3.3		1.7	0.08	0.005	0.13	2.0	2.6			2.4						1.7
n-butane	8.7		3.6	0.21	0.01	0.03	4.6	3.4			5.9						2.6
1-butene	0.01		0.02	0.13	0.003	0.005	0.24	0.13			1.6						0.05
i-butene	0.06		0.08	0.51	0.002	0.01	0.98	0.38			6.6						0.26
t-2-bu	0.01		0.06	0.16	0.001	0.01	0.78	0.24			2.3						0.07
c-2-bu	0.01		0.03	0.12	0.0005	0.003	0.47	0.18			1.6						0.04
i-pentane	2.0		0.77	0.04	0.004	0.09	1.2	1.1			1.2						0.80
vent	**RD07**	**RD07A**	**RD08N**	**RD08NA**	**RD08kr**	**RD08krA**	**RD080**	**RD11L**	**RD11U**	**RD110**	**BTM1**	**BTM1A**	**BTM10**	**BTM102**	**BTM103**	**BTM104**	**BTM2**
n-pentane	3.3		1.1	0.07	0.005	0.06	2.1	1.2			1.7						0.91
isoprene	bdl		0.002	0.03	0.0003	0.001	0.02	0.01			0.13						0.003
1,3-budi	bdl		bdl	0.02	0.0002	bdl	bdl	bdl			0.03						bdl
t-2-pte	0.12		0.01	0.06	0.001	0.01	0.21	0.08			0.81						0.01
c-2-pte	0.05		0.003	0.03	0.0004	0.002	0.09	0.03			0.39						0.003
n-heptane	0.44		0.18	0.03	0.001	0.01	0.58	0.41			0.93						0.11
n-octane	0.05		0.05	0.03	0.0007	0.004	0.18	0.21			0.57						0.05
n-nonane	0.03		0.46	0.02	0.0003	0.001	0.08	0.15			0.44						0.10
n-decane	0.003		0.02	0.01	0.0001	0.0002	0.02	0.11			0.60						0.01
2,3-DMB	0.12		0.04	0.003	0.0002	0.004	0.09	0.08			0.08						0.06
2-MPT	0.64		0.22	0.01	0.001	0.02	0.53	0.43			0.33						0.28
3-MPT	0.26		0.10	0.02	0.001	0.001	0.22	0.17			0.11						0.17
cpt	0.70		0.25	0.01	0.0004	0.003	0.40	0.27			0.22						0.20

vent	RD07	RD07A	RD08N	RD08NA	RD08kr	RD08krA	RD08o	RD11L	RD11U	RD11o	BTM1	BTM1A	BTM1o	BTM1o2	BTM1o3	BTM1o4	BTM2
benzene	5.3		3.2	4.3	0.13	0.46	5.5	33			21						1.3
toluene	0.16		0.27	0.66	0.005	0.13	1.6	0.53			7.6						0.27
EtB	0.01		0.08	0.08	0.0003	0.02	0.09	0.10			3.6						0.06
m/*p*-X	0.03		0.11	0.28	0.001	0.05	0.34	0.36			3.0						0.17
o-X	0.01		0.05	0.13	0.0004	0.01	0.18	0.15			1.1						0.07
styrene	0.001		0.01	0.005	0.00003	0.001	0.01	0.01			0.14						0.003
i-PrB	0.001		0.02	0.01	bdl	0.001	0.003	0.02			0.71						0.01
n-PrB	0.002		0.01	0.04	0.0001	0.002	0.003	0.02			0.27						0.01
m-EtT	0.01		0.05	0.11	0.0001	0003	0.01	0.08			0.54						0.05
p-EtT	0.002		0.02	0.05	0.00003	0.001	0.01	0.06			0.38						0.02
o-EtT	0.004		0.02	0.04	0.00005	0.001	0.01	0.05			0.25						0.02
1,3,5-TMB	0.01		0.02	0.05	0.0001	0.001	0.01	0.11			0.38						0.02
1,2,4-TMB	0.01		0.08	0.17	0.0001	0.003	0.02	0.13			1.0						0.05
1,2,3-TMB	0.01		0.05	0.07	0.0001	0.001	0.01	0.14			0.55						0.04

Values in parentheses denote overrun of the upper measurement range.
[1]*the "A" add denotes samples taken from the depth of 0.8–1 m (below the ground level), while "a" and "o" denote nearby vents; "r" – repeated measurement; "P" – pyrometamorphic zone, "S" – sulfur-mineralized vent.*
[2]*Abbreviations explained under* ***Table 1****; ViCl – vinyl chloride; furan, pyridine and DMDS were analyzed but were below their detection limits.*
[3]*Notable (>100 ppm) enrichment given in bold.*

Table 2.
Results of the pFTIR and GC gas analyses of the "Marcel" mine BCWH in Radlin (RD, second gas study) and a heap in Bytom (BTM).

vent [1]	SWC1	SWC1r	SWC1oP	SWC1oSW	SWC1oB	SWC2	SWC2o	SWC2o2	SWC2o3	SWC3	SWC3A	RCH1	RCH1A	ZBB1	ZBB1A	ZBB2	ZBB2o	ZBB3
T [°C]	45	45	100	100	45	180	180	65	65	43	300	30	49	150	210	86	86	100
pFTIR																		
main components, vol. %																		
H_2O^2	2.47	3.91	4.38	4.29	3.98	6.42	6.37	6.24	6.37	4.43	4.50	7.28	7.81	25.63	22.56	25.53	21.52	21.13
CO_2	bdl	bdl	bdl	bdl	bdl	bdl	bdl	bdl	bdl	bdl	bdl	0.11	0.79	38.31	33.95	34.75	25.77	38.41
inorganics, ppm																		
CO	92	**143**	**166**	**164**	**144**	**1090**	**1070**	**1060**	**1040**	**172**	**176**	21	21	**1220**	**12600**	**938**	**952**	**26700**
N_2O	bdl	bdl	bdl	bdl	bdl	1.3	1.3	1.3	1.2	bdl	bdl	bdl	2.6	bdl	bdl	bdl	bdl	bdl
NO	12	15	bdl	bdl	19	bdl	bdl	bdl	bdl	bdl	bdl	**100**	bdl	bdl	bdl	bdl	bdl	bdl
NO_2	bdl	bdl	bdl	bdl	bdl	bdl	bdl	bdl	bdl	bdl	bdl	bdl	bdl	bdl	66	bdl	bdl	31
NH_3	1.5	3.1	4.1	3.1	3.0	bdl	bdl	bdl	bdl	2.0	1.4	12	bdl	8.3	bdl	bdl	bdl	8.3
SO_2	bdl	15	46	bdl	14	79	59	58	53	39	47	**129**	**172**	bdl	bdl	bdl	**123**	bdl
HCl	0.48	0.73	2.9	2.7	0.94	8.4	8.3	8.3	8.3	3.7	3.9	0.71	1.2	4.3	1.2	5.6	5.9	4.9
CCl_4	0.57	0.03	0.06	bdl	0.11	bdl	bdl	bdl	bdl	0.34	0.34	bdl	bdl	6.0	8.5	8.5	8.2	4.0
HF	0.19	bdl	bdl	1.1	bdl	bdl	bdl	bdl	bdl	bdl	bdl	bdl	bdl	bdl	bdl	bdl	bdl	bdl
SiF_4	bdl	0.08	2.4	2.5	0.18	48	48	47	47	3.0	3.5	0.42	bdl	25	15	28	26	10
AsH_3	0.04	0.09	0.40	bdl	0.28	bdl	bdl	bdl	bdl	0.38	0.37	bdl	bdl	0.88	**2.9**	0.19	bdl	**2.9**
aliphatic and aromatic hydrocarbons and their derivatives, ppm																		
methane	bdl	bdl	bdl	bdl	bdl	bdl	bdl	bdl	bdl	bdl	bdl	9.8	12	**1110**	**740**	**1130**	**1130**	**880**
propane	**192**	72	bdl	bdl	85	34	37	37	39	**208**	**215**	11	12	**233**	bdl	**257**	**277**	bdl
hexane	bdl	bdl	bdl	bdl	bdl	bdl	bdl	bdl	bdl	bdl	bdl	bdl	bdl	38	69	39	40	**64**
ethene	bdl	bdl	bdl	2.3	bdl	26	26	24	8.5	bdl	bdl	bdl	bdl	18	19	21	23	21
DCM	30	19	11	28	20	**181**	**191**	**191**	**189**	36	39	**173**	**143**	92	**142**	46	76	82
1,1-DCE	12	bdl	bdl	bdl	bdl	**103**	**102**	**102**	**104**	bdl	bdl	11	2.4	bdl	bdl	bdl	bdl	bdl

vent[1]	SWC1	SWC1r	SWC1oP	SWC1oSW	SWC1oB	SWC2	SWC2o	SWC2o2	SWC2o3	SWC3	SWC3A	RCH1	RCH1A	ZBB1	ZBB1A	ZBB2	ZBB2o	ZBB3
1,2-DCE	bdl	bdl	bdl	bdl	bdl	bdl	bdl	bdl	bdl	bdl	bdl	bdl	bdl	bdl	59	bdl	bdl	60
1,1,1-TCE	bdl	bdl	1.8	bdl	bdl	**492**	**447**	**444**	**436**	bdl	bdl	63	bdl	23	bdl	26	65	bdl
vent	**SWC1**	**SWC1r**	**SWC1oP**	**SWC1oSW**	**SWC1oB**	**SWC2**	**SWC2o**	**SWC2o2**	**SWC2o3**	**SWC3**	**SWC3A**	**RCH1**	**RCH1A**	**ZBB1**	**ZBB1A**	**ZBB2**	**ZBB2o**	**ZBB3**
1,2-DCP	bdl	bdl	bdl	bdl	bdl	bdl	bdl	bdl	bdl	bdl	bdl	54	67	61	bdl	**214**	**114**	bdl
1,1-DCEe	7.2	bdl	27	21	bdl	**287**	**274**	**272**	**272**	30	33	**206**	**221**	**166**	**130**	**191**	**176**	56
VC	bdl	bdl	bdl	bdl	bdl	bdl	bdl	bdl	bdl	bdl	bdl	bdl	bdl	**323**	**145**	**365**	**299**	**97**
ClB	9.5	bdl	bdl	51	bdl	19	22	39	24	8.3	7.7	bdl	24	bdl	73	bdl	bdl	67
cumene	bdl	bdl	bdl	bdl	bdl	66	64	62	50	bdl	bdl	42	39	90	16	**128**	92	**153**
phenol	4.9	bdl	0.45	9.7	bdl	bdl	bdl	bdl	bdl	bdl	bdl	bdl	bdl	20	11	34	bdl	36
o-cresole	2.4	1.6	5.6	2.5	1.4	13	13	12	13	6.2	7.1	56	52	30	22	39	63	16
heterocyclic organic compounds, ppm																		
THF	bdl	bdl	bdl	bdl	bdl	bdl	bdl	bdl	bdl	bdl	bdl	bdl	bdl	36	bdl	32	21	bdl
py	**178**	**202**	19	**168**	**206**	bdl	bdl	bdl	bdl	**231**	**232**	bdl	bdl	bdl	bdl	bdl	bdl	bdl
tph	bdl	**146**	26	bdl	**149**	**260**	**204**	**200**	**194**	**151**	**141**	bdl	bdl	**496**	bdl	**448**	**282**	bdl
other organic compounds, ppm																		
fm	1.8	0.21	7.1	5.7	0.31	7.4	7.5	7.5	9.4	2.8	3.1	1.6	1.9	bdl	1.2	2.8	bdl	2.2
DMS	bdl	bdl	bdl	bdl	bdl	bdl	bdl	bdl	bdl	bdl	bdl	23	17	65	**153**	60	**148**	**120**
DMDS	**380**	**289**	bdl	16	**293**	bdl	bdl	bdl	bdl	**322**	**321**	7.1	12	bdl	bdl	bdl	bdl	bdl
GC – additional compounds																		
vent	**SWC1**	**SWC1r**	**SWC1oP**	**SWC1oSW**	**SWC1oB**	**SWC2**	**SWC2o**	**SWC2o2**	**SWC2o3**	**SWC3**	**SWC3A**	**RCH1**	**RCH1A**	**ZBB1**	**ZBB1A**	**ZBB2**	**ZBB2o**	**ZBB3**
CH_3Cl	0.001						0.06			0.01	0.04	0.001	0.001	2.0	0.21			
COS												0.003	0.01	2.1	0.45			
ethyne	0.001						0.0004			0.02	0.24	0.0001	0.004	0.02	0.02			

vent	SWC1	SWC1r	SWC10P	SWC10SW	SWC10B	SWC2	SWC20	SWC202	SWC203	SWC3	SWC3A	RCH1	RCH1A	ZBB1	ZBB1A	ZBB2	ZBB20	ZBB3
propene	bdl						0.01			0.65	1.6	0.001	0.001	1.6	0.03			
i-butane	0.002						0.001			0.15	0.06	0.01	0.03	0.53	0.03			
n-butane	0.004						0.002			0.55	0.27	0.01	0.13	2.0	0.07			
1-butene	bdl						0.001			0.05	0.18	0.0002	0.0002	0.17	0.02			
i-butene	bdl						0.002			0.09	0.37	0.0003	0.0005	0.24	0.01			
t-2-bu	0.001						0.001			0.09	0.20	0.0001	0.0003	0.28	0.02			
c-2-bu	0.001						0.001			0.06	0.14	0.0001	0.0001	0.16	0.01			
i-pentane	0.003						0.001			0.10	0.04							
vent	**SWC1**	**SWC1r**	**SWC10P**	**SWC10SW**	**SWC10B**	**SWC2**	**SWC20**	**SWC202**	**SWC203**	**SWC3**	**SWC3A**	**RCH1**	**RCH1A**	**ZBB1**	**ZBB1A**	**ZBB2**	**ZBB20**	**ZBB3**
i-pentane	0.003						0.001			0.10	0.04	0.002	0.01	0.34	0.01			
n-pentane	0.002						0.001			0.29	0.14	0.005	0.03	0.96	0.03			
isoprene	bdl						bdl			0.003	bdl	0.001	0.00005	0.01	0.001			
1,3-budi	bdl						0.0001			0.01	0.04	bdl	bdl	bdl	0.01			
t-2-pte	0.0002						0.0004			0.04	0.04	0.0002	0.001	0.09	0.01			
c-2-pte	0.0001						0.0003			0.02	0.02	0.0001	0.0002	0.04	0.002			
n-heptane	0.001						0.001			0.12	0.07	0.001	0.01	0.33	0.01			
n-octane	0.0002						0.001			0.10	0.05	0.0001	0.001	0.18	0.004			
n-nonane	0.0003						0.0003			0.08	0.03	0.00004	0.001	0.07	0.003			
n-decane	0.0003						0.0003			0.06	0.02	0.00004	0.001	0.04	0.0002			
2,3-DMBu	0.02						bdl			0.01	bdl	0.0001	0.001	0.02	0.001			
2-MPT	0.26						0.001			0.05	0.03	0.001	0.004	0.15	0.003			
3-MPT	9.9						0.0002			0.02	0.01	0.0003	0.001	0.07	0.001			
cpt	0.0005						0.0002			0.04	0.01	0.0005	0.003	0.10	0.002			

vent	SWC1	SWC1r	SWC1oP	SWC1oSW	SWC1oB	SWC2	SWC2o	SWC2o2	SWC2o3	SWC3	SWC3A	RCH1	RCH1A	ZBB1	ZBB1A	ZBB2	ZBB2o	ZBB3
benzene	0.0001						0.03			0.50	1.7	0.003	0.46	12	33			
toluene	0.01						0.005			0.22	0.41	0.002	0.002	2.7	0.01			
EtB	0.0004						0.001			0.03	0.04	0.001	0.001	0.09	0.004			
m/*p*-X	0.001						0.002			0.07	0.15	0.002	0.002	0.34	0.01			
o-X	0.001						0.001			0.03	0.06	0.001	0.001	0.13	0.005			
styrene	bdl						bdl			0.003	bdl	0.00002	0.0003	0.02	0.0004			
i-PrB	0.0002						bdl			0.002	0.004	0.00002	0.0002	0.003	0.001			
n-PrB	0.0004						0.0004			0.01	0.01	0.0001	0.0003	0.02	0.001			
m-EtT	0.001						0.0002			0.01	0.02	0.0003	0.001	0.04	0.003			
p-EtT	0.0004						0.0001			0.004	0.01	0.0001	0.001	0.02	0.001			
o-EtT	0.0002						0.0001			0.01	0.01	0.0001	0.0004	0.02	0.001			
1,3,5-TMB	0.001						bdl			0.01	0.01	0.0001	0.0004	0.01	0.001			
1,2,4-TMB	0.001						0.0004			0.02	0.03	0.0005	0.001	0.05	0.01			
1,2,3-TMB	0.0004						0.0002			0.01	0.01	0.0002	0.001	0.03	0.004			

Values in parentheses denote overrun of the upper measurement range.
[1]*The "A" add denotes samples taken from the depth of 0.8–1 m (below the ground level), while "a" and "o" denote nearby vents; "r" – repeated measurement; "P" – pyrometamorphic zone, "S" – sulfur-mineralized vent.*
[2]*Abbreviations explained under **Table 1**; ethane, furan and acetic acid were analyzed but were below their detection limits.*
[3]*Notable (>100 ppm) enrichment given in bold.*

Table 3.
Results of the pFTIR and GC gas analyses of a BCWH in Świętochłowice (SWC), "Starzykowiec" heap of the "Chwałowice" mine in Rybnik (RCH), and "Ruda" heap in Zabrze-Biskupice (ZBB).

8.3; SO_2, 120, 31; 671, 25; 388, 160; 532, 202; 79, 40; 123, 123; HCl, 11, 2.4; 6.5, 0.58; 6.2, 3.6; 19, 6.1; 8.4, 3.0; 5.9, 3.8; CCl_4, –, –; 6.6, 6.6; 11, 6.5, 1.1, 1.1; 0.57, 0.15; 8.5, 6.8; HF, 0.62, 0.62; 0.03, 0.03; 0.12, 0.12; 1.1, 0.36; 1.1, 0.46; –, –; SiF_4, 6.3, 1.3; 31, 3.5; 21, 16; 3.2, 1.7; 48, 4.6; 28, 19; AsH_3, 0.17, 0.09; 1.7, 0.38; 1.7, 0.75; 3.4, 1.7; 0.40, 0.20; 2.9, 1.1; CH_4, 262, 107; 2950, 16; 3470, 1238; 819, 491; –, –; 1130, 984; ethane, –, –; 30, 30; 142, 142; 281, 281; –, –; –, –; propane, 42, 15; 729, 15; 601, 275; 694, 410; 215, 77; 277, 255; hexane, 11, 1.8; 152, 51; 139, 73; 608, 225; –, –; 69, 48; ethene, 6.6, 3.2; 79, 6.9; 12, 6.0; 84, 27; 26, 13; 23, 20; DCM, 141, 36; 368, 69; 126, 47; 84, 46; 191, 51; 142, 82; 1,1-DCE, 7.2, 7.2; 17, 8.9; –, –; 16, 16; 104, 67; –, –; 1,2-DCE, –, –; 77, 77; 39, 39; 38, 38; –, –; 60, 59; 1,1,1-TCE, 99, 46; 417, 26; 36, 25; 59, 11; 492, 150; 65, 34; 1,2-DCP, 31, 31; 56, 56; 384, 159; 568, 226; –, –; 214, 114; 1,1-DCEe, 77, 21; 347, 67; 195, 123; 28, 17; 287, 66; 191, 132; vinyl chloride, –, –; –, –; 416, 235; chlorobenzene, 39, 24; 186, 15; 77, 44; 206, 81; 51, 18; 73, 70; cumene, 34, 16; 399, 5.7; 126, 97; 264, 81; 66, 81; 153, 60; 153, 76; phenol, 29, 10; 7, 4.2; 51, 16; 67, 41; 9.7, 2.8; 36, 23; *o*-cresol, 46, 5.8; 66, 3.6; 65, 36; 81, 54; 13, 5.3; 63, 30; furan, 0.14, 0.14 (no records for other sites); THF, 1.3, 0.41; 177, 177; 96, 12; 293, 261; –, –; 36, 29; thiophene, 192, 146; 173, 38; 556, 332; 689, 241; 260, 143; 496, 397; formaldehyde, 12, 3.7; 13, 1.2; 5.5, 2.0; 17, 2.3; 9.4, 3.0; 2.8, 1.9; acetic acid, –, –; –, –; 9.1, 9.1; 50, 14; –, –; –, –; DMS, 62, 21; 893, 28; 401, 69; 1540, 534; –, –; 153, 101; DMDS, 104, 28; 37, 21; –, –; –, –; 380, 194; –, –; ad pyridine, 7.1, 7.1; 86, 7.3; –, –; –, –; 232, 144; –, – [ppm]. As compared to these vents, the one at the RCH site. The geometric mean concentrations for the whole range are: H_2O 9.5, CO_2 3.83 [vol.%], CO 350, NO 20, NO_2 54, N_2O 0.51, NH_3 7.5, SO_2 72, HCl 2.5, CCl_4 2.3, HF 0.26, SiF_4 4.5, AsH_3 0.53, methane 201, ethane 106, propane 58, hexane 29, ethene 11, DCM 54, 1,1-DCE 17, 1,2-DCE 53, 1,1,1-TCE 37, 1,2-DCP 127, 1,1-DCEe 60, vinyl chloride 165, chlorobenzene 30, cumene 36, phenol and *o*-cresol and THF 13, furan 0.14, thiophene 200, formaldehyde 2.2, acetic acid 13, DMS 65, DMDS 39, and pyridine 29.

5. Discussion

In general, the data provided for additional vents from additional BCWH probed allows to enlarge the span of the maximum observed values of only some compounds. They include (with excess in parentheses) CO (10x), SO_2, 1,1,1-TCE (12x), 1,1-DCEe (2.5x), cumene (2x), formaldehyde (3x), and pyridine (21x). Higher than previously observed geometric mean values are also observed for NO_2, CCl_4, ethane (2x), propane, ethene, and thiophene. This is clearly seen in the case of the latter two compounds, with more frequent positive determinations than within the previous studies. Similar levels of geometric means are found for HCl, AsH_3, chlorobenzene, phenol, and DMDS. As formerly observed, concentration ranges are usually extremely variable. Cumene is a good example of a compound with very high maximum but very low geometric mean. So is true, though less clearly, for, e.g., *o*-cresol. Some compounds often show large contents but single records. For many BCWHs there are large discrepancies between the geometric mean values and maximums, while for less number of the objects studied the amounts emitted are at very steady level. Some constituents, like vinyl chloride and even methane, may show very high concentrations (>100 ppm) but may be "absent" (below detection limits) at other BCWHs or vents. As explained in the former papers, this results from very high dynamics of the local combustion processes. The *ex situ* GC values obtained are, again, usually much lower than those observed by *in situ* FTIR, thus confirming their uncertain and, possibly, semi-quantitative value. On the other hand, two compounds not observed within the previous GC data are now determined: CH_3Cl (chloromethane or methyl chloride) and cyclopentane.

At the time of the BCWH gas analyses the author could not find paper showing the usage of FTIR for environmental studies. Stockwell et al. [27] used this method to measure H_2O, CO_x, NO_x, HCl, SO_2, NH_3, methane, acetylene, ethene, propane, formaldehyde, formic acid, methanol, acetic acid, HCN, furan, glycolaldehyde, and HONO (the latter also initially reported in [6]) in biomass emissions, though in a Fire Lab at Missoula Experiment. A more *in situ* type of work, engaging airborne FTIR, is by Yokelson et al. [28] who measured African savanna fires, with 14 compounds analyzed.

It is noteworthy that numerous organic and organo(semi)metallic compounds (or similar ones) detected in the BCWHs exhausts are also detected in volcanic fumaroles (mainly via GC, or modeled, as summarized by Wahrenberger [29]) or algal emissions (by GC–MS; [30]). Examples of interesting species include CO_2, COS, CS_2, S_2, S_8, SO_2, AsH_3, HCl, HF, HBr, $CHCl_3$, NO_2, propanal, methanol, acetaldehyde, 1,1,2-trichlorotrifluoroethane, hexafluoropropene, tetrachloroethene, vinyl chloride, *i*-butene, hexane, octane, octane, butadiene, benzene, toluene, α-pinene, *i*- and *n*-propanol, methylacrolein, MEK, acetone, 1,4-dioxane, dimethyldifluorosilane, thioformaldehyde, ethylthiophene, trimethylborane, methylphosphine, and uncertain [*N*-(phenyl-2-pyridinylmethylene)benzeneamine-*N*,*N*′]-irontricarbonyl and silver benzoate; geosmin, cyclopentane, cyclohexane, acetic acid, acetamide, glucopyranose, dibutyl phthalate, cholest-5-en-22-one, benzaldehyde, hydrazine, 8-amino-2-naphthalenol, ethanethioimide, thiourea, 1,3-oxathian-2-one, tetrahydro-2,5-dimethylthiophene, 6-methylbenzo[*b*]thiophene, 3,3,5,5,-tetramethyl-1,2,4-trithiolane, thiirane, $C_2H_7O_2B$ borane, trimethylsilane, butytrimethylsilane, or undecanoic acid 11-chloro- and 11-fluorotrimethylsilyl esters.

Acknowledgements

This work was financed by the NCN (Narodowe Centrum Nauki, or National Science Centre) grant no. 2013/11/B/ST10/04960.

Author details

Łukasz Kruszewski
Institute of Geological Sciences, Polish Academy of Sciences, Warszawa, Poland

*Address all correspondence to: lkruszewski@twarda.pan.pl

References

[1] Nasdala, L., Pekov, I.V., 1993. Ravatite, $C_{14}H_{10}$, a new organic mineral species from Ravat, Tadzhikistan. European Journal of Mineralogy, 6, 669–705.

[2] Cebulak, S., Smieja-Król, B., Tabor, A., Misz, M., Jelonek, I., Jelonek, Z., 2005. Oxyreactive Thermal Analysis (OTA) – A Good and Cheap Method of the Estimation of Coal Self-Ignition in Bingsteads – Preliminary Research Results. Geological Publishing Houses (Geological Institute, Warsaw, Poland), 135–138 [in Polish].

[3] Sokol, E.V., Maksimova, N.V., Nigmatulina, E.N., Sharygin, V.V., Kalugin, V.M., 2005. Combustion Metamorphism. Publishing House of the SB RAS, Novosibirsk, Russia [in Russian, with fragments in English], 312 pp.

[4] Stracher, G.B., 2007. The origin of gas-vent minerals: isochemical and mass-transfer processes. In: Stracher, G. B. (ed.), Geology of Coal Fires: Case Studies from Around the World, v. 18. The Geological Society of America, Reviews in Engineering Geology, USA, 91–96.

[5] Colman, J.J., Swanson, A.L., Meinardi, S., Sive, B.C., Blake, D.R., Rowland, F.S., 2001. Description of the Analysis of a Wide Range of Volatile Organic Compounds in Whole Air Samples Collected During PEM-Tropics A and B. Analytical Chemistry, A-I. American Chemical Society, 73(15), 3723–3731.

[6] Kruszewski, Ł., Fabiańska, M.J., Ciesielczuk, J., Segit, T., Orłowski, R., Motyliński, R., Moszumańska, I., Kusy, D. 2018. First multi-tool exploration of a gas-condensate-pyrolysate system from the environment of burning coal mine heaps: An in situ FTIR and laboratory GC and PXRD study based on Upper Silesian materials. Science of the Total Environment, 640-641, 1044–1071.

[7] Wickenheiser, B., Michalke, K., Hensel, R., Drescher, C., Hirner, A.V., Rutishauser, B., Bachofen, R., 1998. Volatile compounds in gases emitted from the wetland bogs near Lake Cadagno. In: Peduzzi, R., Bachofen, R., Tonolla, M. (Eds), Lake Cadagno: a meromictic Alpine lake. Documenta dell'Istituto italiano di idrobiologia, 63, 137–140.

[8] Craig, P.J. (ed), 2003. Organometallic Compounds in the Environment, 2nd ed. John Wiley & Sons Ltd., Chichester, West Sussex, UK, 415 pp.

[9] Khoury, J., Lewis, R., Sattler, A., Linsley, B., Magers, K., Lee, J., 2008. Analysis of volatile arsenic compounds in landfill gas. WEF/A&WMA Odors and Air Emissions 2008, Water Environment Federation, 630–641.

[10] Kruszewski, Ł., 2013a. Komora termiczna jako narzędzie do zrozumienia procesów przeobrażenia materiałów odpadowych [Thermal chamber as a tool for understanding the processes of trasformation of waste materials; presented on: 4th Meeting of the Users of the Bruker Systems]. IV Spotkanie Użytkowników Systemów Firmy Bruker, Poznań 03–04.10.2013, 59–60 [poster #6; in Polish].

[11] Kruszewski Ł., 2013b: Supergene sulphate minerals from the burning coal mining dumps in the Upper Silesian Coal Basin, South Poland. International Journal of Coal Geology, 105, 91–109.

[12] Kruszewski, Ł., Fabiańska, M.J., Segit, T., Kusy, D., Motyliński, R., Ciesielczuk, J., Deput, E., 2020. Carbon-nitrogen compounds, alcohols, mercaptans, monoterpenes, acetates, aldehydes, ketones, SF6, PH3, and other

fire gases in coal-mining waste heaps of Upper Silesian Coal Basin (Poland) – a re-investigation by means of in-situ FTIR extrernal database approach. Science of the Total Environment, 698, 134274.

[13] Kruszewski Ł. 2006. Oldhamite-periclase-portlandite-fluorite assemblage and coexisting minerals of burnt dump in Siemianowice Śląskie – Dąbrówka Wielka area (Upper Silesia, Poland) – preliminary report. Mineralogia Polonica, 28, 118–120.

[14] Kruszewski Ł. 2008: Apatite-ellestadite solid solution and associated minerals of metacarbonate slags from burning coal dump in Rydułtowy (Upper Silesia). Mineralogia Special Papers, 32, 100.

[15] Kruszewski, Ł., Ciesielczuk, J., Misz-Kennan, M., 2014a. Mineralogy of some metacarbonate rocks from burned coal-mining dump in Przygórze (Lower Silesian Coal Basin) and its analogy to "olive" rocks from the Hatrurim Formation. 4th Central-European Mineralogical Conference - CEMC 2014 (Proceedings of the international symposium CEMC 2014; ISBN 978–80–210-6832-2), p. 77.

[16] Kruszewski, Ł., Ciesielczuk, J., Misz-Kennan, M., Fabiańska, M., 2014b. Chemical composition of glasses and associating mineral species in various pyrometamorphic rocks from coal-mining dumps of the Lower Silesia. Mineralogia – Special Papers, 42, 70–71.

[17] Kruszewski, Ł., 2012. Unique chloride assemblage of exhalative origin from burning coal mining dump in Radlin (Rybnik Coal Area, S Poland). Mineralogia – Special Papers 40, 90–92.

[18] Fabiańska M., Ciesielczuk J., Kruszewski Ł., Misz-Kennan M., Blake D.R., Stracher G., Moszumańska I. 2013: Gaseous compounds and efflorescences generated in self-heating coal-waste dumps – A case study from the Upper- and Lower Silesian Coal Basins (Poland). International Journal of Coal Geology, 116-117, 247–261.

[19] Kruszewski, Ł., Sierny, W., 2019. Radlin coal fire heap: thiosulfate- and dithionate-bearing alkaline mineralization; Cu, Fe, As, and P mineralization; and second worldwide occurrence of tsaregorodtsevite. Mineralogia – Special Papers, 49, 54.

[20] Fabiańska, M., Ciesielczuk, J., Nádudvári, Á., Misz-Kennan, M., Kowalski, A., Kruszewski, Ł., 2019. Environmental influence of gaseous emissions from self-heating coal waste dumps in Silesia, Poland. Environmental Geochemistry and Health, 41(2), 575–601.

[21] Lewińska-Preis, L., Szram, E., Fabiańska, M., Nádudvari, Á., Misz-Kennan, M., Abramowicz, A., Kruszewski, Ł., Kita, A., 2020. Selected ions and major- and trace elements as contaminants in coal-waste dump water from the Lower- and Upper Silesian Coal Basins (Poland). International Journal of Coal Science and Technology, DOI: 10.21203/rs.3.rs-35061/v1 (preprint).

[22] Kruszewski, Ł., Ciesielczuk, J., and Misz-Kennan, M., 2012. What have meteorites to do with coal fires? A case of Upper and Lower Silesian Coal Basins. Mineralogia – Special Papers 40, 28–29.

[23] Kruszewski Ł., Fabiańska, M., Ciesielczuk, J., Segit, T., 2017. Coal fires – Titan – interstellar medium – life: what do they have in common? Potential gaseous bio-precursors in burning mining heaps. Life Origins 2017 conference (Early Earth and ExoEarths: origin and evolution of life), Book of Abstracts, 51–52.

[24] Kruszewski, Ł., Matlakowska, R., 2018. Burning coal-mining waste heaps

versus biological activity – a potential model for survivability in a harsh environment?; Life Origins, COST Workshop 2018, Bertinoro, Italy, March 19–23 (poster, unpublished).

[25] Myhre, G., Shindell, D., Breón, F.-M., Collins, W., Fuglestvedt, J., Huang, J., Koch, D., Lamarque, J.-F., Lee, D., Mendoza, B., Nakajima, T., Robock, A., Stephens, G., Takemura, T., Zhang, H. 2013. Antrhropogenic and natural radiative forcing. Chpt. 8, in: Stocker, T. F., Qin, D., Plattner, G.-K., Tignor, M., Allen, S.K., Boschung, J., Nauels, A., Xia, Y., Bex, V., Midgley, P.M. (eds), Climate Change 2013: The Physical Science Basis. Contribution of the Working Group I to the Fifth Assessment Report of the Intergovernmental Panel on Climate Changes, Cambridge University Press, Cambride, UK, and New York, NY, USA, www.ghghprotocol.org, 640–659.

[26] ESRL, 2019. Earth System Research Laboratory GMD (Global Monitoring Division), www.esrl.noaa.gov/mgd/dv/iadv (accessed on 08.03.2019).

[27] Stockwell, C.E., Yokelson, R.J., Kreidenweis, S.M., Robinson, A.L., DeMott, P.J., Sullivan, R.C., Reardon, J., Ryan, K.C., Griffith, D.W.T., Stevens, L., 2014. Trace gas emissions from combustion of peat, crop residue, domestic biofuels, grasses, and other fuels: configuration and Fourier transform infrared (FTIR) component of the fourth Fire Lab at Missoula Experiment (FLAME-4). Atmospheric Chemistry and Physics, 14, 9727–9754.

[28] Yokelson, R.J., Bertschi, I.T., Christian, T.J., Hobbs, P.V., Ward, D.E., Hao, W.M., 2003. Trace gas measurements in nascent, aged, and cloud-processed smoke from African savanna fires by airborne Fourier transform infrared spectroscopy (AFTIR). Journal of Geophysical Researchm 108(D13), 8478.

[29] Wahrenberger, C.M., 1997. Some aspects of the chemistry of volcanic gases. Ph.D. thesis, Swiss Federal Institute of Technology, Zurich, Switzerland, 233 pp.

[30] Ali, G., 2004. Identification of volatile organic compounds produced by algae. Egyptian Journal of Phycology, 5, 71–81.

Chapter 14

Evaluation of Particulate Matter Pollution in Micro-Environments of Office Buildings—A Case Study of Delhi, India

Saurabh Mendiratta, Sunil Gulia, Prachi Goyal and Sanjeev Kumar Goyal

Abstract

High level of particulate matter in an office building is one of the prime concerns for occupant's health and their work performance. The present study focuses on the evaluation of the distribution pattern of airborne particles in three office buildings in Delhi City. The study includes the Assessment of PM_{10}, $PM_{2.5}$ and PM_1 in the different indoor environments, their particle size distribution, I/O ratio, a correlation between pollutants their sources and management practices. The features of buildings I, II, and III are old infrastructure, new modern infrastructure, and an old building with good maintenance. The results indicate that the average concentrations of PM_{10}, $PM_{2.5,}$ and PM_1 are found in the range of 55–150 μg m^{-3}, 41–104 μg m^{-3} and 37–95 μg m^{-3}, respectively in Building I, 33–136 μg m^{-3}, 30–84 μg m^{-3} and 28–73 μg m^{-3}, respectively in Building II and 216–330 μg m^{-3}, 188–268 μg m^{-3} and 171–237 μg m^{-3}, respectively in Building III. The maximum proportion of the total mass contributed by $PM_{0.25–1.0}$ i.e., up to 75%, 86%, and 76% in the meeting room of Building I, II and III, respectively. The proportion of ultrafine particles was found higher in the office area where the movement was minimum and vice versa. The higher I/O indicates the contribution of the presence of indoor sources for ultra-fine and finer particles. Further, possible strategies for indoor air pollution control are also discussed.

Keywords: ultrafine particulate matter, size segregated particles, distribution pattern, indoor sources, indoor/outdoor ratio, office buildings

1. Introduction

Indoor Air Quality (IAQ) refers to the level of air pollutants and thermal (temperature and relative humidity) conditions that affects the health, comfort, and performance of the occupants inside a building. The high concentration of air pollutants indoor is a major concern in Delhi city, which has been many times reported as one of the polluted cities of the world [1]. The major sources in an office building are infiltration of ambient air pollutants; emissions from office equipment like printers, xerox. Etc.; emission of VOCs from building materials, re-suspension of floor dust; emission from cleaning chemicals among them [2]. In addition to the sources, poor ventilation builds the pollutant level indoors [3]. The increasing level of pollutants

 IntechOpen

needs to be managed as it can affect occupant's health, comfort, and work output. Researchers in the past showed evidence that the air within residential and other commercial buildings including offices can be more polluted than the outdoor air even in the largest and most industrialized cities [4–7]. It is also reported that the health risks may be greater due to exposure of air pollutants indoor than outdoor as people spend more time in indoor environment, be it office or at home.

There are no specific criteria pollutants and standards defined to categorize classes of indoor air quality i.e. satisfactory or hazardous. Generally, researchers focus on the CO_2 level and thermal comfort parameters for Indoor air quality, which are indicators of sick building syndrome [8, 9]. There is limited information available on the high exposure of indoor particulate matter in buildings, especially working offices, where people spend almost 8–9 hours daily and are exposed. Limited research is carried out on size segregated PM in indoor air in cities of developing countries where outside PM levels are higher [10–12]. Researchers also observed that fine and ultrafine particles are more harmful than coarse particles irrespective of indoor or ambient environments [13].

In the past, Wargocki et al. [14] have experimented in a typical office environment in which two exposure conditions were produced i.e., with and without emission source where the same office staff worked for 5 h in each condition. The productivity of the staff was found 6.5% less in poor air quality and experienced significantly increased incidences of headache i.e., a symptom of sick building syndrome [15]. Fisk [16] concluded relatively strong evidence of relationships among characteristics of buildings and indoor environments, which influence the occurrence of communicable respiratory illness, allergy and asthma symptoms, sick building symptoms, and worker performances. It is also reported that any improvement in IAQ by a factor of 2–7 can increase occupant's productivity in offices.

In the recent past, some studies are carried out to assess the Indoor PM_{10} and $PM_{2.5}$ levels in school buildings and residential buildings in different Indian cities (**Table 1**). The major objectives of these studies were to assess PM exposure on children, who are more sensitive [22, 23, 26]. Kulshreshtha and Khare [21] have found average PM_{10} concentrations in the range of 373–894 μg m^{-3} in winter and 107–199 μg m^{-3} during summer in a residential building in Delhi. The $PM_{2.5}$

Author/ City	Type of Building	Pollutant Concentration	Study Period/ Sampling duration	Key Findings of the Study
Saraga et al. [17]/ Goudi, Athens	Museum	$PM_{2.5}$: 20.3 ± 2.69 μg/m^3 (Summer)	Jun 22 - Jul 2, 2007/ 24 hr	The higher number of occupants and re-suspension of PM leads to elevated fine particles levels.
	Smoker's Office	$PM_{2.5}$: 37.6 ± 27.3 μg/m^3 (Summer)	July 16–22, 2007/ 24 hr	
	Non-Smoker's Office	$PM_{2.5}$: 30.7 ± 6.7 μg/m^3 (Summer)	July 23–27, 2007/ 24 hr	
Razali et al. [18] / Malaysia	School	$PM_{2.5}$: 22 ± 6 μg/ m^3 (Summer) PM_{10}: 35 ± 11 μg/ m^3 (Summer)	June 14–July 1, 2011,/ 8 hr., day time	The classroom location and the movement of students in and out of the classrooms influence the PM concentrations.
Zwoździak et al. [19]/ Wrocław, Poland	School	$PM_{2.5}$: 59.8 ± 21.6 μg/m^3 (Winter) and 13.5 ± 4.1 μg/m^3 (Summer) PM_{10}:	Winter (Dec.09 – Jan. Mar, 10) Summer (Apr.-Jun.,10,)/08 hr.	Fine and coarse particles were generated by indoor sources i.e. dust re-suspension due to children activities

Author/ City	Type of Building	Pollutant Concentration	Study Period/ Sampling duration	Key Findings of the Study
		68.5 ± 21.8 μg/m^3 (Winter) and 43.8 ± 17.9 μg/m^3 (Summer)		
Taneja et al. [20]/ Agra, India	Residential Rural & Urban	PM_{10}: 245 μg/m^3 (rural) and 339 μg/m^3 (urban)	Oct. 04 – Dec. 05/24 hr	The concentrations of $PM_{2.5}$ and PM_{10} were higher inside during the winter. Coarse particles were generated by indoor sources i.e. cooking, burning, etc.
Kulshreshtha & Khare [21]/ Delhi, India	Residential Homes	PM_{10}: 373–894 μg/m^3 (Winter) and 107–199 μg/m^3 (Summer); $PM_{2.5}$: 197–713 μg/m^3 (winter) and 34–60 μg/m^3 (Summer) PM_1:169–623 μg/m^3 (winter) and 23–36 μg/m^3 (Summer)	Winter and Summer season of 2008, Hourly average	The PM concentrations were significantly higher during the winter period. Emission from the kitchen is the dominant source of Indoor particles in small houses with poor ventilation.
Goyal and Khare [22]/ Delhi, India	School	$PM_{2.5}$: 30–160 μg/m^3 (Non Winter); 110–789 μg/m^3 (Winter) $PM_{10:}$ 19.5–110.6 μg/m^3 (Non Winter) and 77.3–713.1 μg/m^3 (Winter)	(Aug., 06, Sep. 06, Apr., 07); (Nov., 06, Dec., 06, Jan.- Feb., 07)/ 6 hr	The results of this study indicated that the concentration of pollutants particularly PM is influenced by the occupant's activity.
Chithra & Nagendra [23]/ Chennai, India	School	$PM_{2.5:}$ 61 ± 29 μg/m^3 (Winter) and 32 ± 16 μg/m^3 (Summer) PM_{10}: 149 ± 69 μg/m^3 (Winter) and 95 ± 61 μg/m^3 (Summer)	Winter (Jan.- Mar. 11) Summer (Apr.- May 11)/24 hr	A strong seasonal variability with poor IAQ was observed during winter. Human activity seems to be an important factor influencing the coarse particle level.
Datta et al. [24]/ Delhi, India	Office & School	$PM_{2.5:}$ 43.8 μg/m^3 (Office) -22.8 μg/m^3 (School) (Summer)	June–July 2015/ 8 hr., day time	The study indicates that the occupant density in the air-conditioned non-residential buildings plays a vital role in controlling indoor air pollution levels inside the building.
Gupta et al. [25]/ Delhi, India	Office	$PM_{2.5}$: 116.5 ± 67 μg/m^3 (Winter)	Jan.- Feb., 18	A higher concentration of $PM_{2.5}$ in the building could be due to its maximum proximity to urban busy roads and poorly maintained HVAC ducting system, which may lead to infiltration and more leakages of $PM_{2.5}$ from outdoors.

Table 1.
Indoor air quality in different types of buildings.

concentrations in the range of 197–713 μg m^{-3} in winter and 34–60 μg m^{-3} in summer while PM_1 found in the range of 169–623 μg m^{-3} in winter and 23–36 μg m^{-3} in summer, respectively in a low and medium-income group house where emissions from kitchen are generally high due to Indian cooking style. The results indicate a higher level of PM concentration during the winter season. However, the level of $PM_{2.5}$ and PM_{10} in western cold countries is comparatively very less [19, 25]. Recently, Gupta et al. [25] have reported $PM_{2.5}$ concentration level as 116 ± 67 μg m^{-3} in one of the office buildings in Delhi during the winter period. The higher level of $PM_{2.5}$ might be due to the penetration of ambient $PM_{2.5}$, which is generated from nearby high traffic roads.

The high level of indoor PM (PM_{10}, $PM_{2.5,}$ and $PM_{1.0}$) in the office building is an emerging issue in view of its adverse effects on working productivity and health. There is a need to assess it comprehensively at different locations in the city for different types of office buildings.

The present study is an attempt to assess the size of fractioned PM in different sections of the office buildings. The study has monitored the different size PM in different building sections like staff cabin, meeting/conference room, office halls, accountant room, and outside building in Delhi city. PM monitoring is carried out in three different types of buildings. The particle size distribution plot of each monitoring location has been carried out and compared to further correlate with the sources. The correlation between sites for a particular size PM is calculated along with the Indoor/outdoor ratio.

2. Materials and methods

2.1 Site description

Three different types of office buildings are considered in the present study to assess the PM levels in the indoor environment. The buildings are named Building I,

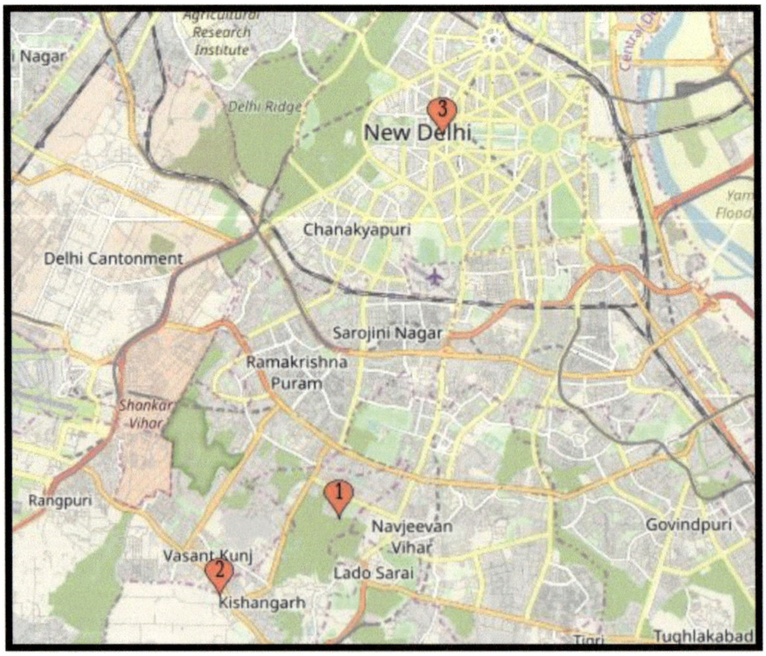

Figure 1.
Delhi's map showing site I, II, and III.

Locations	Direction wrt Centre of Delhi	Types of Building/ Features/ Old or New (approx. age)	Surrounding Landuse features
Building I/ Site I	South Delhi	Typical office building, lots of old files, close cabins, poor ventilation, congested place, approach, 30–40 years old, etc.	Located in the institutional area, Medium density traffic road outside building, residential area on one side of the building, and green forest area on the other side
Building II/ Site II	South Delhi	Modern office building, clean and spacious Hall, New infrastructure, 1–2 years old building, no open files, located at the lower ground floor, Office is part of a shopping mall having eateries, offices, coffee shops and retail shops etc.	High-density traffic road, Car parking outside the building, covered by Residential area on three sides and open land on one side
Building III/ Site III	Central Delhi	Old and very well maintained building hall, high movement of people for meeting/discussions, no public dealing office, daily cleaning activities through cleaning reagents, etc.	Smooth traffic movement outside, commercial area nearby and lots of green areas

Table 2.
Building types and surrounding features at site I, II, and III in Delhi.

Building II and Building III located at sites I, II and III, respectively. Sites I and II represent South Delhi, while site III represents Central Delhi, as shown in **Figure 1**. Building I is a traditional office building with old infrastructure, Building II is a newly constructed office space (with Modern Infrastructure) and Building III is an old Building, but maintained very well. The details of the buildings and surrounding site features are described in **Table 2**.

3. Instrumentation and monitoring protocol

PM monitoring was carried out using a laser aerosol monitor (GRIMM Aerosol Technik Gmbh & Co. KG, Ainrig, Germany, Mini-LAS Model 11R) [27]. The instrument captures every single particle ranging from 0.25 to 32 μm and classifies it into 31 size range channels. The instrument was calibrated before monitoring. The data were recorded and stored at every 6-second interval. The monitor provides concentration levels at the cut of point of $PM_{1.0}$, $PM_{2.5}$, and PM_{10} which are generally monitored for health exposure and from a regulatory compliance point of view.

Monitoring was carried out for 1 day each at all the three sites in December 2018 (Winter) on different dates. At each location of the building at all three selected sites, 15 minutes of measurement was recorded. The monitor was placed in the center of each room (about 1 m above the floor), which corresponds to the breathing zone of the sitting occupants, and the outdoor monitor was placed at least 1.5 m away from any obstacle at a height of 1 m above the ground. The details of the monitoring protocol followed are summarized in **Table 3**. The photographs of monitoring locations for Buildings I and II are shown in **Figure 2**, whereas photographs were not taken at site III due to security reasons. Based on the discussion with the office staff about their comfort and visualizing the situations of the Heating, Ventilating and Air Condition (HVAC) system, in each compartment of all three buildings, building I am categorized as poor ventilated, however, buildings II and III as good ventilated. Kulshreshtha and Khar [21] have correlated the

Location	Monitoring Details	Monitoring Locations
Building I	Monitoring was carried out in the second week of December 2018, during 10 am to 3 pm, monitoring was carried out for 15 minutes at each of the selected locations.	1. Halls A-I mainly occupied by staffs (Vol.315–350 m^3 of each) 2. Account department (Vol. 84 m^3) 3. Meeting rooms (150 m^3) 4. Conference room (294 m^3) 5. Common area at entrance (360 m^3) 6. Outdoor air in front of the entrance gate. PM monitor was placed at an average height of 1 m above ground. Poor Ventilation System
Building II	Monitoring was carried out in the first week of December 2018, during 11 am to 3 pm, monitoring was carried out for 15 minutes at each of the selected locations.	1. Halls mainly occupied by clerical/ technical staffs, (Vol. 210–280 m^3) 2. Staff Cabin (Vol. 27m^3) 3. Conference (315 m^3) 4. Meeting room (215 m^3) 5. Cafeteria/Pantry (60 m^3) 6. A common indoor area at the entrance (20 m^3 and Reception (60 m^3) 7. Outdoor air at entrance. PM monitor was placed at an average height of 1 m above ground. Good Ventilation System
Building III	Monitoring was carried out during the last week of December 2018 from 10 am to 1 pm for one day. The monitoring was carried out for a period of 15 minutes each at the selected locations.	1. Meeting Room/Hall (Vol. 12000 m^3) 2. Common Indoor area. (3000 m^3) PM monitor was placed on the table of an average height of 1.2 m in the meeting room and chair of height 0.40–0.45 m in the common area Good Ventilation System

Note: Vol. – Volume of indoor space where monitoring was carried out. These volumes are calculated based on tentative measures of length, width, and height of indoor compartments.

Table 3.
Monitoring protocol adopted in each building.

ventilation parameters with the comfort level of occupants in a residential building where poor indicate inadequate ventilation and a high potential for complaints and Good indicate satisfaction for all occupants.

4. Results and discussion

4.1 Status of indoor PM_{10}, $PM_{2.5}$, and PM_1 concentration

Generally, the particles are monitored in terms of PM_{10} (particles having aerodynamic diameter ≤ 10 μm), $PM_{2.5}$ (particles having aerodynamic diameter ≤ 2.5 μm), and PM_1 (particle having aerodynamic diameter ≤ 1.0 μm) for regulatory as well as health exposure assessment in ambient as well as in the indoor environment. Additionally, the particles are defined as Ultrafine (<1 μm), Fine or accumulation mode (1 to 2.5 μm) and Coarse particle (> 2.5 μm) as described by Tiwary & Williams [28]. A similar assessment was carried out in the present study as well to evaluate the level of these particles. The monitored data of PM_{10}, $PM_{2.5}$, and $PM_{1.0}$ concentrations were analyzed statistically and are summarized in graphical form in **Figures 3** to **5** for Buildings I to III, respectively.

Figure 2.
Photographs showing monitoring location in building I and II (Note: Building III photographs not taken due to security reason).

The average concentrations of PM_{10}, $PM_{2.5}$, and $PM_{1.0}$ in Building I were found to be 101 μg m^{-3} (range 55–150 μg m^{-3}), 72 μg m^{-3} (range 41–104 μg m^{-3}) and 64 μg m^{-3} (range 37–95 μg m^{-3}), respectively. The concentration of particulate matter was found higher in the account's department compartment and Halls E and

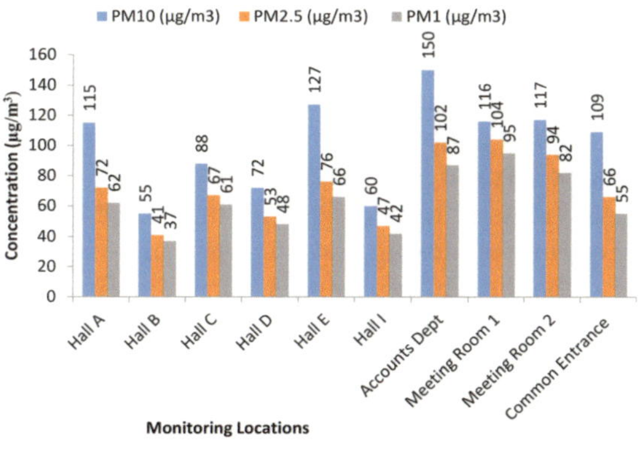

Figure 3.
Average PM concentration in different indoor rooms at building I.

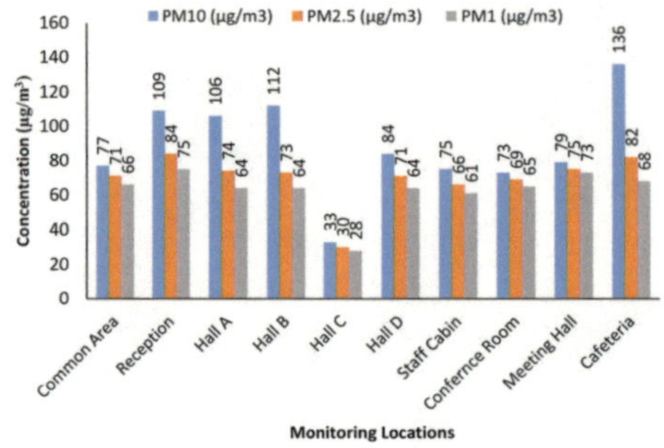

Figure 4.
Average PM concentration in different indoor rooms at building II.

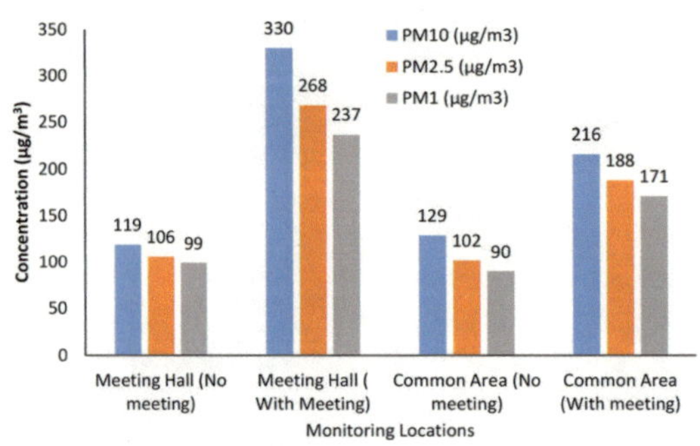

Figure 5.
Average PM concentration in different indoor rooms at building III.

Hall A, which might be due to higher activities and deposition of particles on files that move here and there on daily basis along with the staff.

At Building II, the average levels of PM_{10}, $PM_{2.5}$, and PM_1 were 88 µg m^{-3} (range 33–136 µg m^{-3}), 70 µg m^{-3} (range 30–84 µg m^{-3}) and 63 µg m^{-3} (range 28–73 µg m^{-3}), respectively. The concentrations of all three fractions of PM were found higher at the reception area and in the cafeteria/ pantry area, which is directly correlated with the high activity area. The high level of PM_{10} at reception and halls A and B (next to the reception area) might be due to the high movement of staff and visitors in the office compared to other office areas.

At Building III, the concentrations of PM_{10}, $PM_{2.5}$, and PM_1 were found in the range of 119–129 µg m^{-3}, 102–106 µg m^{-3} and 90–99 µg m^{-3}, respectively during non-meeting hours, however, these values during meeting hours were found to be high as 216–330 µg m^{-3}, 188–268 µg m^{-3} and 171–237 µg m^{-3}, respectively. This difference might be due to the penetration of PM due to the opening and closing of doors from the entrance gate to the lobby area and then the lobby gate to outside due to the high movement of people. The meeting was going during the monitoring and about 60–70 persons were present in the meeting hall.

Each of the compartments of respective buildings varied notably in dimension, number of doors, frequency of closing and opening, and the number of units of air filtration vents as described in **Tables 2** and **3**. The combination of these variables provided highly variable ventilation conditions and huge differences in indoor PM concentrations. A higher proportion of ultrafine particles also indicates the possibility of bio-aerosols in indoor spaces, which needs to be assessed and managed from a health impact point of view.

Further, the correlation between the size of the room/halls (indoor volume, m^3) and size segregated particulate concentrations were estimated for Building I and II.

The correlation coefficient (r^2) values for PM_{10}, $PM_{2.5}$ and PM_1 were estimated to be −0.35, −0.55, −0.54, respectively at Building I and − 0.19, −0.28, −0.28 at Building II. The negative correlation means larger halls/rooms increase the dispersion of particles, which results in low concentrations. It is also important to note that particulate concentrations at Building I (Old infrastructure and poor ventilation) have a good negative correlation with the size (volume) of the indoor compartments/rooms as compared to Building II (Modern infrastructure and good ventilation). It might be due to the impact of a good ventilation system, which dominated the impact of room size. Further, the fine and ultra-fine particles have a good correlation with the size of the room compared to coarser particles.

4.2 Particle size distribution in indoor work environment

The particles in the atmosphere may be primary or secondary, solid, or liquid depending upon their formation/sources. In the air, particles remain in suspended form for a longer time depending upon their sizes, which vary from very ultra-fine particles (nm) to coarse fine particles (μm). In literature, it is reported that ambient air particles below 2.5 μm are called fine particles which are further divided into transient nuclei (<0.1 μm) and accumulation range (0.1–2.5 μm). The fine particles are mainly generated due to primary emissions (controlled combustion activities, bio-aerosols, secondary aerosol, room air freshener, room cleaner spray in Indoor environments etc.). The particles in the size range of 2.5–100 μm are called coarse particles and are generated from wind-blown dust, sea spray etc. [29, 30].

In the present study, particle size between 0.25 μm to 32 μm is monitored at different 31 intervals. The fraction of different sized particle mass is compared between different indoor work environments and then with the ambient air. The fraction of total mass (%) contributed by different size range particles are described in **Tables 4-6** and **Figure 6**.

In Building I, the maximum mass was contributed by particles of size range 0.25–1.0 μm, in the range of 33–55% in Halls (Staff sitting area with half-sized individual cabin). These values for meeting rooms were even higher, being in the range of 60–75% (empty room during monitoring). The second dominant particle size range was 2.5–10 μm, which contributed 27–40% of the total mass in Halls, 12–21% in meeting rooms, however, the contribution at the common building entrance was 38%. The proportion of particle size 10–32 μm was between 11 and 24% (except Hall D, 5% only), 4–7% in meeting rooms, and 18% at the common entrance gate. In ambient air the mass contribution by particles of size 0.25–1.0 μm, 2.5-10 μm and 10-32 μm was found as 7%, 48%, and 41% respectively, which seems to be opposite to the trend of mass distribution in the different indoor environment except for the common entrance area.

In Building II, which is a modern office and located in the lower ground floor of a shopping mall (no direct opening in the ambient environment), the trend of particle size distribution was more or less similar with more percentage of ultrafine particles (0.25–1.0 μm); in the range of 36–64% in Office Halls, 82–86% in conference/meeting rooms, 30% at Main entrance of the Mall. In this building, the pantry area is near to the office staff sitting area and where the dominant particle size range was 2.6–10 μm with 40% of the total mass.

In Building III, a similar trend was observed for the meeting hall and common Indoor lobby area during non-meeting hours. However, during meetings, the proportion of ultrafine particles decreased from 78–64%, whereas particles of 2.6–10 μm increased from 13–22%. This indicates the re-suspension of particles due to the movement of people in the indoor environment. In the meeting hall, approx. 60–70 people were present during the meeting, which enhanced the particle concentrations even in the presence of sufficient ventilation systems.

Particle Size Range (µm)	Hall A	Hall B	Hall C	Hall D	Hall E	Hall I	Accounts Department	Meeting Room 1	Meeting Room 2	Common Entrance	Outside (Ambient Air)
0.25–1.0	41	48	55	55	33	50	43	75	60	35	7
1.1–2.5	8	6	6	8	7	7	8	9	10	9	5
2.6–10	40	27	27	31	35	27	32	12	21	38	48
10–32	11	18	12	5	24	14	17	4	7	18	41

Table 4.
Proportion of mass (%) contributed by different size particles in the different indoor compartment of building I.

Particle Size Range (µm)	Hall A	Hall B	Hall C	Hall D	Common Area (Lobby)	Staff Cabin	Conference Room	Meeting Hall	Cafeteria	Reception	Main Entrance	Ambient Air
0.25–1.0	38	36	55	64	78	70	82	86	30	53	30	18
1.1–2.5	8	7	7	10	9	8	8	5	9	9	13	7
2.6–10	31	37	12	21	11	18	8	7	40	28	47	31
10–32	22	19	26	4	1	3	0	1	21	9	10	44

Table 5.
Proportion of mass (%) contributed by different size particles in different indoor compartments of building II.

Particle Size Range (μm)	Meeting Hall (No meeting)	Common Area (No meeting)	Meeting Hall (With Meeting)	Common Area (With meeting)	Ambient Air
0.25–1.0	78	62	64	73	39
1.1–2.5	6	10	9	8	8
2.6–10	13	25	22	16	35
10.1–32	3	3	5	3	19

Table 6.
Proportion of mass (%) contributed by different size particles in different indoor compartments of building III.

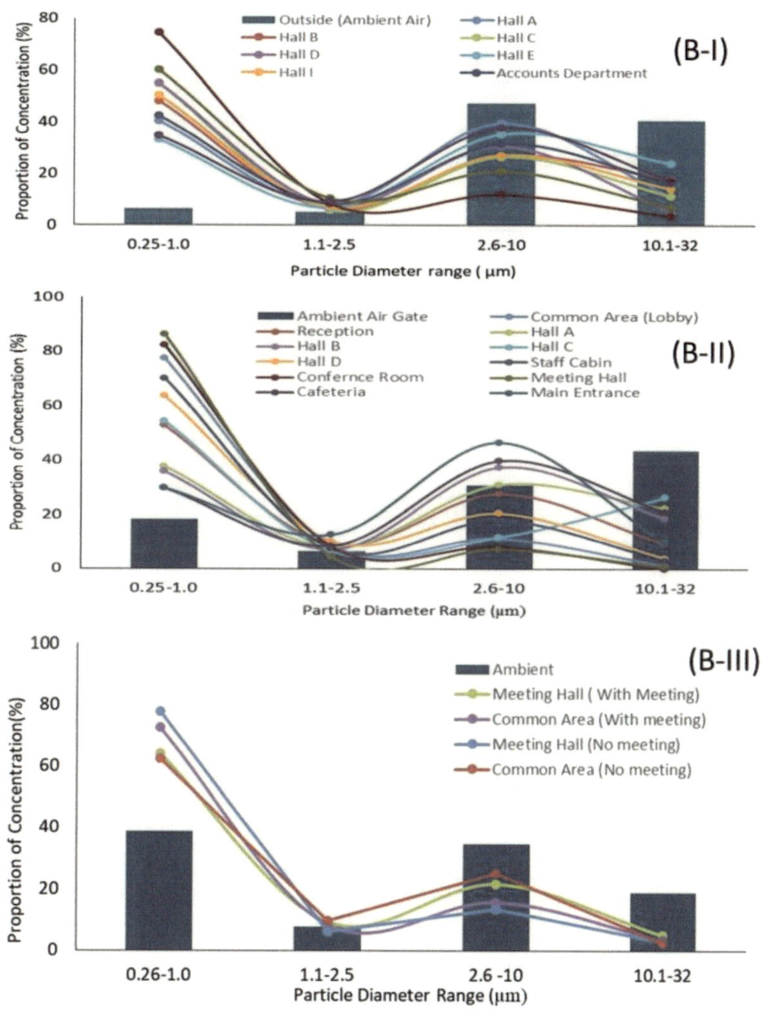

Figure 6.
Proportion of mass (%) contributed by different size particles in building I, II, and III.

It is observed that the proportion of finer particles is maximum in the indoor environment where the activity level is minimum (meeting rooms), followed by staff sitting area and then common building entrance (high people movement). It indicates that ambient air particles are more influenced by windblown dust particles from road and construction dust and natural dust. The particle size distribution indoor indicates the possibility of accumulated particles and bio-aerosol which are generally found in the range of fine particle size (diameter < 1 μm).

Norhidayah et al. [31] also found a dominant particle size range 0.3–0.5 μm in an office building in Malaysia and reported printing and photocopier machines as the major source of particles which is supported by work carried out by Massey and Taneja [13]. They have found that photocopier and printer machines generated accumulation phase particles i.e., 0.25–1.0 μm, and air freshener and cleaner generate particles of size 1 μm. Similarly, Tang et al. [32] reported a significant increase in fine and ultra-fine particle concentration in 43 out 62 office's rooms. They reported the average size of emitted particles in the range from 0.23 and 20 μm.

4.3 The ratio of Indoor/Outdoor (I/O) PM Concentrations

The I/O ratio of a pollutant is generally calculated to evaluate the possibility of intrusion of outdoor pollution inside the building. In the present study, the I/O ratio of size segregated PM (range 0.25–1.0 μm, 1.1–2.5 μm, 2.6–10 μm, and 10.1–32 μm) is calculated for each compartment of each building where monitoring was carried out as shown in **Figure 7**.

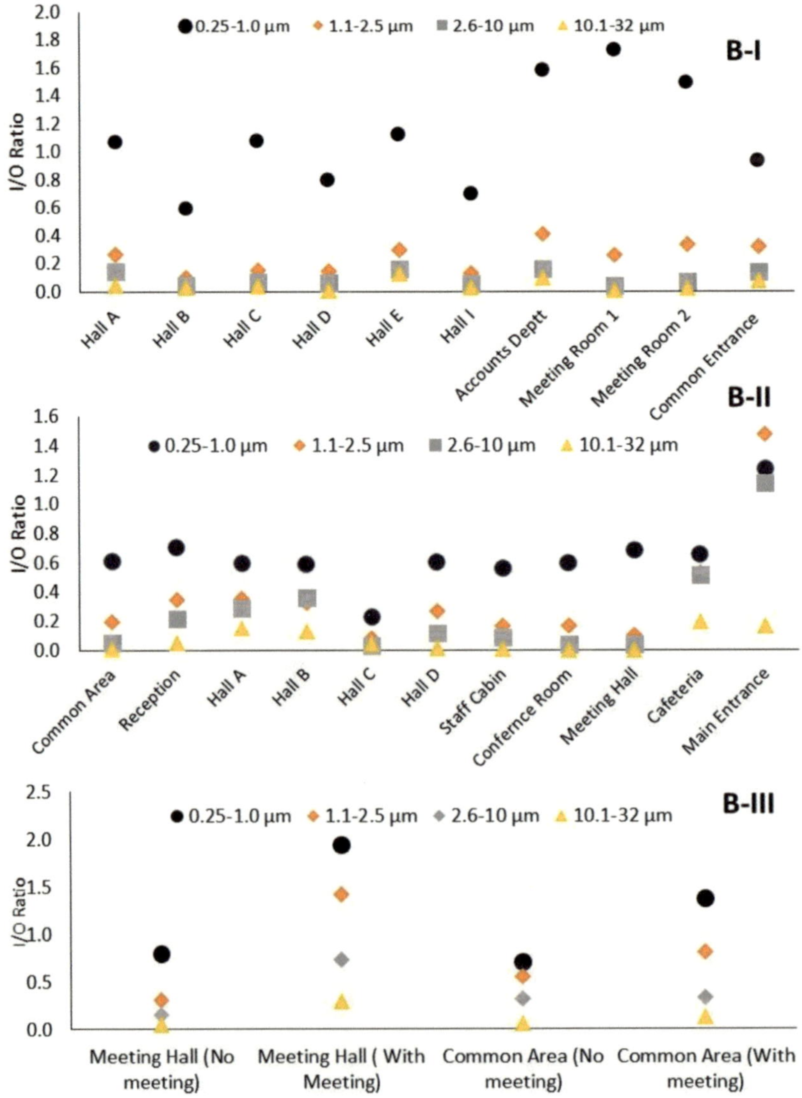

Figure 7.
Indoor/outdoor (I/O) ratio of size segregated PM in building I, II, and III.

In Building I, the I/O of ultrafine particles (0.25-1 μm) was found higher as compared to fine and coarse sized particles in all indoor compartments of the building. The I/O ratio of ultra-fine particles in the meeting/conference room (non-active area) was found maximum in the range of 1.6–1.7 when compared to other compartments of the building where office staff movement was more (0.5–1.1). Secondly, particles of size 1.1–2.5 have a higher I/O ratio in the range of 0.1–0.4. The I/O of coarse sized particles in the range of 10.1- 32 μm is lowest in all building compartments (up to 0.1).

In Building II, the I/O of ultrafine particles was found higher in the range of 0.6–1.2 (except Hall C of 0.2) compared to fine and coarse sized particles in all indoor compartments of the building. The particle size of 1.1–2.5 μm and 2.6-10 μm was more or less similar in the range of 0.1–0.4 except the cafeteria/pantry and main entrance. The values at the cafeteria/pantry were 0.5 for both sizes ranged particles and 1.2 and 1.5 for the main entrance area. The I/O of coarse sized particles in the range of 10.1- 32 μm is lowest in all building compartments (up to 0.2).

The I/O ratio pattern of Building III is more or less similar to Building I during non-meeting hours which is found to be 0.8, 0.3, 0.2, and 0.1 in the meeting hall for particle sizes of 0.25–1.0 μm, 1.1–2.5 μm, 2.6-10 μm and 10.1–32 μm, respectively. However, during meeting hours, these I/O values were found to be higher as 1.9, 1.4, 0.7 and 0.3, respectively. The occupants found the I/O ratio in Building III higher because of the opening and closing of the door many times during the monitoring period.

It is inferred that fine and ultrafine particles have higher I/O at all three sites, which might be due to the presence of Indoor sources and/or poor ventilation. Building II and III are well maintained, ventilated, and have modern infrastructure compared to Building I. This is reflected by low I/O values in all indoor compartments of Building II compared to Building I for ultrafine and fine particles. At Building III, there were no open files on desks, no cafeteria activities like Building II, however, still I/O was found >1 for the finer particles. Based on the discussions, it was found that regular cleaning of the tables, chairs and other areas was carried out through cleaning spray in the meeting room, which might generate fine aerosols. Similar observations were found by Goyal and Kumar [33], they found that I/O ratio for PM_{10}, $PM_{2.5}$ and $PM_{1.0}$ varied from 0.37–3.1, 0.2–3.2 and 0.17–2.9 respectively, at a commercial building in Delhi city. In one of the office buildings in Delhi, the I/O ratio of $PM_{2.5}$ was found to be 0.28–1.07 $\mu g\ m^{-3}$, which indicates that indoor $PM_{2.5}$ sources exist in the building apart from infiltration from outdoors [25]. The findings in developed countries also indicate average I/O ratios of $PM_{2.5}$ between 0.4 and 0.9 in an office room in Beijing and Xi'an cities in China [34, 35] and 0.62 ± 0.14 in Milan, Italy [36].

The analysis indicates that fine and ultrafine particles are dominantly generated from indoor activities at the monitoring location, which is not directly connected with outdoor gate (e.g. reception area, common entrance area etc). High I/O ratio for ultrafine and fine particles neglect the hypothesis of intrusion of outside PM in a mechanical ventilated building as coarse particles do not have such trend in I/O ratio.

5. IAQ management approach

Adequate and properly designed ventilation systems are the most effective strategies for achieving IAQ objectives. Smart planning of building uses and internal layout may help prevent many unnecessary IAQ problems. The mixed-use buildings having common facilities like Xerox facilities, pantry area among others should be properly ventilated and disconnected from the main office sitting area by the air-filter. There should be proper storage spaces for the old office records. The partitioning of the layout may affect the effectiveness of air distribution resulting in stagnant zones with poor air quality, which needs to be taken care of by architectural planning and ventilation engineering. Housekeeping is important in preventing IAQ problems as it keeps dust levels down and removes dirt, which could otherwise become sources of contamination, including mold growth. The cleaning schedule should be arranged according to occupancy patterns and activity levels. Daily cleaning of surfaces and vacuuming of floors is advisable for areas with high occupancy or which are in constant use during the day. The use of eco-friendly or non-toxic chemicals for cleaning also improves IAQ.

Numerous studies are available that strongly suggest that foliage plants in offices may improve health and reduce discomfort symptoms [37, 38] Kobayashi et al. [39] tested more than 20 plants to improve indoor air quality. Gawrońska, & Bakera [40] concluded that Spider plants (*Chlorophytum comosum* L.) phytoremediation particulate matter from indoor air. Torpy & Zavattaro [41] tested *Chlorophytum comosum* (Spider Plant) and *Epipremnum aureum* (Pothos) and concluded that indoor green plants can significantly reduce particulate matter concentration and hence improve Indoor Environment Quality (IEQ).

6. Conclusion

The study has focussed on the assessment of size segregated particulate matter (PM) in different indoor environments of three office buildings located in different parts of Delhi city. The PM concentrations were found higher in the indoor environment where activities were high but had poor ventilation. The levels of PM in the old building were found higher compared to the newly built office building having the modern infrastructure and well-maintained activities/files etc. The presence of people and activities generated re-suspended particles greater than 2.5 $\mu g\ m^{-3}$, which is noticed when compared PM concentration in the common area, reception area with office cabin area and meeting room with and without meeting hours. The indoor/outdoor ratios were greater for ultrafine and fine particles than coarser particles, which indicates presence of sources of finer particles indoors in all three buildings. Further, the meeting room/conference hall has a higher portion of ultra-fine particles of the total PM concentration. Further, correlation between room size (Indoor volume) and size segregated PM concentration found good negative correlation with finer particles in both buildings. This helps the indoor air quality managers to decide the suitable technology for the improvement of IAQ in different compartments of an office building.

Currently, the country does not have any IAQ standards nor have any monitoring protocol for Indoor air quality assessment. Therefore, it is suggested that country should come out with regulatory framework for IAQ assessment in different types of buildings. The findings of the present study suggest that any proposed IAQ standards should cover ultrafine (PM_1) and fine particle ($PM_{2.5}$) instead of coarser particles especially in office buildings.

Authors' contribution

Saurabh Mendiratta: Monitoring and Original writing, Sunil Gulia: Methodology, Data Analysis, Review, and Re-writing, Prachi Goyal: Monitoring and review, S.K. Goyal: Concept, Methods and Review.

Declaration of conflicting interests

The author(s) declared no potential conflicts of interest with respect to the research, authorship, and/or publication of this article.

Author details

Saurabh Mendiratta, Sunil Gulia*, Prachi Goyal and Sanjeev Kumar Goyal
CSIR-National Environmental Engineering Research Institute, Delhi Zonal Centre, Naraina, New Delhi, India

*Address all correspondence to: s_gulia@neeri.res.in; sunilevs@gmail.com

References

[1] Gulia S, Nagendra SS, Khare M and Khanna I. Urban air quality management-A review. *Atmos Pollut Res* 2015;16(2):286–304.

[2] Cheng, Y. H. (2017). Measuring indoor particulate matter concentrations and size distributions at different time periods to identify potential sources in an office building in Taipei City. *Building and Environment*, *123*, 446–457.

[3] Dorizas, P. V., Assimakopoulos, M. N., Helmis, C., & Santamouris, M. (2015). An integrated evaluation study of the ventilation rate, the exposure and the indoor air quality in naturally ventilated classrooms in the Mediterranean region during spring. Science of the Total Environment, 502, 557–570.

[4] Rohra, H., Tiwari, R., Khare, P., & Taneja, A. (2018). Indoor-outdoor association of particulate matter and bounded elemental composition within coarse, quasi-accumulation and quasi-ultrafine ranges in residential areas of northern India. *Science of The Total Environment*, *631*, 1383–1397.

[5] Srivastava, A., & Jain, V. K. (2003). Relationships between indoor and outdoor air quality in Delhi. *Indoor and Built Environment*, *12*(3), 159–165.

[6] Zhao, J., Birmili, W., Wehner, B., Daniels, A., Weinhold, K., Wang, L., ... & Hussein, T. (2019). Particle Mass Concentrations and Number Size Distributions in 40 Homes in Germany: Indoor-to-outdoor Relationships, Diurnal and Seasonal Variation. *Aerosol and Air Quality Research*, *20*(3), 576–589.

[7] Khan, S. A. R., Zhang, Y., Kumar, A., Zavadskas, E., & Streimikiene, D. (2020). Measuring the impact of renewable energy, public health expenditure, logistics, and environmental performance on sustainable economic growth. *Sustainable Development.*

[8] Tsai, D. H., Lin, J. S., & Chan, C. C. (2012). Office workers' sick building syndrome and indoor carbon dioxide concentrations. *Journal of occupational and environmental hygiene*, *9*(5), 345–351.

[9] Yau, Y. H., Foo, Y. W., & Mohyi, M. H. H. (2008). A preliminary study on HVAC systems and thermal comfort in a tropical university building in Malaysia. *International Journal of Mechanical and Materials Engineering*, *3*(2), 160–175.

[10] Goel, S., Patidar, R., Baxi, K., & Thakur, R. S. (2017). Investigation of particulate matter performances in relation to chalk selection in classroom environment. *Indoor and Built Environment*, *26*(1), 119–131.

[11] Majumdar, D., Gajghate, D. G., Pipalatkar, P., & Chalapati Rao, C. V. (2012). Assessment of airborne fine particulate matter and particle size distribution in settled chalk dust during writing and dusting exercises in a classroom. *Indoor and Built Environment*, *21*(4), 541–551.

[12] Khan, S. A. R., Yu, Z., Sharif, A., & Golpîra, H. (2020). Determinants of economic growth and environmental sustainability in South Asian Association for Regional Cooperation: evidence from panel ARDL. *Environmental Science and Pollution Research*, 1–13.

[13] Li, N., Georas, S., Alexis, N., Fritz, P., Xia, T., Williams, M. A., ... & Nel, A. (2016). A work group report on ultrafine particles (AAAAI) why ambient ultrafine and engineered nanoparticles should receive special attention for possible adverse health outcomes in humans. *The Journal of allergy and clinical immunology*, *138*(2), 386.

[14] Wargocki, P., Wyon, D. P., Baik, Y. K., Clausen, G., & Fanger, P. O. (1999). Perceived air quality, sick building syndrome (SBS) symptoms and productivity in an office with two different pollution loads. *Indoor air*, *9* (3), 165–179.

[15] Wyon, D. P. (2004). The effects of indoor air quality on performance and productivity. *Indoor air*, *14*(1), 92–101.

[16] Fisk, W. J. (2000). Health and productivity gains from better indoor environments and their relationship with building energy efficiency. *Annual review of energy and the environment*, *25* (1), 537–566.

[17] Saraga, D., Pateraki, S., Papadopoulos, A., Vasilakos, C., & Maggos, T. (2011). Studying the indoor air quality in three non-residential environments of different use: a museum, a printery industry and an office. *Building and Environment*, *46* (11), 2333–2341.

[18] Razali, N. Y. Y., Latif, M. T., Dominick, D., Mohamad, N., Sulaiman, F. R., & Srithawirat, T. (2015). Concentration of particulate matter, CO and CO2 in selected schools in Malaysia. *Building and environment*, *87*, 108–116.

[19] Zwoździak, A., Sówka, I., Krupińska, B., Zwoździak, J., & Nych, A. (2013). Infiltration or indoor sources as determinants of the elemental composition of particulate matter inside a school in Wrocław, Poland?. *Building and Environment*, *66*, 173–180.

[20] Taneja, A., Saini, R., & Masih, A. (2008). Indoor air quality of houses located in the urban environment of Agra, India. *Annals of the New York Academy of Sciences*, *1140*(1), 228–245.

[21] Kulshreshtha, P., & Khare, M. (2011). Indoor exploratory analysis of gaseous pollutants and respirable particulate matter at residential homes of Delhi, India. *Atmospheric Pollution Research*, *2*(3), 337–350.

[22] Goyal, R., & Khare, M. (2009). Indoor–outdoor concentrations of RSPM in classroom of a naturally ventilated school building near an urban traffic roadway. *Atmospheric Environment*, *43*(38), 6026–6038.

[23] Chithra, V. S., & Nagendra, S. S. (2012). Indoor air quality investigations in a naturally ventilated school building located close to an urban roadway in Chennai, India. *Building and Environment*, *54*, 159–167.

[24] Datta, A., Suresh, R., Gupta, A., Singh, D., & Kulshrestha, P. (2017). Indoor air quality of non-residential urban buildings in Delhi, India. *International Journal of Sustainable Built Environment*, *6*(2), 412–420.

[25] Gupta, A., Goyal, R., Kulshreshtha, P., & Jain, A. (2020). Environmental Monitoring of PM 2.5 and CO 2 in Indoor Office Spaces of Delhi, India. In *Indoor Environmental Quality* (pp. 67–76). Springer, Singapore.

[26] Habil, M., & Taneja, A. (2011). Children's exposure to indoor particulate matter in naturally ventilated schools in India. *Indoor and Built Environment*, *20*(4), 430–448.

[27] GRIMM, The Ultimate New Model 11-R Mini Laser Aerosol Spectrometer (Mini-LAS). http://www.envitech-bohemia.cz/files/008-indoor/grimm/01-mini-las/mini-las-en.pdf. Accessed on 9th April 2020.

[28] Tiwary, A., & Williams, I. (2018). *Air pollution: measurement, modelling and mitigation*. CRC Press.

[29] Harrison, R. M. (1999). Measurements of concentrations of air pollutants. In *Air pollution and health* (pp. 63–81). Academic Press.

[30] Massey, D. D., & Taneja, M. (2011). Emission and formation of fine particles from hardcopy devices: the cause of indoor air pollution. *Monitoring, Control and Effects of Air Pollution*, 121–134.

[31] Norhidayah, A., Aui, S. H., Ismail, N., Sukadarin, E. H., & Jalil, M. E. A. (2016). Indoor particle size distribution in office. *ARPN J Eng Appl Sci*, *11*(11), 7161–7165.

[32] Tang, T., Hurraß, J., Gminski, R., & Mersch-Sundermann, V. (2012). Fine a nd ultrafine particles emitted from laser printers as indoor air contaminants in German offices. Environmental Science and Pollution Research, 19(9), 3840-3849.

[33] Goyal, R., & Kumar, P. (2013). Indoor–outdoor concentrations of particulate matter in nine microenvironments of a mix-use commercial building in megacity Delhi. *Air quality, atmosphere & health*, 6(4), 747–757.

[34] Shi, S., Chen, C., & Zhao, B. (2017). Modifications of exposure to ambient particulate matter: Tackling bias in using ambient concentration as surrogate with particle infiltration factor and ambient exposure factor. Environmental pollution, 220, 337–347.

[35] Kalimeri, K. K., Bartzis, J. G., Sakellaris, I. A., & de Oliveira Fernandes, E. (2019). Investigation of the PM2. 5, NO2 and O3 I/O ratios for office and school microenvironments. Environmental research, 179, 108791.

[36] Sangiorgi, G., Ferrero, L., Ferrini, B. S., Porto, C. L., Perrone, M. G., Zangrando, R., ... & Bolzacchini, E. (2013). Indoor airborne particle sources and semi-volatile partitioning effect of outdoor fine PM in offices. Atmospheric environment, 65, 205–214.

[37] Deng, L., & Deng, Q. (2018). The basic roles of indoor plants in human health and comfort. *Environmental Science and Pollution Research*, *25*(36), 36087–36101.

[38] Moya, T. A., van den Dobbelsteen, A., Ottele, M., & Bluyssen, P. M. (2019). A review of green systems within the indoor environment. *Indoor and Built Environment*, *28*(3), 298–309.

[39] Kobayashi, K. D., Kaufman, A. J., Griffis, J., & McConnell, J. (2007). Using houseplants to clean indoor air.

[40] Gawrońska, H., & Bakera, B. (2015). Phytoremediation of particulate matter from indoor air by *Chlorophytum comosum* L. plants. *Air Quality, Atmosphere & Health*, *8*(3), 265–272.

[41] Torpy, F., & Zavattaro, M. (2018). Bench-study of green-wall plants for indoor air pollution reduction. J. Living Archit, 5(1), 1–15.

Chapter 15

Correlation between Air Quality and Wastewater Pollution

Karzan Mohammed Khalid

Abstract

Recently, air pollution is a universal problematic concern which adversely affects global warming and more importantly human body systems. This chapter focuses on the importance of air quality, and indicates the negative effects of emissions originated from both municipal and industrial wastewaters to atmosphere. More importantly, the improvements in wastewater treatment plants to eliminate the crisis of emissions on environment and human health is also clarified. Urbanization and distribution of industrials in urban areas influence the air pollution via releasing pollutants and contaminants to environment. The pollutant emissions from wastewaters are volatile organic compounds, Greenhouse gases and other inorganic pollutants (heavy metals) which are causes to many reactions through atmosphere, then products detriment whole environment and living organisms including human. Moreover, contaminants are also released into air from influents of municipal wastewaters and they are considered as the main resources of most threatened infections in human and other animals. As conclusion, because of the persistently development urbanization and industrialization as the wastewater pollutant sources, the environmental technology regarding wastewater treatments must depend on prevention of emissions to air before thinking on cost and good quality effluents.

Keywords: air pollution, municipal wastewater, pollutant emission, infectious agents, industrial wastes, promised techniques

1. Introduction

The world recognizes air pollution as detrimental issue that significantly affects public health. There has been intensive studies and documentation of the effects of air pollution around the world [1, 2]. Sustainable development in any society provides a good living standard for the individuals. Also, these include social progress and equality, environmental protection, conservation of natural resources, and stable economic growth [3]. Industrial and transportation emissions and their burden in regional and global harm on health, climate and vegetation have been well studied in last few decades [4].

Health effects due to air pollution are a big concern for the World Health Organization. Air pollution does not only cause toxicological effects on human health, it has also significantly degraded the environment in the last years [5, 6]. Now a day, wastewater treatment plants (WWTPs) are definitely known as one of the most crises on air quality and availability of gases, chemical pollutants and biological contaminants in environment directly resourced from sewage wastewaters [7].

Moreover, municipal wastewater drastically increased, and due to household waste contents and draining to trunk cannels close to the urban areas, they adversely affect human health rather than environmental damage.

Water pollution is a problematic on humanity and the aquatic life, and increase catalyzes climatic changes [8]. For instance, various human activities as well as the release of greenhouse gases by industries greatly contributes to global warming, planet temperature enhancement, and lowering of atmospheric air quality.

Sustainability of environment among different societies is importantly developed as an initiation of living standard improvement for individuals. And It aimed to solve the challenges faced to environment, economy and society) without effects on human and environment in the future. Sustainability is also important to progression and equality of social, environmental safety, preservation of natural resources and economic growth [3].

The main contributors of air pollution and their cooperativeness cause to increase risks on air quality. For instance, With the growth of population, there is also a growth in demand for gas, oil, and other energy sources. This has also increased the number of refineries and petroleum wastewater treatment plants [9]. The pollutants are mostly chemicals which present in items used by individuals, chemicals containing preservative compounds, dyes, hydrocarbons, proteins as nutrients, etc. In last few decades, the demand on synthetic chemical products increased and products easily delivered to homes, due to advertisement and evolution in lifestyle such as; internet availability and easily contact in society. On the other hand, more urbanization around the world leads to increase wastes per individuals, however, this is different among various countries while still considered as one of the most reasons of developed more and more liquid and soil wastes (municipal wastewater).

Fortunately, along all society wastewater treatment plants (WWTPs), purification seems to be familiar and properly applicable to remediate municipal and industrial wastes and there are good understandings in this aspect. However, the wastewater treatment systems are variously performed within different society, while the all of them are targeted for one reason of improvement air quality and human health. Thus, all organization including WHO and governments hardly work to be far from wastewater emission impacts and improvement of air quality.

This chapter is aimed to better understanding in adverse impacts of wastewater effluents on air quality via emission process which is directly and indirectly affects human health via respiratory and skin diseases. It is also aimed to keep air emission in line level in recent technologies, however the risks associated with exposure to emissions from WWTPs are uncertain and require more research, stronger regulatory frameworks and safer design consideration.

2. Wastewater pollutants emitted to air

Generally, the presence of high concentrations of pollutants in atmosphere are results from unsustainable regional policy and lack of affordable green technology transfer [10]. There are different pollutants and contaminants emissions, the diversity of chemical pollutants leads to classify emissions according to their etiological agent within different types of wastewaters. The design of constructed sewage channels also affects the emission rate into the atmosphere. Open wastewaters are more efficiently exhaust emissions than close box or underground constructed wastewaters, as a result of abiotic effects which leads to worm the water and stimulate more volatilize and release. The followings are the air pollutants which are also originated from wastewaters effluents and easily releasable:

2.1 Hydrocarbons

Hydrocarbon pollutants are defined as one of the great serious emissions that effects all life forms [11]. Aliphatic and aromatic hydrocarbons are released into the air from industrial outlets rother than solid wastes which are directly emitted into the air, particularly from petroleum industries [12]. Recently, in developed countries industrial wastes undergo several processes of purification such as; conversion, separation and treatment, while during processing hydrocarbon emission persistently occur and adversely affect air quality. Despite of the above occurrence of emission, transportation of refined or purified products through tanks and pipelines can also leakage to water bodies and additional hydrocarbon emission occurs through wastewater treatment plants [12]. According to Aljuboury et al., the effluents/outlets from petroleum industries containing pollutant products with easily emission [13]. Therefore, auxiliary emissions arise when volatile organic compounds are stripped off from the contaminated wastewater in aeration basins, drains, and ponds which are all considered as indirect emissions [12]. The pollution via hydrocarbons sometimes due to the accidents during over sea transportations, when crude oil and gasses leak and release on the surface of water body. Finally, these pollutants directly and indirectly reach human, animals and plants and adversely affect them [12].

2.2 Volatile compounds

Volatile compounds are chemical substances; they have low boiling points and are immediately released into the air after contact. The concentration and identity of volatile compounds in wastewater and their emission to air varies according to the wastewater resources, transport system, characteristics of the employed treatment plant and the weather (physical) conditions. The aeration process and mechanism involved in oxygen diffusion in wastewater treatment plant states transfer characteristics between air and wastewater. And several organic substances in wastewater are either adsorbed, biodegraded or volatilized [14].

The emission of volatile organics from municipal wastewater plants is the main problem for wastewater treatment systems. Different types of pollutants (solvents and chemicals) that originated from municipal wastewater considered as a major source of VOCs. They are also presence in gaseous forms and leads bad odors/toxicity, they are crises on natural environment and air pollution resulted in the availability of VOCs [15]. In addition, VOCs also released during the composting of different organic wastes [16]. According to He and his colleagues, volatile organic compounds are also released during bio-drying of municipal solid waste. Biodegradation of wastes causes the production of these compounds in composting sites. During the process of biological decomposition, a huge quantity of VOCs is released from the organic matrix as well [17]. Because of the close relationship between wastewater streams and landfills and high occurrence of leaking from solid wastes of landfills to water streams, it is important to discuss the efficiency of landfills in VOCs emissions. Gases are also produced in landfills when household chemical products are vaporized in the landfill sites [18]. Landfills in many countries closed to municipal wastewaters, therefore solid wastes from this site certainly drained to wastewaters. The abandoned landfills had volatile organic compounds over the permissible limits, and their release to wastewater and directly for atmosphere is estimated over the permissible levels. Benzene, toluene, ethylbenzene, and xylene were the major volatile organic compounds detected in the air [19].

Petroleum as highly pollutant in environment contains high concentration of VOCs. Controlling the release of volatile organic compounds into the air is a big

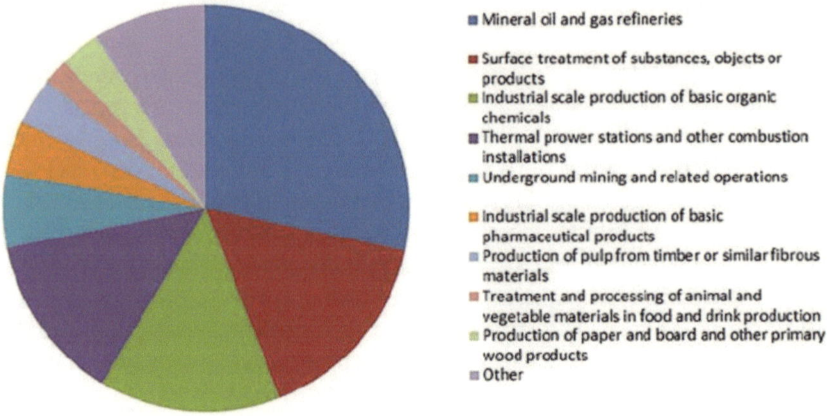

Figure 1.
Non-methane volatile organic compound emission shares from European industrial facilities in 2012 [22].

challenge for petroleum and the oil refining industries [12]. Malakar and Saha (2015) had concluded that, High concentrations of VOCs are derived from streams effluents of petroleum industries and refineries [20]. Moreover, the applicable fossil fuels in treatment plants (desalination) are noticed to release about 16,000 tons of VOCs. Thus, there may be a constitution fuel for desalination process [21]. Despite the anthropogenic or municipal wastes, emissions of VOCs from industrial wastes are also defined, and different industrial sectors are presented in the **Figure 1**.

2.3 Greenhouse gases

Municipal wastewater treatment plants are known to be one of the minor sources of greenhouse gases that are distributed in the atmosphere. Generally, there are three major sources of greenhouse gases (methane, carbon dioxide, and nitrous oxide) which are easily and frequently emitted into space, they are also found to cause indirect emissions from energy generation process [23, 24]. Aerobic biological treatment plants produce very large amounts of greenhouse gases as they require a large amount of energy to carry out various processes. The quantities of the resulted gasses depend on the influent of the wastewater, off-site treatments, and treatment processes in WWTPs [25]. According to United State Environmental Protection Agency (USEPA) in 2018, the three gasses emission in United States of America were ~ 81%, 10% and 7% for CO_2, methane and N_2O, respectively. With the remained 3% of emitted Fluorinated gases [26] (**Figure 2**).

Effect of wastewater treatments coming out of refineries and petrochemical industries is definitely problematic to environment and human health. Nevertheless, they are also known to cause large-scale emission of greenhouse gases into the atmosphere. According to Li et al., (2016), in United States of America around 0.40% of the total greenhouse gasses are emitted by wastewater treatment plants of refineries and petroleum industries [27]. Logistical pollutants which are frequently rely on fossil fuel consumption are considered as main reason of CO_2 and GHG emissions, and they are classified as significant contributor which affects environmental sustainability [28, 29].

2.4 Airborne microbial contaminants

Another critical issue that negatively affect air quality is the availability of microorganisms in atmosphere termed as microbial air pollution. The emission of

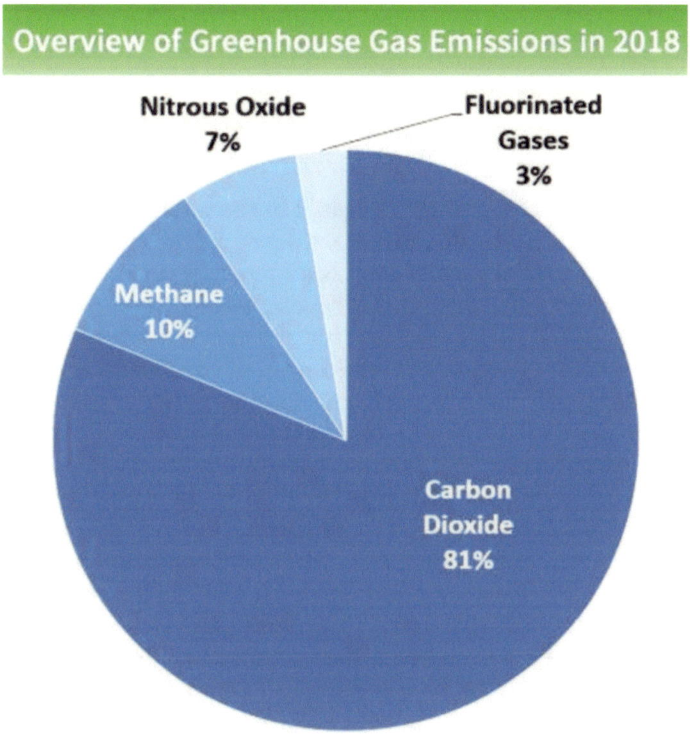

Figure 2.
Gas emission in 2018 in US (EPA).

bio-aerosols from wastewater to environment resulted from pollution of main sewage streams by human excreta (urine and feces) which containing a lot of microorganisms specifically bacteria (gram negative bacteria) [30]. The most common bacterial types released from municipal wastewaters are mostly include mesophilic pathogenic bacteria and psychrophiles, among them; *S. aureus*, Coliform bacteria, *Pseudomonas fluorescens* [31]. *Salmonella* sp., *Shigella* sp., *Pseudomonas aeruginosa*, *Clostridium perfringens*, *Bacillus anthracis*, *Listeria monocytogenes*, *Vibrio cholerae*, *Mycobacterium tuberculosis*, *Streptococcus faecalis*, *Proteus vulgaris* [32]. It is worthy to note that, many pathogenic microorganisms died and removed, while still some of them can survive in sewage sludge for the period of months [33]. Wastewater treatment plants are also known to release aerosols into the atmosphere and cause health issues to people working with the plants and living in the surrounding areas. The droplets from WWTPs documented to carry ten to thousand times more bacteria into the atmosphere than the sources of water pollution. The release and emission of microorganisms mainly depend on the temperature, wind velocity, humidity, smog, and other factors. Importantly, the available humidity triggers microorganism's proliferation, as it eliminates the solar efficiency to eradicate microorganisms [34]. The presence and absence of microorganisms in wastewater also related to the quality and climate of the site, even in purified wastewaters depend on the employed method of purification [35]. During purification processes, the microorganisms can get atmosphere via aerosols, particularly when the wastewater undergo aeration process by using air diffuser and biological bioreactor chamber [36].

The basic nutrients in wastewater (N, C and P) directly affects microbial life, due to this, their availability in wastewaters and any other sites leads to increase microbial activity. Furthermore, the presence of abundant microbial community in environment commonly sensed around the world, particularly in highly polluted zones. Microorganisms either contaminate atmosphere by themselves or involved in degradation of chemical compositions and finally a part of produced pollutants in the form of gasses (Volatile compounds) emitted into the air.

2.5 Nitrogen oxides and sulfur oxides

There are various processes that take place in the wastewater treatment plants. These processes lead to the production of oxides of sulfur and oxides of nitrogen. Nitrous oxide emission from wastewaters is known as a problematic contaminant which needs to be addressed. During this several years, dramatic increase of N2O was noticed. However, N2O is threatened but the emitted value is less than other chemical pollutants [37]. Particularly, the design of the sewers and their operational conditions potently facilitate N_2O emission into surrounded environment. Domestic wastewater which is originated from household activities by human also known to contain high concentrations of different nitrogen forms rother than phosphor and other chemical pollutants. The plants which take more efforts in nitrogen removal emit a less amount of nitrous oxide into the air [37]. The released nitrous oxide in atmosphere interact with other VOCs to make products such as tropospheric ozone [38]. The flux of atmospheric GHGs directly increases with the input nutrients, and it is also different from one type of wastewater to another one [23]. For instance, N_2O emissions are related to nitrification and denitrification processes which is triggered by some particular microorganisms. Denitrification of NO_3 and NO_2 as a result of metabolism of Nitrobacter more N_2O emitted into the air (**Figure 3**).

Combustion of fuel to run these plants also leads to the large-scale production of oxides of sulfur and nitrogen [10]. Continuously, sulfur dioxide (SO_2) emission is a detrimental issue in many developing countries, especially from influents of coal fired power plants and the coal industries. It is a pollutant that is directly emitted from the source and released into air, unlike tropospheric ozone which is indirectly produced from combination of chemical pollutants in atmosphere. The consumption of coal in human activities in form of wood, dung and crop residues for domestic energy at home also contributes to ambient SO_2 concentrations and adversely affect the children and adults who are exposed to high levels of this pollutant [39]. Sulfur oxide may create acidic forms of sulfuric and sulfurous as a result of presence vaporized water, then acid rain precipitation occurs [40]. On the other hand, the precipitated acid rain causes disturbance fresh water and vegetation on earth. Both nitrogen oxides and sulfur dioxide considered as the biggest sources of pollutants from desalination plants for wastewaters [41]. Desalination plants are known as producers of NOx and SOx with values about 60,000 and 200,000 tons per year, respectively [21].

2.6 Heavy metals

Heavy metals are listed at the top of inorganic pollutant with wide range of negative effects on organisms, plants, and human [42]. Heavy metals released into the environment via different routes such as industries, domestic, mining activities, agricultural activities and etc. [43]. Heavy metals are not degradable and accumulated in living systems [44]. Thus, air pollution by heavy metals is considerable even at low concentrations and the long-term cumulative is threatened for human health [45].

The effects on human health and the environment from exposure to the three most common heavy metal pollutants (mercury, lead and cadmium) include:

Mercury is known as one of the toxic forms of metals that can harm different systems from human body (the brain, heart, kidneys and lungs), and lowering immune response against foreign objects of all ages. Moreover, childhoods affected in central nervous system and causes to less able to think and learn [46]. The mercury cycle is particularly important for understanding how it can reach atmosphere from different sources within wastewater, specifically illustrates how this metal that can be methylated. There are many compounds which are widely used in industrial processes and big demand on this metal observed around the world. Inorganic mercury that concentrated at the bottom muds of the water was methylated by anaerobic bacteria of the genus Desulfovibrio (in **Figure 4** below).

Collaboration among each of atmosphere, apical aerobic water layer and anaerobic sediment is seriously affecting Hg cycle. Some anaerobic microorganisms at the bottom, can convert free mercury to methylated forms that can be transported to water and the atmosphere such as Desulfovibrio as an example of bacteria. Methylated mercury processed through biomagnification. The produced volatile elemental mercury (HgO) is easily distributed into upper oxygenated water body

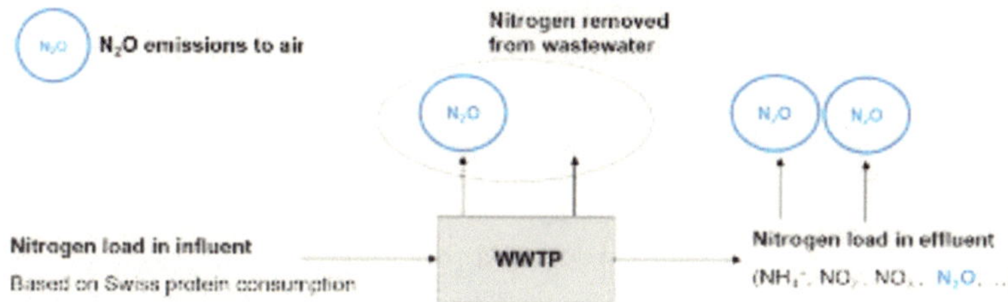

Figure 3.
Nitrous oxide emission to atmosphere from wastewater treatment plants.

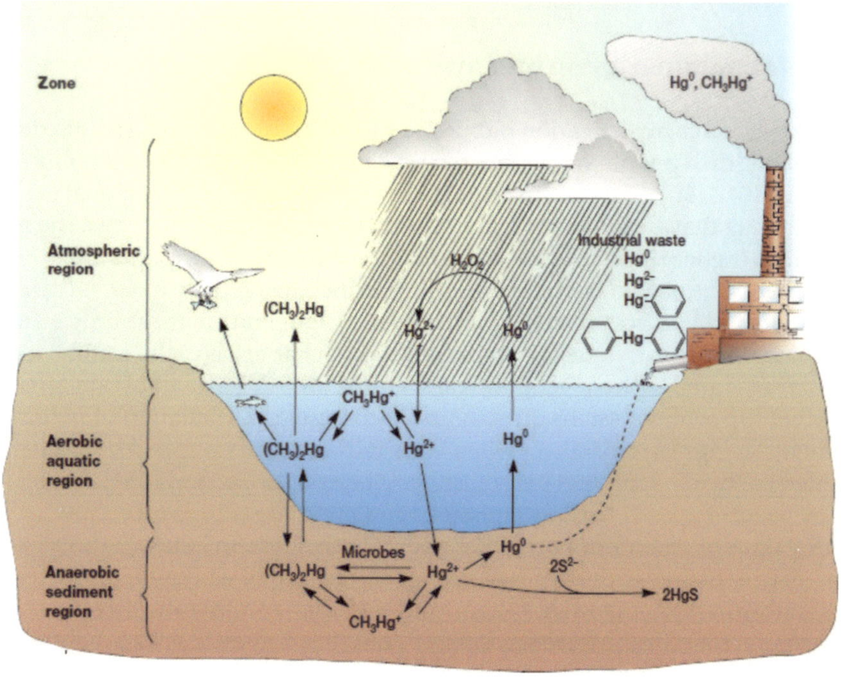

Figure 4.
The mercury cycle.

and even atmosphere, and then to waters and the atmosphere. Ionic mercury can react with sulfide in anaerobic sediment and resulted in less soluble HgS.

The produced methylated mercury is characterized as volatile and lipid soluble, and the mercury concentrations increased in the food chain (by the process of bio-magnification) and also increased in atmosphere. For example, the directly emitted Hg into atmosphere is evaluated by 2500 tons/year, which is accounting for about 31% of the overall emissions [47].

Lead as a non-essential heavy metal causes to late neuro-developmental in children, even at trace levels of exposure. Other effects include cardiovascular, renal, gastrointestinal, hematological and reproductive effects. Children six years old and under are most at risk. Currently, the thresholds of Pb are unknown. Lead frequently reaches human, animal and plant bodies and accumulates as a toxic substance [48].

Cadmium as a toxic metal and it is presence can make many problematic issues including: pulmonary irritation, kidney disease and cancer, bone weakness and prostate. Among the Cd sources, food and cigarette smoke are the common which easily exposure for the general population. About 90% of Cd exposure from dairy sources into the environment and to people who are not smoking.

Cadmium (Cd) as a toxic metal to plants, animals and microorganism. It may cause lowering growth rate of plants and many serious diseases for human as a result Cd accumulation, mainly in the kidney and liver of vertebrates and in aquatic invertebrates and algae. Severe toxic effects on fish, birds and other animals may include death or fetal malformations [49].

The availability of Pb, Hg and Cd was studied by Du and his team in Heilongjiang City- China, samples were collected from 27 WWTPs with intervals monthly during 2015. And the results ensured the removal of heavy metals from wastewater is highly affective as they released to environment which finally adversely affects humanity [7].

3. Effects of polluted air on human health

According to the investigation published by the World Health Organization, the number of died people was estimated by 7 million as a result of air pollution exposure in 2012 [50]. This number indicates one out of eight total global deaths and confirming that air pollution is the world's largest health risk. Thus, the mortality rate much higher than that caused by malaria and AIDS. Air pollution is not only the concern of one nation or country, while it is increasing daily along with increase urbanization and industrialization. The universal cooperation is the only solution to overcome this critical issue (air pollution) which has crisis on humanity, and air is a natural resource without geopolitical boundaries [4]. Individuals are affected by different types of emissions directly and indirectly via inhalation of pollutants and climate change (for instance; when solar radiation gets trapped by gaseous and suspended particulate matters in atmospheric layers, respectively [51].

Many people exposed to these emissions and microorganisms may show unhealthy signs of respiratory problems and digestive system issues [34]. Bio-aerosols are known to contain various types of microorganisms that can cause disorders of the respiratory system, digestive system, and skin. Bio-aerosols also affect the quality of air in the surrounding. Moreover, it was found that domestic sewage containing animal and human excreta contains the highest amounts of microorganisms. They are usually treated and released by municipal wastewater plants which cause various micro-organisms to enter the atmosphere [52]. The spread of microorganisms in atmosphere depends on the weather and season [53].

Chronic obstructive pulmonary disease, acute lower respiratory illness, ischemic heart disease and lung cancer have been contributed to most air pollution related cases and even deaths. Inhalation of fine particles from air (particulate matter) and produced ozone are detected as the origin of those diseases [50]. All studies agreed on the presence of correlation between green technology improvement and environmentally sustainable. For instance, Khan and his colleagues (2020) have prepared two important hypotheses, which are "(1. Greater environmental performance reduces the health expenditure) and (Country environmental performance has a positive correlation with economic growth)" [28].

4. How to keep air pollution in line from wastewater emissions?

The previous wastewater treatment plants were afforded only to obtain large purified effluents from wastewaters and cost-effective protocol, while no consideration about the emissions as a result of biological reactions. Now a day, global attempts considered to increase environmental sustainability which is derived from greenhouse gasses (GHGs), organic and inorganic compounds that directly emitted into the atmosphere and air quality disruption.

To minimize GHG emissions from wastewater treatment plants, the following recommendations can improve the practical systems with lowering gasses emissions. To minimize N_2O emissions, biological wastewater treatment plants should be operated at high solid retention times (SRT) to preserve low ammonia and nitrite concentrations. Moreover, big bioreactors are suggested to dispose of systems able to large volume loading buffer and to decrease the risk of transient oxygen depletion. The emissions N_2O can be reduced (if nitrous oxide stripping by aeration is limited since microorganisms would have more time to consume it) [19]. On the other hand, application of anammox processes can be used to remove ammonia. On the basis of the metabolism of anammox bacteria, N_2O is not directly produced [54], and therefore, it is considered as a promised process to emission N_2O in the WWTPs as a constitute of the conventional nitrification–denitrification processes.

Methane (CH_4) emissions can be reduced into a minimum value by properly covering sludge tank and sludge disposal tanks, to prevent gas leakages and emitted CH4 captured by hoods which could be undergo through burning with excess biogas in a torch [51]. Methane gas usually produced within WWTP itself and from it is sources. Methane is mainly undergo oxidation by approximately 80% in the activated sludge tanks, which could be exploited to further decline methane emissions to atmosphere from WWTPs [55].

The SRT as a promised invention applied to the biological reactor which triggers GHG emissions to atmosphere. The activated sludge system, especially when SRT value seems to be high, improves biomass endogenous respiration, which stimulates the COD oxidization to CO_2 and reduces the produced sludge. The lower sludge production indicates CH4 decline, thus any reduction of CO_2 release correlated to its combustion [56].

5. Conclusions

Air quality is not less important than any pandemic diseases throughout the world, due to uncontrolled emissions from anthropogenic wastes. The reports indicated that wastewaters are considered as one of the detrimental sources of pollutants in environment specifically within atmosphere. And the emissions from wastewater to air directly increase with urbanization and industrialization, which are known as the main sources of wastewater pollution. Any additions of VOCs and GHGs in air directly and indirectly harm environment and human health. The priority of wastewater treatment plant systems must be changed from cost effective and good effluent quality to emission prevention then follow other aspects.

Conflict of interest

It is my pleasure to do further studies on environmental technologies, particularly bioremediation, phytoremediation and air quality in advance laboratories. Hope to get offers from different parts of the world to do whatever I want in my specialization.

Author details

Karzan Mohammed Khalid
Faculty of Science, Soran University, Soran-Erbil, Kurdistan Region of Iraq

*Address all correspondence to: karzan.khalid@soran.edu.iq

References

[1] Kelly FJ, Fussell JC. Air pollution and public health: emerging hazards and improved understanding of risk. Environmental geochemistry and health. 2015 Aug 1;37(4):631-649.

[2] Genc S, Zadeoglulari Z, Fuss SH, Genc K. The adverse effects of air pollution on the nervous system. Journal of toxicology. 2012 Oct;2012.

[3] Ilin I, Kalinina O, Iliashenko O, Levina A. Sustainable urban development as a driver of safety system development of the urban underground. Procedia Engineering. 2016 Jan 1;165:1673-1682.

[4] Piqueras PM. Urban Air Pollution: Wastewater Treatment Sources and Impacts on Agriculture. University of California, Riverside; 2017.

[5] Ghorani-Azam A, Riahi-Zanjani B, Balali-Mood M. Effects of air pollution on human health and practical measures for prevention in Iran. Journal of research in medical sciences: the official journal of Isfahan University of Medical Sciences. 2016;21.

[6] Lovett GM, Tear TH, Evers DC, Findlay SE, Cosby BJ, Dunscomb JK, Driscoll CT, Weathers KC. Effects of air pollution on ecosystems and biological diversity in the eastern United States. Annals of the New York Academy of Sciences. 2009 Apr;1162(1):99-135.

[7] Du P, Zhang L, Ma Y, Li X, Wang Z, Mao K, Wang N, Li Y, He J, Zhang X, Hao F. Occurrence and Fate of Heavy Metals in Municipal Wastewater in Heilongjiang Province, China: A Monthly Reconnaissance from 2015 to 2017. Water. 2020 Mar;12(3):728.

[8] Palmate SS. Modelling spatiotemporal land dynamics for a trans-boundary river basin using integrated Cellular Automata and Markov Chain approach. Applied geography. 2017 May 1;82:11-23.

[9] Helmy Q, Kardena E. Petroleum oil and gas industry waste treatment; common practice in Indonesia. Journal of Petroleum & Environmental Biotechnology. 2015 Jan 1;6(5):1.

[10] Piqueras P, Vizenor A. The rapidly growing death toll attributed to air pollution: A global responsibility. Policy Brief for GSDR. 2016:1-4.

[11] Srivastava M, Srivastava A, Yadav A, Rawat V. Source and Control of Hydrocarbon Pollution. InHydrocarbon Pollution and its Effect on the Environment 2019 Nov 27. IntechOpen.

[12] Ragothaman A, Anderson WA. Air quality impacts of petroleum refining and petrochemical industries. Environments. 2017 Sep;4(3):66.

[13] Aljuboury DA, Palaniandy P, Abdul Aziz HB, Feroz SA. Treatment of petroleum wastewater by conventional and new technologies-A review. Glob. Nest J. 2017 Nov 1;19:439-452.

[14] Sree U, Bauer H, Fuerhacker M, Ellinger R, Schmidt H, Puxbaum H. Hydrocarbons emissions from a municipal wastewater treatment pilot plant in Vienna. Water, air, and soil pollution. 2000 Nov 1;124(1-2):177-186.

[15] Pagans E, Font X, Sánchez A. Emission of volatile organic compounds from composting of different solid wastes: abatement by biofiltration. Journal of hazardous materials. 2006 Apr 17;131(1-3):179-186.

[16] He P, Tang J, Zhang D, Zeng Y, Shao L. Release of volatile organic compounds during bio-drying of municipal solid waste. Journal of Environmental Sciences. 2010 Jan 1;22(5):752-759.

[17] Font X, Artola A, Sánchez A. Detection, composition and treatment of volatile organic compounds from waste treatment plants. Sensors. 2011 Apr;11(4):4043-4059.

[18] Scheutz C. Attenuation of methane and trace organics in landfill soil covers (Doctoral dissertation, Environment & Resources DTU. Technical University of Denmark).

[19] Kim KH, Shon ZH, Kim MY, Sunwoo Y, Jeon EC, Hong JH. Major aromatic VOC in the ambient air in the proximity of an urban landfill facility. Journal of Hazardous Materials. 2008 Feb 11;150(3):754-764.

[20] Malakar S, Saha PD. Estimation of VOC emission in petroleum refinery ETP and comparative analysis with measured VOC emission rate. The IJES. 2015;4(10):20-29.

[21] Cipollina A, Tzen E, Subiela V, PapapetrouM,KoschikowskiJ,SchwantesR, Wieghaus M, Zaragoza G. Renewable energy desalination: performance analysis and operating data of existing RES desalination plants. Desalination and Water Treatment. 2015 Sep 11;55(11):3120-3140.

[22] Ojala S, Koivikko N, Laitinen T, Mouammine A, Seelam PK, Laassiri S, Ainassaari K, Brahmi R, Keiski RL. Utilization of volatile organic compounds as an alternative for destructive abatement. Catalysts. 2015 Sep;5(3):1092-1151.

[23] Chen GC, Tam NF, Wong YS, Ye Y. Effect of wastewater discharge on greenhouse gas fluxes from mangrove soils. Atmospheric environment. 2011 Feb 1;45(5):1110-1115.

[24] Campos JL, Valenzuela-Heredia D, Pedrouso A, Val del Río A, Belmonte M, Mosquera-Corral A. Greenhouse gases emissions from wastewater treatment plants: minimization, treatment, and prevention. Journal of Chemistry. 2016 Jan 1;2016.

[25] Das S. Estimation of greenhouse gases emissions from biological wastewater treatment plants at Windsor.

[26] US-EPA. United State Environmental Protection Agency. 2018.

[27] Li QF, Han J, Elgowainy AA. Industrial Wastewater Treatment in GREET® Model: Energy Intensity, Water Loss, Direct Greenhouse Gas Emissions, and Biogas Generation Potential. Argonne Natl Lab, 1-13.

[28] Khan SA, Zhang Y, Kumar A, Zavadskas E, Streimikiene D. Measuring the impact of renewable energy, public health expenditure, logistics, and environmental performance on sustainable economic growth. Sustainable Development. 2020 Jan 6.

[29] Zhang Y, Khan SA, Kumar A, Golpîra H, Sharif A. Is tourism really affected by logistical operations and environmental degradation? An empirical study from the perspective of Thailand. Journal of Cleaner Production. 2019 Aug 1;227:158-166.

[30] Kusnierz M, Swierczek E. Infrastruktura krytyczna a niezawodność systemu odprowadzania i oczyszczania ścieków w niekorzystnych warunkach pogodowych. Infrastruktura i Ekologia Terenów Wiejskich. 2014(I/1).

[31] Sánchez-Monedero MA, Aguilar MI, Fenoll R, Roig A. Effect of the aeration system on the levels of airborne microorganisms generated at wastewater treatment plants. Water Research. 2008 Aug 1;42(14):3739-3744.

[32] Klapec T, Cholewa A. Zagrożenia dla zdrowia związane ze stosowaniem nawozów organicznych i organiczno-mineralnych. Medycyna Ogólna i Nauki o Zdrowiu. 2012;18(2).

[33] Michalska-Szymaszek M. The influence of temperature on survival, phenotype and antygen properties of Escherichia coli O157 strains isolated from water and clinical material. Roczniki Panstwowego Zakladu Higieny. 2010 Jan 1;61(2):213-219.

[34] Vantarakis A, Paparrodopoulos S, Kokkinos P, Vantarakis G, Fragou K, Detorakis I. Impact on the quality of life when living close to a municipal wastewater treatment plant. Journal of environmental and public health. 2016 Jan 1;2016.

[35] Kołwzan B, Jadczyk P, Pasternak G, Głuszczak J, Pawlik M, Krawczyńska M, Klein J, Rybak J. Ocena stanu sanitarnego powietrza w otoczeniu wybranej oczyszczalni ścieków. Ochrona środowiska. 2012;34(2):9-14.

[36] Breza-Boruta B. Ocena mikrobiologicznego zanieczyszczenia powietrza na terenie oczyszczalni ścieków. Woda-środowisko-obszary wiejskie. 2010;10:49-57.

[37] Law Y, Ye L, Pan Y, Yuan Z. Nitrous oxide emissions from wastewater treatment processes. Philosophical Transactions of the Royal Society B: Biological Sciences. 2012 May 5;367(1593):1265-1277.

[38] Piqueras P, Vizenor A. The rapidly growing death toll attributed to air pollution: A global responsibility. Policy Brief for GSDR. 2016:1-4.

[39] Osborne SP, Radnor Z, Strokosch K. Co-production and the co-creation of value in public services: a suitable case for treatment?. Public management review. 2016 May 27;18(5):639-653.

[40] Josipovic M, Annegarn HJ, Kneen MA, Pienaar JJ, Piketh SJ. Atmospheric dry and wet deposition of sulphur and nitrogen species and assessment of critical loads of acidic deposition exceedance in South Africa. South African Journal of Science. 2011 Apr;107(3-4):01-10.

[41] Cowan MK. Microbiology: A Systems Approach. New York: McGraw-Hill Education; 2015.

[42] Lau CK, Diem MD, Dreyfuss G, Van Duyne GD. Structure of the Y14-Magoh core of the exon junction complex. Current Biology. 2003 May 27;13(11):933-941.

[43] Cao Y, Zhang S, Wang G, Li T, Xu X, Deng O, Zhang Y, Pu Y. Enhancing the soil heavy metals removal efficiency by adding HPMA and PBTCA along with plant washing agents. Journal of Hazardous Materials. 2017 Oct 5;339:33-42.

[44] Rai PK, Tripathi BD. Comparative assessment of Azolla pinnata and Vallisneria spiralis in Hg removal from GB Pant Sagar of Singrauli Industrial region, India. Environmental Monitoring and Assessment. 2009 Jan 1;148(1-4):75-84.

[45] Opaluwa OD, Aremu MO, Ogbo LO, Abiola KA, Odiba IE, Abubakar MM, Nweze NO. Heavy metal concentrations in soils, plant leaves and crops grown around dump sites in Lafia Metropolis, Nasarawa State, Nigeria. Advances in Applied Science Research. 2012;3(2):780-784.

[46] http://www.epa.gov/mercury/about.htm.

[47] Outridge PM, Mason RP, Wang F, Guerrero S, Heimburger-Boavida LE. Updated global and oceanic mercury budgets for the United Nations Global Mercury Assessment 2018. Environmental Science & Technology. 2018 Sep 18;52(20):11466-11477.

[48] http://www.epa.gov/lead/.

[49] http://www.e-b-i.net/ebi/contaminants/cadmium.html. UNEP (2008b)

[50] WHO. 7 Million Premature Deaths Annually Linked to Air Pollution." 2014, March. World Health Organization. 2014 (http://www.who.int/mediacentre/news/releases/2014/air-pollution/en/)

[51] Manoli G, Katul GG, Marani M. Delay-induced rebounds in CO_2 emissions and critical time-scales to meet global warming targets. Earth's Future. 2016 Dec 1;4(12):636-643.

[52] Szala B, Paluszak Z. Wykorzystanie paciorkowców kałowych w mikrobiologicznej ocenie procesu higienizacji kompostowanych osadów ściekowych. Woda-Środowisko-Obszary Wiejskie. 2010;10:179-188.

[53] Chen W, Li Y, Chen CE, Sweetman AJ, Zhang H, Jones KC. DGT passive sampling for quantitative in situ measurements of compounds from household and personal care products in waters. Environmental Science & Technology. 2017 Nov 21;51(22):13274-13281.

[54] Daelman MR, van Voorthuizen EM, van Dongen UG, Volcke EI, van Loosdrecht MC. Methane emission during municipal wastewater treatment. Water research. 2012 Jul 1;46(11):3657-3670.

[55] Kartal B, Kuypers MM, Lavik G, Schalk J, Op den Camp HJ, Jetten MS, Strous M. Anammox bacteria disguised as denitrifiers: nitrate reduction to dinitrogen gas via nitrite and ammonium. Environmental microbiology. 2007 Mar;9(3):635-642.

[56] Ge H, Batstone DJ, Keller J. Operating aerobic wastewater treatment at very short sludge ages enables treatment and energy recovery through anaerobic sludge digestion. Water Research. 2013 Nov 1;47(17):6546-6557.